Management of Information Technology

Third Edition

Management of Information Technology

Third Edition

Carroll W. Frenzel

COURSE
TECHNOLOGY

ONE MAIN STREET, CAMBRIDGE, MA 02142

an International Thomson Publishing company I(T)P®

Cambridge • Albany • Bonn • Boston • Cincinnati • London • Madrid • Melbourne • Mexico City
New York • Paris • San Francisco • Singapore • Tokyo • Toronto • Washington

This book is dedicated to Alexandra and Jason, with love.

Managing Editor	Kristen Duerr
Associate Product Manager	Lisa Ayers
Production Editor	Nancy Shea
Interior Design and Composition	GEX, Inc.
Cover Design	Ann Turley

© 1999 by Course Technology—I(T)P®

For more information contact:

Course Technology
One Main Street
Cambridge, MA 02142

ITP Europe
Berkshire House 168-173
High Holborn
London WCIV 7AA
England

Nelson ITP Australia
102 Dodds Street
South Melbourne, 3205
Victoria, Australia

ITP Nelson Canada
1120 Birchmount Road
Scarborough, Ontario
Canada M1K 5G4

International Thomson Editores
Seneca, 53
Colonia Polanco
11560 Mexico D.F. Mexico

ITP GmbH
Königswinterer Strasse 418
53227 Bonn
Germany

ITP Asia
60 Albert Street, #15-01
Albert Complex
Singapore 189969

ITP Japan
Hirakawacho Kyowa Building, 3F
2-2-1 Hirakawacho
Chiyoda-ku, Tokyo 102
Japan

Trademarks

Course Technology and the Open Book logo are registered trademarks and CourseKits is a trademark of Course Technology. Custom Editions is a registered trademark of International Thomson Publishing.

I(T)P® The ITP logo is a registered trademark of International Thomson Publishing.

Some of the product names and company names used in this book have been used for identification purposes only and may be trademarks or registered trademarks of their respective manufacturers and sellers.

Disclaimer

Course Technology reserves the right to revise this publication and make changes from time to time in its content without notice.

ISBN 0-7600-4990-4

Printed in Canada

1 2 3 4 5 6 7 8 9 WC 03 02 01 00 99

Contents

Chapter 4 Information Technology Planning...................................84

PART TWO *TECHNOLOGY AND INDUSTRY TRENDS*

PART THREE *MANAGING APPLICATION PORTFOLIO RESOURCES*

PART FOUR *TACTICAL AND OPERATIONAL CONSIDERATIONS*

PART FIVE *CONTROLLING INFORMATION RESOURCES*

PART SIX   *PREPARING FOR IT ADVANCES*

Chapter 18   People, Organizations and Management Systems

Preface

Management of Information Technology, Third Edition, is written for students and practitioners seeking sound management principles and their applications in the dynamic field of information technology (IT)—integrated computer and telecommunication systems. Since the first edition in late 1991, IT has grown exponentially, creating opportunities and challenges for qualified, energetic individuals. This book teaches principles and applications for students and managers wanting to grow and prosper in today's tumultuous environment.

The future promises continued growth and increased management challenges as organizations install dynamic multiprocessors, giant centralized data repositories, and huge networked systems to support enterprise and interorganizational applications. From 1997 through 2000, global businesses will sell 400 million PCs, install 120 million miles of optical fiber, and launch hundreds of communication satellites into orbit.

During this time, PC manufacturers will double processor speeds and telecommunication suppliers will quadruple the capacity of fiber strands. Data processing capabilities will grow significantly; however, global communication capacity will expand even more rapidly.

Most PCs manufactured in the next four years will be installed in homes and small businesses. According to current estimates, nearly one billion people worldwide will access the Internet by the year 2000. Within the next decade, two billion people now without telephones will acquire phone service, many from new, high-function cellular systems. Telecommunications firms are developing and installing wireline and wireless networks capable of maintaining nearly everyone online all the time regardless of geographic location. We are realizing the information age, an age unprecedented in mankind's history.

Struggling to gain advantage from advancing technology, today's organizations downsize, outsource, and reengineer, attempting to capitalize on IT's enormous capabilities. Frantically scrambling to position themselves for the future, many form alliances, partnerships, mergers, and joint ventures. For business organizations, the importance of information technology's revolution seems impossible to overestimate. To informed and thoughtful observers, it is enormously profound.

Management of Information Technology, Third Edition, focuses on management issues surrounding information technology and lays foundations of management knowledge needed by successful managers. This text teaches sound, proven IT management basics and describes processes and procedures for applying them. It describes what managers must do and how they can best implement these procedures. Successful managers must know not only management principles but also when and how to apply them. This text teaches both.

Because technology's rapid pace has confounding influences on management, this edition emphasizes management basics, addresses environmental changes, and introduces telecommunication industry dynamics. The text discusses important new developments such as intranets, Web technology, Java programming, the 1996 Telecommunications Act, and low-earth-orbit satellite systems, and focuses on management concepts for new architectures like client/servers and intranets. As the information industry develops and introduces new technology, it undergoes rapid change and links more tightly to business. This text highlights these important developments for managers.

Management of Information Technology, Third Edition focuses on the complex topics listed above for both college students and current practioners. To take full advantage of this material, students should already have been introduced to information systems or have experience in programming or system operation and use. Students from fields such as business, economics, law, and military science, practioners in marketing, engineering, procurement, and manufacturing as well as IT personnel will benefit from the material presented in this book.

THIS TEXT'S SCOPE

The text considers information technology from several perspectives—from first line managers to chief executive officers. Experienced *and* new or aspiring managers learn to attain increased benefits from rapidly advancing technology by studying the text's frameworks and principles and applying them effectively.

IT managers are gaining considerable staff responsibility in organizations with widely dispersed technology. When IT lines and staff responsibilities grow, senior IT managers' job scopes also expand and their responsibilities increase. These trends and others demand that IT managers' skills closely resemble those of general managers. General management skills are important, but a keen awareness and knowledge of technical issues and technology trends is vital. IT executives have many weighty responsibilities and need a broad array of finely-tuned management skills to be successful. This text prepares individuals for these responsibilities.

CONTENT OVERVIEW

Part One: Foundations of IT Management

Part One lays the foundations of information technology in organizations and discusses its management issues. Readers gain perspectives on IT management through examples and management frameworks. Part One explains information technology's strategic importance and teaches essential elements of strategizing and strategic plan development. Strategic planning is logically extended to tactical and operational planning models. This part models planning processes, relating tools information, and the interacting of organizations over time.

Part Two: Technology and Industry Trends

Part Two explores information delivery systems, technology trends, and industry dynamics. Microcomputers, mainframe, and supercomputer advances are correlated with semiconductor logic and memory trends. This part discusses advanced programming, telecommunications, and workstation systems. Students gain perspectives on technology's pace and direction that translate into management opportunities. Part Two focuses extensively on telecommunication's value to modern businesses and reveals the importance of information industry dynamics to business managers.

Part Three: Managing Application Portfolio Resources

Organizations' application portfolios are major assets and profound sources of difficulty. Large development backlogs, extended completion dates, and unplanned expenditures plague many managers. Large, growing, and complex databases supporting application programs further complicate managers' work while increasing their opportunities. This part develops procedures and techniques to help managers handle these complex and exasperating circumstances. Part Three focuses on application development and acquisition in centralized and distributed computing environments, teaching managers models for success in this challenging, critical business activity.

Part Four: Tactical and Operational Considerations

The text's fourth section teaches disciplined processes for managing tactical and operational IT activities in centralized and distributed environments. It develops the basis for managing customer expectations and IT's response to user needs. Problem management and avoidance and change management techniques are developed in detail, followed by contingency management and disaster recovery planning. This part teaches computer system and network management, showing how to meet customer needs while conserving resources.

Part Five: Controlling Information Resources

Information technology rapidly infusing into organizations demands careful controls. Part Five relates IT investment and returns to financial control, customer relations, and client expectations. Students learn the need for application system controls and audits and methods to meet these needs. Managers must secure information assets, develop sound business controls, and assure senior executives they have done so. This part teaches managers how to protect and control critically important IT assets.

Part Six: Preparing for IT Advances

Part Six teaches how high-performance managers restructure business processes, introduce new technology, and manage talented people with sensitivity and skill. First-line and senior managers must manage individual transitions skillfully when restructuring. This section teaches how to achieve high morale and discusses ethical and legal considerations in managing employees. Part Six summarizes the text's management principles and practices into cohesive frameworks for business managers. The final chapter unifies this text's themes by teaching how successful CIOs develop policy, introduce technology, facilitate change, and use management systems effectively.

A NOTE TO THE READER

This text's subject matter is broad and comprehensive. Students, managers, and aspiring managers who devote sufficient time to the vignettes, text material, questions, assignments, and references will be rewarded with significant, valuable knowledge. Nevertheless, this book does not pretend to answer all the questions about this complex subject. People and technology are incredibly complicated. Even after extensive study and prolonged experience, wise managers maintain high levels of humility and flexibility. Management can be a highly satisfying profession; I hope this text brings success and increased satisfaction to students and managers.

FEATURES

This book includes several pedagogical features that were designed to help the professor further illustrate the important themes of the text. Each chapter begins with an outline that succinctly details the topics covered. Most chapters then open with a business vignette. These vignettes illustrate, using examples from industry, how sound management practices can be applied to current business situations or to challenging new problems. Each chapter ends with extensive review and discussion questions, as well as sample assignments. Many chapters also contain appendices which expand upon the material in the chapter. These end-of-chapter materials reinforce the key points in the text. A comprehensive bibliography provides students with many current articles and texts to find additional information.

TEACHING SUPPLEMENT

Management of Information Technology, Third Edition includes an Instructor's Manual to provide instructors with materials to help make their classes informative and interesting. The Instructor's Manual includes numerous value-added features, such as teaching tips, discussion topics, sample syllabi and solutions to the end-of-chapter materials.

ACKNOWLEDGEMENTS

My management knowledge and experience was enriched by associations with many wonderful people at IBM, the University of Colorado, and other organizations. I was taught by superiors, colleagues, students, clients, and subordinates alike. With this text, I reflect their input and acknowledge their contributions to my education.

I would like to thank these special people for their thorough reviews of the manuscript for this textbook. Many reviewer comments and suggestions were incorporated in the final text. The following reviewers receive my grateful acknowledgement of their assistance.

Jon W. Beard
University of Tulsa

Barbara Grabowski
Benedictine University

Bruce Hartman
University of Arizona

Sam Hijazi
Florida Keys Community College

Gerald L. Isaacs
Carroll College

Wade Jackson
University of Memphis

Jack Klag
Colorado Technical University

Jack M. Ligon
American University

Mary Anne Robbert
Bentley College

Chetan S. Sankar
Auburn University

Madjid Tavana
LaSalle University

—Carroll Frenzel
CWFRENZEL@aol.com

Management of Information Technology

Third Edition

Part
One

Foundations of IT Management

Modern information and telecommunication systems create enormous opportunities for skillful managers and their organizations. Managing information technology successfully demands general management expertise focusing on strategy development, long-range planning, and business controls. Successful IT managers are suburb generalists with outstanding technical and human resource management skills. This part includes chapters on Management in the Information Age, Information Technology's Strategic Importance, Developing the Organization's IT Strategy, and Information Technology Planning.

Managing information technology is an exceptionally challenging task. Developing sound IT strategies and plans that support the organization's goals and objectives is a vital first step toward successful IT management.

1

Management in the Information Age

The typical large business 20 years hence will have fewer than half the levels of management of its counterpart today and no more than a third of its managers. Its structure and its management problems and concerns will bear little resemblance to the typical manufacturing company, circa 1950, that textbooks still present as the norm. Instead it is far more likely to resemble organizations that neither the practicing manager nor the management scholar pays much attention to today: the hospital, the university, the symphony orchestra. For like them, the typical business will be knowledge-based, an organization composed largely of specialists who direct or discipline their own performance through organized feedback from colleagues, customers, and headquarters. For this reason, it will be what I call an information-based organization.[1]

INTRODUCTION

Electronic technology designed to process and transport data and information has been developing at an exceptional rate for more than four decades. This information technology "revolution" has significantly affected employees, managers, and their organizations. It has created countless opportunities and challenges for millions of individuals and organizations. In particular, the challenges facing managers responsible for introducing and using this technology have been especially high. In our information-based society, managers must learn to maximize advantages offered by information technology while avoiding the many pitfalls associated with rapid technological change.

Important information, broadly disseminated by new technology, radically alters the balance of power among institutions, governments, and people. Power bases dependent on information are being built, transformed, and destroyed as critical information flows around the globe with little restriction. Information technology (IT) has altered the way many people do their jobs and has changed the nature of work in industrialized nations.[2] Management practice has been greatly affected. To succeed in this rapidly changing environment, managers must be fluent in new practices and techniques.

The technological revolution that we experience today developed over the past 45 years. Many individuals now entering the workplace have directly experienced this technological phenomenon and perhaps take it for granted. Employees completing their careers have seen the whole spectrum of events unfold during their lifetimes. For many people, the growth of information technology has been an unmitigated blessing. "New" technology was a major part of their formal education. It is essential in their employment and an important ingredient of their future success. For others, information technology has been a complicating factor, creating apprehension or outright fear. To a considerable degree, information technology has brought change to nearly everyone.

Information technology has become increasingly vital for creating and delivering products and services in most nations. Spending by U.S. companies on information technology has increased steadily since 1972, exceeding spending on basic industrial equipment since 1982. By 1994, IT spending exceeded basic industrial equipment spending by a factor of three. By 1996, IT, including computing and telecommunications, became the largest U.S. industry, generating revenues of $866 billion and creating 6.2 percent of the nation's goods and services according to a study sponsored by the American Electronics Association and NASDAQ.[3]

For individuals managing information technology within a larger enterprise, the rapid pace of innovation brought unprecedented growth of job opportunities fueled by an ever-increasing need for skilled managers. Senior executives who can manage complex IT tasks are in great demand. This demand is driven by the growth of valuable computer applications such as decision-support systems and expert systems, and by computer-aided design, computer-aided manufacturing systems, electronic imaging, and multimedia and Internet applications. Demand for skilled executives has also been created by rapid advances in telecommunications, which have spawned local- and wide-area networking, electronic data interchange and electronic commerce, and interorganizational systems. The future for skilled managers of information technology is promising and challenging.

James Sutter, 14-year veteran vice president and general manager of information systems (IS) at Rockwell International Corporation (defense and space), is an example of a highly successful IT manager. He manages an IS budget of about $400 million for a company focused on productivity improvements. He consolidated IT outposts in a single data center, reducing computer operations personnel by 50 percent, but still spends 15 percent of software development resources on client/server applications. His goal is to fund new projects from productivity improvements, thus coping with difficult times in the defense business. Rockwell is one of the Premier 100 companies as measured by *Computerworld's* productivity index.[4]

This introductory chapter sets the scene for the management task in the field of information technology. It describes how organizations use information and why they now critically depend on it. It analyzes the foundations of information technology and relates them to the challenges of management. This chapter describes challenges and issues facing IT executives and reveals critical success factors or conditions necessary for management success.

Individuals within modern organizations have visions or expectations of what technology can do for them, but they may appreciate less the impact technology will have on them. This chapter teaches a superior management system that helps managers deal with expectations and cope with organizational and personal change. It builds solid foundations in general management principles that strong IT managers need. This text teaches how to apply these management principles to the complex tasks of introducing, implementing, and managing technological change.

HOW ORGANIZATIONS USE INFORMATION

Modern organizations use resources of several types to benefit the people they serve. The most common resources upon which organizations depend to carry out their missions are money, skilled people, information, physical property, and time. Managers' primary tasks are to combine these resources, optimize their value in discharging the organization's mission, and obtain superior returns for the organization's stakeholders. Thus, asset management is a managers' chief task.

Although these assets are important to nearly all firms, their proportionate value is not the same for all organizations. For example, physical property is very important to an energy company or mining operation but much less important to a financial institution. Time, i.e., rapid response, is especially important in the service industry or in new product development in highly competitive industries; however, passage of time, sometimes long periods of time, is important to forest products firms or lending institutions. Information is important to all organizations, but some, such as newspaper publishers or credit reporting agencies, are information centric—information is their lifeblood. But, skilled people are the most important asset of all organizations, regardless of form or purpose.

In today's information-intensive environment, the creative combination of information and people is a powerful force for achieving superior performance. High-performance organizations that attain or exceed challenging goals, satisfy and expand established markets or develop important new ones, and create superb value for owners, employees, and customers are likely to employ empowered, motivated workers supported by well-developed information systems. The leverage of information and people is so powerful that managers in high-performance organizations devote considerable energy to managing information, its delivery system, the people who deliver it, and those who use it. The combination of skilled people and advanced information technology has revolutionized the concept of management.

Indeed, information technology permanently and irrevocably alters the social and organizational structure of the workplace and transforms the nature of work for managers and nonmanagers alike. Adopting new information technology involves more than installing impersonal, technical appendages to current activities. Advanced technology redistributes knowledge among all employees and undermines traditional managerial authority based on privileged access to information. In today's environment, a new portrait of authority must be developed that nurtures knowledge workers. It must be based heavily on participative management techniques.

From the employee's perspective as well, adopting new information technology is not an insignificant activity. The need for more information to work in today's increasingly knowledge-based environment implies increased worker responsibility, and it demands more intellectual effort from workers, too. Because much information is shared and collectively used, a heightened awareness of joint responsibility for and ownership of organizational success develops among workers, yet these conditions conflict with the traditional notion of authority. Thus, today's knowledge workers and their managers need to establish new relationships.

Transformations in work content and social and organizational relationships at work do not primarily result from the routine automation of work-related tasks. In contrast with automation through mechanical devices, the introduction of electronic information technology not only requires data and instructions for its use, but it supplies useful data about the task it is automating. Information generated about the task to which automation is applied provides new insights into the task itself and gives opportunities for further technology applications.

For example, order-entry systems with caller-ID (the caller's name and phone number is available when the phone rings) permit the operator to respond with the caller's name, a nice personal touch in many sales situations. Using the phone number to recall information about a customer's prior orders, the system lets the operator portray a more intimate knowledge of the customer's preferences. This greatly improves rapport with the customer and may also permit the operator to make suggestions leading to increased sales. Many advantages stem from using system-generated data to tilt the ordering function toward sales. The operator gains important new responsibilities, the firm adds another sales interface, employee-supervisor relations change, and the order-entry department plays a new role.

All these effects are additions to many other broader and more common benefits made possible by sophisticated order-entry systems, such as rapid sales analysis, optimized inventory levels, responsive pricing or marketing strategies, and enhanced information for manufacturers and suppliers. To benefit from information-generating automation, however, organizational structures and social relationships within the firm must change.

In contrast with earlier automation efforts, applications of computerized information technology have a two-fold nature. They can supplement human activities in routine operations, thereby improving productivity and increasing operational quality. Simultaneously, they can provide important information about the process, which can be used to improve it. These properties enable subsequent technology implementations and human activities to be based on a deeper knowledge of the processes and their results.[5] Thus, evolving, more pervasive technology implementations lead to improved processes, greater benefits from information technology, increased responsibility for technology implementors and users, and constantly shifting social and organizational relationships within the firm and between it and others outside. For this reason, workers at all levels face a constantly changing environment.

In addition to changing the traditional notion of managerial authority through more broadly dispersed information, information technology encourages employees to become more specialized and offers them performance feedback from many more sources. As this transformation occurs, managerial hierarchy gives way to a more interdependent, less formal structure. To focus on people and obtain their commitment, participative management, decentralization, and self-managing teams develop. Drucker described this transformation in the opening paragraph of this chapter.

To capture the many important advantages of information technology, however, a firm must have strategies and plans to adopt and use technology, and individuals and departments to develop and support them. In most firms, specialized IT departments develop these strategies and plans and also implement major portions of them. Because information technology is usually widely distributed, the IT department also guides and assists other parts of the firm in its implementation and use. In most firms, the IT department is critical.

INFORMATION TECHNOLOGY ORGANIZATIONS

Most firms today have an information technology organization with a status similar to that of marketing, accounting, engineering, or manufacturing. With information technology broadly dispersed through client/servers, intranets, and other forms of distributed processing, some thought that centralized IT organizations and centralized processing would disappear. This is highly unlikely, however, because IT depends on a collection of specialized knowledge and skills, important to the firm and not available more generally. In addition, development of coordinated plans, maintenance and development of infrastructures, and greatly increasing staff responsibilities require firms to have a strong and vital IT function.[6]

Like other functions in a firm, the IT organization has departments that have specific responsibilities and report through department managers to the senior IT manager. Traditionally, the senior IT manager reported to the finance organization; however, many alternate reporting relationships evolved as IT responsibilities grew to encompass the entire firm. In recognition of this broadened responsibility, the senior IT executive is frequently called the "chief information officer," or CIO.[7] This text more fully discusses various approaches to IT organizational alignment later.

In many firms today, the IT organization operates as a business within a business, supporting all other functional units in a variety of ways. Table 1.1 illustrates some of these important functional units and the IT applications that typically support them.

TABLE 1.1   Functions and IT Applications Supporting Them

Functions	Applications Supported
Product development	Design automation, parts catalog
Manufacturing	Materials logistics, factory automation
Distribution	Warehouse automation, shipping and receiving
Sales	Order entry, sales analysis, commission accounting
Service	Call dispatching, parts logistics, failure analysis
Finance and accounting	Ledger, planning, accounts payable
Administration	Office systems, personnel records

The rather conventional and easily recognized examples in Table 1.1 represent only a small fraction of the total portfolio of computerized applications. In many firms, applications developed by functional areas for specific purposes are also an important part of the application portfolio. A medium-size corporation may have an application portfolio holding several thousand computer programs.[8]

Today's senior IT managers and most IT departments have both line and staff responsibilities. Line responsibilities are to directly accomplish the objectives of the enterprise. For example, Chevrolet manufacturing at General Motors is a line responsibility. On the other hand, the personnel department at General Motors performs a staff responsibility. Staff activities help line functions accomplish the primary objectives of the enterprise.[9]

To emphasize the distinction further, the purpose of line organizations is to implement the mission of the firm. Product development, manufacturing, sales, and service are examples of line organizations. The purpose of staff organizations is to help line organizations accomplish their missions. Personnel, finance and accounting, and legal counsel are examples of staff organizations. In most firms today, IT is a hybrid organization because it has both line and staff responsibilities.

IT organizations contain core activities that represent their line responsibilities and other activities that discharge various staff responsibilities. For example, IT functions usually have departments to build computer applications, operate centralized computer facilities, develop and maintain IT plans and strategies, and provide other kinds of technical service to the firm. IT organizations also have people or departments to handle staff responsibilities. For example, there may be an IT group that supports computer operations in manufacturing, or a department to support telecommunications in the company, or persons who develop internal technical standards for the firm.[10]

The firm's culture and the industry it's in are key factors that govern the activities and responsibilities of the IT function. For example, in firms that depend heavily on information technology, the IT technical support role may be broad and comprehensive. In others, however, the programming group, computer operators, or other technical specialists may handle technical support. Structures vary with corporate and departmental needs, and they evolve as the firm adopts new technology or reengineers to improve operations. Subsequent portions of this text discuss in much greater detail the role of the IT organization and its relationship to the larger organization.

MANAGING INFORMATION TECHNOLOGY

Information technology is the term that describes the organization's computing and communications infrastructure, including computer systems, telecommunication networks, and multimedia (combined audio, text, and video) hardware and software. Another commonly used term is *information management*. It refers to business-related data and communications standards and operations in the firm. Still another term, *information resource management*, refers to the activities of investing in and managing people, managing technology and data, and establishing policy regarding these assets' use. In this text, *information technology management* encompasses these three terms and includes the tasks of managing the infrastructure, the standards, and operations; making technology-related investments; and recommending appropriate corporate policy.

The extensive task defined above evolved over several decades from more narrowly defined tasks as information technology itself evolved and as firms deployed it more broadly and more deeply. As all aspects of information technology became increasingly valuable and critical to organizational success, the task of managing these assets became more difficult and encompassed broader responsibilities. As managers' jobs expanded over the years, they also changed character, from a technology orientation toward general management tasks. Today, IT managers must understand technology and its trends but must be outstanding generalists as well. Larger and more important jobs for skilled managers have resulted from the evolution of technology and its management.

The Evolution of IT Management

Electronic information processing has been used in business processes for more than four decades and has evolved through several identifiable phases.[11] In the late 1950s and throughout the 1960s, routine business data handling was automated by punched cards, electronic accounting machines (EAM), and physically large but relatively low-power electronic computers. Because these early capabilities were readily applied to accounting and financial activities, the data processing (DP) or electronic data processing (EDP) departments were usually within the accounting or finance functions. In many firms, these early EDP departments struggled in isolation from the remainder of the organization to automate financial applications. In these early years, some employees viewed EDP apprehensively because they feared job loss through automation.

During this period, investments in DP were based on traditional budgeting processes and, in many cases, handled as overhead. Attempting to recover costs through charge-outs, some served by DP were dissatisfied: they were paying for an organization that only marginally served their needs. At the same time, however, firms found EDP essential to their operations. Typically, as EDP managers grappled with immature technology, they slighted some managerial dimensions of their jobs. Sometimes, just keeping the systems running was a full-time job.

A decade later, terminals were being connected to mainframes, and database management systems were introduced to handle the large business data files that accumulated. During the 1970s, as computers began to handle functions other than finance and accounting, emphasis changed from providing data to creating information, and the responsible function became a centralized entity more likely to be called information systems (IS). Information infrastructures began to emerge within the firm and massive reports of financial and production data proliferated as IS tried to assist managers with information systems designed for their use. During this time, decision-support systems (DSS) began to emerge. In this era of management information systems (MIS) much confusion remained over what management information really was.

As the firm's technology investments increased and applications multiplied, IS managers were motivated to concentrate more on efficient operations—keeping large systems fully loaded around the clock became an important goal. To satisfy managers of client organizations, IS began to concentrate on plan alignment with them. During this time, the role and contribution of IS was refined as it became obvious that the role was expanding and the contribution growing.

During the 1980s, telecommunications and networking flourished with the introduction of distributed data processing, office systems, and personal computers. With the proliferation of incompatible systems and distributed operations, firms discovered infrastructure fragmentation and experienced the accompanying loss of operational discipline. Recognizing the potential for enormous gain from information systems, firms searched for competitive advantage through information systems development and business transformations. During this time, the function began to be called information management, information resource management, or information technology (IT) management. As telecommunications and computers merged into one discipline, the leader of the combined function began to be called the "chief information officer" (CIO).

Strategic planning, competitive advantage, organizational learning, and the role and contribution of IT were central issues facing most organizations struggling with advancing technology and restructuring to take advantage of it. Communication between IT and other executives and functions remained an issue as IT-induced changes swept the firm. Getting functional managers involved in using IT to reshape business processes and gain efficiency and effectiveness became important tasks for senior executives.

As technology advances at a torrid pace in the 1990s, firms depend on it to streamline structures and to link electronically to suppliers and customers. Business process reengineering, downsizing, outsourcing, and restructuring take on added meaning as firms emphasize quick response and flexible infrastructures to improve effectiveness. Executives embrace networked organizations as they decentralize operations while maintaining centralized coordination through IT systems. Today, IT receives top-level attention because of its increasingly important role in corporate strategy and policy.

As information technology penetrates more deeply into all business processes, IT executives and other senior executives struggle with technical and policy issues—systems integration is difficult because open systems are not necessarily as open as advertised—and internal standards and governance policies must be rigorously established.[12] Outsourcing considerations, application program make-*vs*.-buy decisions, and powerful network technologies such as client/servers and the Internet greatly complicate life for CIOs and their peers. New information technology capabilities and many new, superior information systems create tremendous opportunities and daunting challenges that today's corporate executives must routinely and effectively manage.

Types of Information Systems

New information systems based on Internet technology, data warehousing concepts (very large databases of operational data), or interorganizational demands add to earlier, more familiar types of systems commonly discussed in the IT literature and found in most organizations. These include transaction processing systems (TPS), decision support systems (DSS), office automation systems (OAS), management information systems (MIS), and expert systems.

Transaction processing systems handle routine information items, usually manipulating data in some useful way as it enters or leaves the firm's databases. Order-entry programs are an example of a TPS. Orders arriving at a manufacturer are a transaction that leads to a workorder on the plant floor or to a shipping order at the warehouse. The plant workorder system and the shipping system may themselves be transaction processing systems. Many transaction processing systems are online, meaning that many users interact with the database simultaneously, performing updates or retrievals from workstations at their desks. TPS are the most common form of information systems.

Management information systems provide a focused view of information flow as it develops during the course of business activities. This information is useful in managing the business. Labor reporting programs, inventory transaction reports, sales analyses, and purchase order systems are some of the many management information systems. Reports, generated for managers' use, are the usual product of management information systems. Decision support systems (DSS) are analytic models used to improve managerial or professional decision making by bringing important data to managers' attention. In many cases, these systems use the same data as management information systems, but DSS refine the data, making it more useful to managers. According to Keen, however, "the term DSS is now so vague as to be indistinguishable from executive information system, end-user computing, expert systems, and even personal computing."[13]

Office automation systems provide electronic mail, word processing, electronic filing, scheduling, calendaring, and other kinds of support to office workers. They became very popular with the introduction of personal computers. End-user computing, a form of distributed data processing usually implemented through PCs connected to servers, places computational capability into the hands of users to develop or execute programs. Programs created and used this way serve a wide variety of purposes.

Many other types of systems are designed for specific purposes. For example, engineering design systems enable skilled engineers to design complex computer chips by manipulating design algorithms and laying out millions of circuits on a chip while obeying numerous electrical ground rules. Chip design today is impossible without systems of this type. Some programs, termed expert systems, are fashioned to augment human reasoning in complex analyses. Many applications of expert systems exist. IBM's Deep Blue chess-playing program is an expert system.

Some of these systems are operational in nature, supporting the business in the short term. Others aid managers in making intermediate- or long-term decisions or have other long-term uses. These systems evolved as information technology itself evolved and as IT professionals and others sought to apply technology to support business activities. Computer and communications technology continues to evolve today, challenging professionals and managers to apply it in new, innovative ways to further organizational goals and objectives.

IT MANAGEMENT CHALLENGES

For several important reasons, information technology managers have many challenging tasks to perform. Although many IT managers established brilliant performance records, others fared less well. Mediocre performance and lack of success dominated the careers of some IT managers whose efforts to manage rapid technological change effectively fell short.

Many information technology managers find themselves and their organizations in untenable positions. They spend large sums implementing new technologies that promise high potential value to the organization and carry commensurately high risk. They support clients who request increased services, while senior executives, observing slowly rising or level productivity, question increasing expense levels.[14] Unmanaged demand for services leads to expanding work backlogs. As expectations for information technology rise in many organizations, senior IT managers struggle to meet growing challenges.

IT managers, trained in technology but lacking the general management skills that their organizational roles demand, find that their jobs require knowledge of people management and organizational considerations as well as programming or hardware expertise. They and their organizations are constantly scrutinized as they strive to improve productivity in return for the resources they consume. Furthermore, in many cases, mergers, acquisitions, and reorganizations of all kinds threaten the stability of their position.

Executives expect to restructure and streamline their organizations around information technology, but the technology itself causes structural and personal dislocations and generates high expectations within the organization and among its managers. To address this situation, IT managers must use general management skills that let them contend successfully with these complex issues. Their skills must include managing expectations and coping with business and structural changes, while firing the productivity engine for the firm. In short, today's IT managers must be technological leaders as well as superb generalists. They must capitalize on their technological expertise while solidifying and strengthening their platforms of managerial experience and skill.

Growth and Complexity

The computer-communication era began several decades ago. It is now entering an explosive growth phase that dwarfs earlier remarkable advances. In the next three years, according to industry analysts, consumers around the globe will install more than 100 million PCs, telecommunications firms will lay millions of miles of optical fibers, and, if current plans materialize, satellite system developers will launch 1000 or more communications satellites into orbit.[15] One billion people, now without telephones, will acquire phone service during the next 10 years. Fueled by advancing technology, political change, competition, and consumer demand, the world's electronic information infrastructure is drawing growing numbers of people from around the globe and evolving rapidly for those currently within its domains.

Today's development of the information infrastructure is the most complex human activity ever undertaken—its use and application is also monumentally complex. Although the application of modern computer and communication systems greatly benefits organizations of all kinds, complexities inherent in technology adoption and utilization raise management challenges to new heights. Opportunities for individuals expertly prepared to manage these new challenges are equally unprecedented.

Pervasiveness

Computer and telecommunication systems are widely dispersed and deeply infused in modern organizations. Nearly all have installed intelligent workstations at a rapid pace, and, today, most professional employees use an individual workstation or terminal.[16] Workstations for office and professional workers in most organizations are interconnected via local area networks that connect to other nets or to large computer data stores through servers that function as message-switching, message-processing systems for network users. These internets or information highways also unite the firms' sites nationally and internationally and support electronic commerce by linking firms to their customers and suppliers. Future activities based on these systems will grow rapidly, engaging increasing numbers of people around the world.

Applications and Data

Advanced computer and telecommunication systems enable large, sophisticated, and very valuable application programs to operate on databases that grow rapidly in size and importance. The acquisition and maintenance of these vast program and data resources demand very careful attention from many members of the firm's senior management team. Valued at $1 trillion or more in the U.S., these programs and data stores are expensed during development and not recorded on the corporate balance sheet. But they and other intellectual assets are the foundation of information-based organizations.[17]

Computer Network Operations

Most contemporary organizations are critically dependent on skillfully managed computer network operations. Today, network-centric firms process hundreds of revenue producing transactions per second that must be executed promptly and flawlessly. For these firms, the "information system" is in series with the firm's critical activities—the drive shaft of its operations. Service disruption for even a few seconds has serious consequences for the firm's financial health and reputation. This kind of operation demands extremely challenging management performance.

Controls and Environmental Factors

Knowledge-based organizations rely on rigorous, dependable control mechanisms. Ineffective control or loss of control in highly valuable, highly automated operations permits rapid error propagation. Careful attention to business controls of all types must neutralize external and internal threats to sophisticated human and machine business operations. The control issue grows larger as networked systems and internetworking increase.

The social and political environment surrounding technological evolution is also critically important. Technological advances considerably enabled internationalization of business and growing international competition, which altered the way firms conduct their affairs. Governmental actions in many countries also shape global business enterprise and increase the rate of change in the business sector. Business strategy and planning takes place against this moving backdrop, greatly increasing management difficulty and complexity for the firm's executives, including its IT executive.

Competitive Considerations

Today, information technology offers great competitive value and significance for most organizations. Firms expect information technologists and their organizations to provide the tools for capturing and maintaining competitive advantage. Consequently, many information technology organizations and their managers are on their firm's critical path to success. If not operating effectively, they may limit their parent organization's long-term performance. Obviously, this poses a very special challenge for the firm and its executives.

As technical development continues, information technology applications will find important new uses and become involved in more complex processes. Future information and communications systems will engage human and organizational activity ever more broadly and more deeply, becoming more sophisticated and more important to competitive business operations. For many reasons, managing information technology has important people considerations, too, because it demands highly skilled technicians and managers and because its use impacts people in many ways.

People and Organizations

Information technology impacts organizations, managers, and workers at every level as it alters the nature of work in industrialized societies. Not all consequences are favorably perceived—many threaten or intimidate managers and workers alike. These personal and organizational dislocations further heighten the management challenges accompanying technology introduction.

To cope successfully with these challenges, effective managers rely upon a discerning awareness and a keen appreciation of the social phenomena; they continuously fine-tune their people-management skills. High-performance organizations always find ways to maximize human productivity. They recognize that people are the key to capturing the benefits of technological advances in today's information age. Effective executives understand this extremely well.

IT MANAGEMENT ISSUES

The task of managing information technology remains challenging for a variety of important reasons, some of which were discussed earlier. Additional challenges for IT managers arise from the organization's sociology and culture—the backdrop against which information technology evolves within the organization. For the most part, these are managerial issues rather than technical concerns. Resolving them requires comprehensive management skills rather than rich technical skills.

Because these matters are so important, considerable research has been conducted to determine critical issues facing IT managers, their peers, and their superiors. CSC Index, Inc., Datamation, Digital Equipment Company, University researchers, and several consulting firms have developed lists of critical issues, ranked them by importance, and analyzed the considerations behind them. Of particular significance is the finding that some issues remained prominent for many years.

Although the relative ranking of these issues varies somewhat depending on the researcher and the time of the research, some key issues remain critical over long periods. For example, aligning IT and corporate goals, reengineering business processes, and information architecture have been important to IT and other executives for more than a decade. Researchers also consistently find three issues that are important for most organizations: 1) using IT to improve productivity and quality; 2) creating or maintaining competitive advantage through information technology; 3) redesigning business processes to better support company strategy. Other important issues include strategic planning, obtaining positive returns on IT investments, and IS human resources.[18] For IT managers, these findings portend more realignment and restructuring, and managing within tight financial constraints.

Surveying 603 senior IS executives, the CSC Consulting Group identified the top five issues: 1) aligning IS and corporate goals; 2) instituting cross-functional systems; 3) organizing and using data; 4) implementing business reengineering; and 5) improving the IS human resource.[19] Many researchers found reengineering business processes a critical issue.

According to the CSC Consulting Group, the main goals of reengineering are to increase productivity, improve customer satisfaction, and improve quality. Additional goals of reengineering are to reduce costs, reduce cycle time, and increase revenue. Reengineering is consistently at the heart of business effectiveness—improving business effectiveness remains a key goal for IT managers.

Supporting reengineered business processes raises new management issues, too. Today's managers, for example, are concerned with introducing and managing networked systems and wireless technologies, installing and managing extranets and other interorganizational systems, and dealing with security and privacy issues. All these activities affect the work that people do. Skillful managers must pay attention to the sensitive issues that result.

It should be no surprise that the concerns noted above are critical to IT managers and their firms because information technology is the heart of many businesses today. Thus, organizations using and supporting IT must deal with important issues of strategy development, organizational change, goal alignment, and using current technology and introducing new technology to achieve superior business results. These key issues apply to organizations around the world as well as those in the U.S.

In addition to skills, IT managers must have effective management systems and processes to assist them in dealing effectively with issues and challenges inherent in their responsibilities. Understanding these systems and processes and their application and use within the firm are necessary conditions for success. This text devotes considerable attention to the development and use of superior, comprehensive management systems and processes.

What is the fundamental meaning in all of this? The following section briefly describes the evolution of IT management. It puts today's issues and concerns into historical perspective and constructs the logic and rationale for developing a superior framework for tomorrow's IT managers.

THE MATURATION OF IT MANAGEMENT

Over the past four decades, as information technology moved from supporting mainly accounting activities to enabling modern operations in nearly all facets of businesses, IT managers have changed from mostly technical specialists to more sophisticated generalists. IT managers' issues and concerns have moved from the specific to the general, from technical to managerial, as IT matured from the technology orientation of the 1960s and 1970s to the business orientation of the late 1980s and 1990s.

Telecommunications, competitive advantage, organizational learning, the role and contribution of IS, and acting as change agents dominated IT managers' thinking in the late 1980s. Now, near the turn of the century, after restructuring and reengineering, many organizations are investing in another wave of sophisticated networking technologies to make important information more readily available internally and to bring the firm ever closer to external customers and suppliers.

Although many firms are introducing Internet-based systems and other sophisticated applications, managers are concentrating on improving business effectiveness through other means, too. To reduce costs, improve internal efficiency, capture economies of scale, and obtain critical new skills, corporations and their IT organizations continue to reorganize, downsize resources, and outsource some activities.[20]

In the late 1990s as the power of information technology becomes more critical to most organizations, planning, strategizing, adopting standards, and establishing IT policy become increasingly important. In part, the rising prestige of the CIO position reflects this new emphasis for IT managers. As managing information technology becomes a mature discipline in tomorrow's organizations, governance—establishing rules of conduct in the firm—will become a more dominant concern for IT executives.

Although large, broad issues are important to today's IT managers, many of the old challenges remain. For example, the typical IT organization still struggles in its relationship with senior executives in other departments as it copes with distributed architectures, user-developed systems, and rapid changes in the business environment. As IT executives grapple with these responsibilities, they experience some of the same old problems: work backlogs, mergers and reorganizations, employee training and retraining, and managing with constrained resources. To function successfully in this challenging environment, IT executives need a keen awareness of business issues and refined general management skills.

MANAGING MATURE IT ORGANIZATIONS

As IT management becomes a more mature discipline, IT managers must develop more sophisticated, more mature models of behavior. They must be knowledgeable about technology and its trends, but this knowledge alone is not sufficient. Broad-based business experience is critical for IT managers but, by itself, is insufficient for success. Tomorrow's successful IT manager needs solid business skills *and* a sound understanding of technology, its trends, and its implications.[21] In addition, IT managers need models of behavior and frameworks of business management to help guide their actions.

In an important work, Paul Strassmann presents a model of Information Management Superiority premised on the idea that IT management only has value within the context of business management. He states that "the benefits of investments in information technology can be assessed only as seen from the standpoint of a business plan."[22] His model depicts information management superiority being sustained by five reinforcing and interacting ideas.

A. *Governance*. Governance, or information politics, is used not only to exercise authority but as an art for achieving corporate consensus. It guides how individuals and groups cooperate to achieve business objectives.

B. *Business Plan Alignment*. IT business plans must be congruent with the organization's business plans or the worth of IT plans will be suspect.

C. *Process Improvement*. All IT and business activities should be regularly scrutinized to identify areas where improvements, however small, can be made.

D. *Resource Optimization*. Managers must always question whether money, space, time, or people can be used more effectively to further corporate goals.

E. *Operating Excellence*. All operational details of the business must be performed in a superior fashion. Quality in the business process must be an overriding consideration throughout the business.

According to Strassmann, information management superiority is achieved by constant management interaction among these five activities. As an example of how this works, consider a manufacturing process improvement. Suppose, for instance, that inspection of arriving chemicals will be outsourced to an independent testing lab, and, therefore, the plant no longer needs its receiving inspection program for chemicals. This business process improvement alters business plans for manufacturing and for IT. Later, the change in chemical inspections may be expanded to include all incoming supplies. Thus, the policy for receiving inspection at this plant has changed.

Strassmann's model helps explain the developing maturity of IT management. Thirty years ago, EDP managers were concerned with improving their operations, getting more work through CPUs, and keeping systems running around the clock. They were concerned with IT resource optimization—Strassmann's point D—other management tasks were secondary for many. Twenty years ago, optimizing hardware systems reached a fine art and IT concentrated on improving IT processes—point C in Strassmann's model. IT managers gave more attention to applications development. During the 1980s, IT executives concentrated on end-user systems, teaching organizations about technology, and developing strategies closely aligned with the business strategy—Strassmann's point B.

In the past 10 years, spurred by international competition and other forces, American firms focused intensely on quality in the business process evidenced by the Baldrige quality award competitions. IT organizations joined the effort to improve operating excellence—point E in Strassmann's model.

Today, as IT management continues to mature, executives are concentrating on the rules governing technology dispersal and are establishing policies regarding distributed computing, Internet activities, portable computing (notebooks and laptop computers), and wireless networks. Policy making, such as establishing authority for technology acquisition and deployment, is now more critical than owning and operating large computer systems. Because governance, Strassmann's point A, causes managers to concentrate on larger, more important policy issues, IT executives increasingly focus on their staff responsibilities and more willingly delegate routine line activities to information servicing firms. The importance of Strassmann's model will become clearer as we explore other topics later in this text.

Thus, as information technology penetrates more deeply into organizations and plays a more important role in business success, managers are adopting more sophisticated models to help them oversee increasingly vital information assets. Although now using more sophisticated techniques, managers continue to value older, but still important concepts.

IMPORTANT CONCEPTS IN INFORMATION SYSTEMS

Stages of Growth

Early in the evolution of IT, Richard Nolan and Cyrus Gibson found that organizations go through predictable stages of growth as they adopt and implement technology.[23] Their research provided an important conceptual basis for understanding the introduction and assimilation of new information technology. Table 1.2 lists the growth stages they identified.

TABLE 1.2   Stages of Growth

1. Initiation
2. Contagion
3. Control
4. Integration
5. Data Administration
6. Maturity

The following paragraphs briefly summarize the characteristics of these growth stages.

Stage 1. *Initiation*. New technology is initially introduced into the organization, and some users begin to find applications. Use grows slowly as people become familiar with the technology and its applications.

Stage 2. *Contagion*. As more individuals become acquainted with new technology, demand increases and use proliferates. Enthusiasm for the new technology builds rapidly during this stage.

Stage 3. *Control*. During the control stage, the issue of costs versus benefits intensifies and managers become increasingly concerned about the economics of technology.

Stage 4. *Integration*. As systems proliferate and databases grow rapidly, the notion of systems integration becomes dominant. Managers become interested in integrated systems and databases.

Stage 5. *Data Administration*. At this stage, managers are concerned with the value of the data resources. They take action to manage and control the databases and to ensure their effective utilization.

Stage 6. *Maturity*. In this stage, if it occurs, the technology and the management process are integrated into an efficiently functioning entity.

The stages-of-growth concept is useful because it operates at several different levels within an industry or an organization and provides a measure of predictability. Not all firms within an industry are in the same growth stage, nor are all functions within a corporation. Using this framework, executives can observe where in the growth stage their organizations are and can make reasonable assertions concerning future organizational behavior. From observations of the present, thoughtful managers can prepare themselves and their organizations for the future.

The stages-of-growth concept is also useful in information technology planning. It can provide guidance for planning future activities, and it forms the basis for one IT planning model. As new technology is introduced into the organization, the growth stage theory comes into play again. By observing the technology adoption process, managers can implement practices to capitalize on the technology's potential while minimizing the effects of inevitable pitfalls. The text explores these notions in later chapters in connection with IT planning, the introduction of intranets, and other topics.

Although a conceptual rather than quantitative tool, the growth stage hypothesis is important to managers because it provides insight into technology adoption processes. Skillful IT managers use this knowledge and insight along with other tools to improve their performance and the organization's.

Critical Success Factors

Armed with a knowledge of the important challenges facing IT executives and with a model for understanding them, managers are well positioned to describe the factors necessary for success. What actions must IT executives carry out successfully? What management systems and processes are vital for their personal success and their organization's success?

Answers to some of these questions are found in the notion of critical success factors (CSF) developed by Rockart to help executives define their information needs.[24] Critical success factors identify those few areas where things must go right—they are executives' necessary conditions for success. They apply to IT executives, to their subordinate managers, and to other executives in the firm.

Rockart identified four sources or areas where executives should search for critical factors: the industry in which the firm operates, the company itself, the environment, and time-dependent organizational areas. This last source accounts for the possibility that some organizational activity may be outside the bounds of normal operations and require intense executive attention for a short period. In addition, Rockart identified two types of CSFs: the monitoring type and the building type. The monitoring type tracks the ongoing operation. The building type initiates activity designed to improve the organization's functions in some way. Critical success factors apply to IT managers. Useful in IT planning, they will be discussed again in that context.

Critical success factors for IT managers can be determined by seeking answers to the following questions. What conditions are necessary for IT managers' success today? What tasks must be carried out very well in order for managers to succeed? The information needed to answer these questions can be obtained from the challenges and critical issues studied previously and applied in conjunction with the elements of management superiority found in Strassmann's model. Using this approach, the factors necessary for success can be grouped into the several classes shown in Table 1.3.

TABLE 1.3   Critical Areas for IT Managers

1. Business issues

2. Strategic and competitive issues

3. Planning and implementation concerns

4. Operational items

To be successful, the firm's IT management team must take action in these critical areas. If the organization experiences difficulties in any of these areas, managers must set goals and objectives to eliminate the difficulties. Managers must also act to prevent the development of issues. Outstanding managers use a roadmap of critical success factors to assess their posture on these vital topics. The following CSF list serves as a model that successful IT managers might follow.

1. Business Management Issues
    a. Obtain agreement with the firm's executives on how information technology will be managed within the firm.
    b. Operate the IT function within the parent organization's cultural norms.
    c. Attract and retain highly skilled people.
    d. Practice good people management skills.
    e. Use IT to improve productivity and financial returns.

2. Strategic and Competitive Issues (long range)
    a. Develop IT strategies supporting the firm's strategic goals and objectives.
    b. Provide leadership in technology applications to attain advantage for the firm.
    c. Educate the management team about the opportunities and challenges surrounding technology introduction.
    d. Ensure realism in long-term expectations.

3. Planning and Implementation Concerns (intermediate range)
    a. Develop plans supporting the firm's goals and objectives.
    b. Provide effective communication channels so that plans and variances are widely understood.

c. Establish partnerships with client IT organizations during planning and implementation.

d. Maintain realism within the organization regarding intermediate term expectations.

4. Operational Items (short range)

a. Provide customer service with high reliability and availability.

b. Deliver service of all kinds on schedule and within planned costs.

c. Respond to unusual customer demands and to emergencies.

d. Maintain management processes that align operational expectations with IT capabilities.

Not all these items will appear on every IT manager's list of critical success factors. Normally, most critical areas operate smoothly and routine attention maintains high quality operation. Still, in some situations, managers must add temporal factors or company specific factors to the list. In all cases, superior managers remain attentive to those factors necessary for their success.

Information technology managers who can accomplish the tasks outlined above have a good chance of becoming highly successful. They have developed general management skills that prepare them for increased future responsibility. Throughout the remainder of this text, management tools, techniques, and processes will be developed to enable IT managers to accomplish these critical tasks successfully. Achieving success in these key areas is a high priority goal for IT managers.

EXPECTATIONS

Although opportunities available to highly skilled managers are bright indeed, the challenges of the profession remain equally high. IT trends suggest that these challenges will increase dramatically in the coming years, and that successful IT managers must learn to keep pace *now*.

Many challenges arise from the volatility of technology and from evolving technology's impact on the firm's structure and economics. Expectations of the firm's managers at many levels spawn other challenges. IT managers must anticipate these challenges and create thoughtful plans to manage emerging issues to the firm's advantage. Well-prepared managers seek to understand issues and prepare to control their consequences.

For IT managers to succeed within their organizations, they must do more than cope with issues as they arise; they must also take a strong leadership position in formulating and shaping them. Their insight and vision of future technology is a valuable resource. Executives need this insight and vision to develop strategies for the firm that gain competitive advantage. IT managers need to supply the firm's executives with their technological *and* their business vision so that executives can anticipate and prepare for future structural changes well in advance. With the CEO's financial and strategic company goals firmly in mind, IT managers must champion their technological vision from the general manager's perspective while inspiring the executive team with a realistic, practical, and innovative view of the future.

IT management must be at its best when dealing with expectations. Senior executives' expectations include using information technology for competitive advantage and attaining financial payoffs for the firm. Given that corporations spend anywhere from one to five percent of revenue on information technology, these expectations are entirely reasonable. CEOs have every reason to require the IT branch of the organization to conduct its affairs in a businesslike manner and to conform to business practices common to their organizations. It is reasonable that CEOs require their senior IT managers to perform their jobs with executive-level skills.

Many information sources originating both inside and outside the firm influence CEOs. These information sources include trade associations, government agencies, and informal communication with peers in addition to routine communication with the firm's officers and directors. Information is a basis for expectations, and senior IT managers must respond to these expectations in a disciplined and professional manner.

IT managers must have a good understanding of the corporate culture. Corporate culture or corporate philosophy consists of basic beliefs or ideas that guide members of the organization in their behavior within the organization. As Marvin Bower puts it, these behavior patterns describe "how we do things around here."[25] IT managers must also have a clear and realistic view of technology trends and a keen appreciation for the firm's technological maturity. This knowledge is essential for providing the CEO and the executive staff with information upon which reasonable expectations can be built.

Expectations held by the firm's senior executives constitute a yardstick by which IT managers will ultimately be measured. It doesn't matter where the expectations originate or whether they are realistic. But in most organizations, completely fulfilled expectations lead to a satisfactory performance appraisal. Skillful IT managers, those with a general management view of the business, are likely to position themselves and their organizations to maximize this eventuality. Superior IT managers understand the importance of expectations, and their efforts to manage them effectively are proactive.

Less-skilled managers set and cope with expectations less effectively. They frequently find themselves and their organizations overcommitted or operating reactively. Unskilled managers and their organizations create expectations they are unable to fulfill. Through lack of discipline or excess enthusiasm, they sow the seeds of their own demise. Executives rely upon IT managers to set and manage expectations skillfully. They understand that the success of the IT team is integral to their personal success and to the overall success of their organization. In today's climate of rapid technological change, substandard IT management is rarely tolerated, and savvy IT managers strive to position themselves as indispensable executive resources for achieving expectations.

Skills alone, however, are not enough. Tools and processes, together with a management system in which these processes can operate effectively within the corporate culture, are necessary for success. These processes engage various members of the firm in activities ranging from long-term strategic considerations at one extreme to very short-term considerations at the other. In short, successful IT management can flourish only in a supportive environment. Moreover, all players engaged in this activity bear some responsibility for the success of the processes.

The study of information technology management in this text concentrates on business results, attaining efficiency and effectiveness, and achieving and maintaining competitiveness with the external environment. The intended result is improved operations for the firm. This text is structured to focus on business results and is organized as portrayed in Figure 1.1.

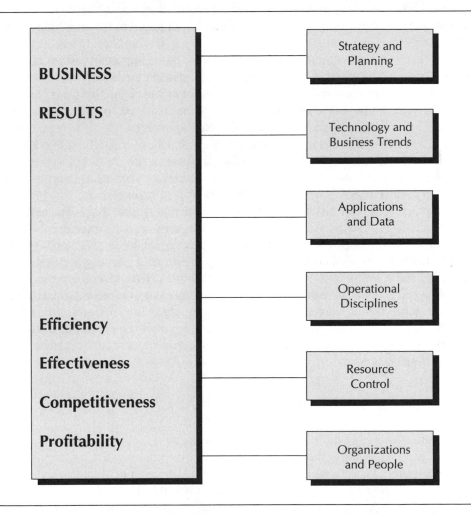

FIGURE 1.1   A Model for the Study of IT Management

Each element on the right side of Figure 1.1 is essential to the success of the firm. Each represents a tangible or intangible asset to be deployed effectively for the firm's benefit. In every area, the IT executive needs reciprocal cooperation with managers throughout the organization to maximize these assets. Each part in this text addresses one of these vital assets.

Part One describes how to develop IT strategic actions and how to plan successful, controlled implementations. Today, competent IT managers are expert in strategy development and planning so they can capitalize on the many opportunities that rapid advances in information and telecommunication systems create. Sound strategizing and effective planning are critical success factors for today's managers.

Part Two discusses advances in computing hardware, operating systems software, storage technology, telecommunication systems, and industry trends. Successful managers need information about technology and industry trends because they use it to prepare themselves, their people, and their organizations for the future. This part provides essential information on these topics.

Software applications and related databases are large, important resources for the firm. IT professionals have special responsibilities to help manage these resources to add value to the firm's business processes and mitigate the effects of technical and business obsolescence. Part Three describes how application resources must be managed to achieve improved business results. This is a critically important task for today's managers.

Part Four describes systematic, disciplined approaches for handling operational IT activities that are so critical to business operation. Success in this activity leads to high levels of customer responsiveness from their information systems. Success in operational activities is mandatory—it is a necessary condition for successful IT management.

Resource control is a basic management responsibility, one that is increasingly important with advanced IT systems. As technology penetrates organizations more deeply, managers must use refined tools, techniques, and processes to discharge their control responsibilities. Part Five defines methods for measuring and controlling investments, ensuring desired returns, and for protecting and controlling vital IT assets such as hardware, data, and programs.

Managing human resources effectively is a manager's most important success factor. Part Six describes sound people management techniques that outstanding managers use and shows how to achieve high morale and maintain high ethical standards in the organization. It also describes how the chief information officer manages the IT management system so the organization achieves the business results it desires.

SUMMARY

Information technology is a powerful force in today's global society. Computer and telecommunication technologies enable important transformations that profoundly affect people, organizations, industries, and nations. Managers in all spheres of endeavor must adjust their behavior to achieve success during these unprecedented times.

This chapter concentrated on the basic ideas surrounding the management of organizations and introduced topics and issues important for the managers of information technology. Subjects introduced in this chapter such as expectations, governance, important current issues, and critical success factors are interwoven throughout the text.

Information technology is an important ingredient in the strategies of nearly all firms today; consequently, IT managers play a vital role in achieving the firm's long-term success. *Management of Information Technology*, 3rd edition, defines and explains the general management principles necessary for success in this endeavor.

The remainder of this text focuses on attaining business success through applying management principles to information processing technology.[26]

Review Questions

1. According to Drucker's thesis, how will information technology alter the future structure and operation of the firm?

2. What are some consequences of disseminating information more broadly to employees?

3. What stages has electronic information processing gone through in the last 45 years?

4. What were the main characteristics of information systems during the 1970s?

5. What is a management information system?

6. What are some factors that cause the challenges for IT managers to remain high?

7. What five major issues did CSC Consulting Group find important?

8. What kind of preparation improves IT managers' chances of success?

9. What changes are occurring as the IT discipline matures?

10. According to Strassmann, on what important topics does Information Management Superiority depend?

11. Why is governance so important today?

12. What six stages of growth did Nolan identify? During which stage does senior management become seriously interested in technology and why?

13. Why is an understanding of stages of growth useful to management?

14. How do necessary conditions for success and sufficient conditions for success differ?

15. Why does the IT manager's list of critical success factors contain statements regarding expectations?

16. What are the causes of the dramatic structural changes predicted by Drucker and others?

Discussion Questions

1. How is Drucker's hypothesis related to changes in the way technology is now used?

2. Discuss the changing nature of authority as important information becomes more widely available in an organization.

3. Information technology has high potential value for most firms. Why does this cause the IT manager's job to be so demanding?

4. Describe the role that IT organizations and their managers play in firms today. Differentiate between line and staff activities.

5. Discuss the evolutionary stages of information technology by describing the characteristics of each. In particular, differentiate the 1980s from the 1990s.

6. Identify the main types of information systems and relate their development to the evolution of information technology.

7. Discuss the challenges of IT management. Which challenge do you think is most difficult and why?

8. What factors cause one to believe that IT management is moving toward maturity? What factors might cause one to reach a different conclusion?

9. Discuss the main ingredients of Strassmann's Information Management Superiority model and relate them to the maturation of information technology.

10. Discuss the significance of the stages-of-growth concept. Can you think of an example of this concept in operation?

11. What personal characteristics would you expect to find in individuals responsible for initiating a technological thrust?

12. Why is it possible that not all parts of a corporation will be at the same stage at the same time? What opportunities does this present management?

13. Some individuals in an organization may be extremely reluctant to embrace a new technology. What special problems does this raise? What tools are available to management to solve this problem?

14. Discuss the relationship between Strassmann's model and the list of critical IT success factors.

15. Discuss the relationship between the ideas of corporate culture and the critical success factor.

16. Discuss the role that expectations play in IT management as described in this chapter. Do you think expectations are as important to managers in other functions in the firm? If so, why?

Assignments

1. Read the article by Peter Drucker (referenced in endnote 1) and write a summary of his thesis, concentrating on its meaning for future organizations and managers. Specifically, how do you think this will affect the managers of IT organizations?

2. Using library reference material, find an IT manager's success story and prepare a synopsis for class presentation. Identify the chief factors contributing to the manager's success.

3. Which of Nolan's stages of growth best describes your school's level of information processing maturity? Prepare your answer by discussing this subject with a knowledgeable individual in your school's administration.

4. Read Chapters 1 and 2, pages 3 through 14, of Paul Strassmann's book, *The Politics of Information Management* (referenced in endnote 22), and prepare a summary of the main points for class discussion.

ENDNOTES

[1] Peter F. Drucker, "The Coming of the New Organization," *Harvard Business Review*, January-February, 1988, 45.

[2] The term "IT" is used throughout this text to mean information technology.

[3] Steve Lohr, "Information Technology Field is Rated Largest U.S. Industry," *New York Times*, November 18, 1997. From a study based on Commerce Department data titled, "Cybernation: The Importance of the High Technology Industry to the American Economy," sponsored by the American Electronics Association and the NASDAQ stock exchange.

4 Premier 100, Supplement to *Computerworld*, September 19, 1994, 25. Sutter is retiring from Rockwell and will join the University of South California to teach information technology strategy.

5 Shoshana Zuboff, *In The Age of The Smart Machine* (New York: Basic Books, Inc., 1988), 10. Activities, events, and objects are translated into and made visible by information when a technology informates (generates information about the process) as well as automates, according to Zuboff.

6 For more detail see Gordon B. Davis, "An International View of the Future of Information Systems," *The DATA BASE for Advances in Information Systems*, Fall 1996, 9.

7 The term "chief information officer" was coined in 1981 by William Synott, senior director of the Yankee Group, an industry consulting group based in Boston.

8 The total number of applications is increasing rapidly as firms purchase programs and employees develop them at their workstations. In most firms, the investments and benefits of all the computer programs are difficult to estimate.

9 Harold Koontz and Cyril O'Donnell, *Principles Of Management* (New York: McGraw-Hill, 1964), 262.

10 One of the most important IT staff responsibilities is to develop the firm's standards and policy guidelines for technology acquisition and use.

11 For more detail on this evolution, see Peter G. W. Keen, *Every Manager's Guide to Information Technology* (Cambridge MA: Harvard Business School Press, 1991), 7.

12 The term "open systems" means an electronic environment that permits hardware, software, and network components from various vendors to co-exist and evolve smoothly.

13 See Keen, note 11, 74.

14 Although spending on information technology remains at high levels, productivity growth in the U.S. has been stagnant or weak. Known as the productivity paradox, this controversial issue still haunts IT managers and other senior executives.

15 Firms laid 19.3 million miles of fiber in 1996 and are projected to lay an additional 83 million miles in the next three years according to Timothy Aeppel, "Why Too Much Stuff Means Little Now," *The Wall Street Journal*, December 12, 1997, A2.

16 Worldwide customer shipments of personal computers are expected to reach 117 million in 2000, of which 62 percent will be outside the U.S. See M. H. Martin, "When Info Worlds Collide," *Fortune*, October, 28, 1996.

17 For a fascinating discussion of intellectual capital, see *Forbes ASAP* April 7, 1997; also see Walter B. Wriston, *The Twilight of Sovereignty* (New York: Charles Scribner's Sons, 1992), 11-12, and "Management of Intellectual Capital," *Long Range Planning*, Vol. 30, June 1997.

18 For a recent example of current issues and trends compiled by University researchers, see James C. Brancheau, Brian D. Janz, and James C. Wetherbe, "Key Issues in Information Systems Management: 1994-95 SIM Delphi Results," *MIS Quarterly*, June 1996, 241.

19 Julia King, "Reengineering Focus Slips," *Computerworld*, March 13, 1995, 6. Also see "Top Concerns of Information Managers," *InformationWeek*, January 30, 1995, 70.

20 Downsizing (reducing numbers of people, middle managers in many cases, and outsourcing moving work to outside specialty firms) are current trends.

21 Current debates about whether CIOs should be technologists or business generalists miss the point: CIOs must combine the skills of both.

22 Paul Strassmann, *The Politics of Information Management* (New Canaan, CT: The Information Economics Press, 1995), 10. Policy is to management what law is to governance according to Strassmann. More about Paul Strassmann, one of the most influential IT management experts in the U.S., can be obtained at www.Strassmann.com/.

[23] Richard L. Nolan and Cyrus F. Gibson, "Managing the Four Stages of EDP Growth," *Harvard Business Review*, January-February 1974, 76. Also Richard L. Nolan, "Managing the Crisis in Data Processing," *Harvard Business Review*, March-April, 1979, 115.

[24] John F. Rockart, "Chief Executives Define Their Own Data Needs," *Harvard Business Review*, March-April, 1979, 81.

[25] Marvin Bower, *The Will to Manage* (New York: McGraw-Hill Book Company, 1966), 22, and *The Will to Lead* (Boston, Harvard Business School Press, 1997), 61-68.

[26] The concepts in this text are applicable to institutions, agencies, government entities, and not-for-profit organizations. The use of terms such as "bottom-line results" and "business results" is intended to encompass the results of all types of organizations, not just profit-making firms.

2 Information Technology's Strategic Importance

Since Bill Gates and Paul Allen founded Microsoft nearly 25 years ago, sales have grown to more than $10 billion and the market values its shares at more than $170 billion—Gates' share is worth more than $40 billion. Now the third largest firm in the U.S. measured by market capitalization, analysts predict that the giant software firm will continue to grow rapidly into the next century.[1]

Microsoft's strategy is to create software that extends beyond the workstation and unites and supports all parts of the emerging national information infrastructure. It wants its software to invade corporate and personal communication systems and occupy the critical junctions of the developing information superhighway. Microsoft intends to replace the middleman in conventional commerce with its electronic commerce software and collect fees for its use in what it calls "friction-free capitalism." It also plans to become a major player in the content revolution by aligning itself with large content providers and by building essential parts of the delivery system.

"This new electronic world of the information highway will generate a higher volume of transactions than anything has to date, and we're proposing that Windows be at the center, servicing all those transactions," says Gates.[2]

In the waning years of the 20th century, an explosion of telecommunications bandwidth is developing in the U.S. (a factor of 100 increase in five years). Microsoft concludes that new telecommunications systems built on cheap and abundant bandwidth will bring enormous changes to society and strengthen the importance of software and its standards. Microsoft hopes to identify and occupy critical segments of these emerging markets and obtain new revenue opportunities. In doing so, however, it confronts the giants of the telecommunications industry.

Microsoft's initial efforts indicate the breadth of its approach to these markets. It now offers *Sidewalk*, an entertainment and activity guide for Seattle and New York; *Expedia*, an online travel service; and *CarPoint*, an information source on car prices and other data for car buyers. It offers *Investor*, for tracking stock and mutual funds online; *Cinemania*, an online movie review service; and *Music Central*, which includes online music clips, concert coverage, reviews, and interviews.[3] These initial steps are part of a much larger plan to collect fees on transactions valued at nearly $2 trillion in the U.S. economy.

Microsoft's purchase of WebTV Networks Inc. in early 1997 is a critical first step in putting more users online using Microsoft products. WebTV allows TV watchers to navigate the World Wide Web, and send and receive e-mail messages from their TV sets. Set manufactures worry that this move gives Microsoft an edge toward putting Windows technology into future television sets and capturing revenue from a wide variety of applications made available with digital TV. In a stunning announcement in June 1997, Microsoft said it would invest $1 billion in Comcast, the nation's fourth largest cable company, to help it build future PC/TV hardware and software applications. Microsoft is also considering investing up to $1 billion in other cable companies. In addition, its alliances with major content producers and satellite broadcasters will give it a tremendous edge in this evolving business.

By embarking on this tack, Microsoft is gambling that its technological approach is superior to the telephone industry's because it already provides multimedia capability, and better than the cable TV industry's because of its technological lead and its freedom to adopt fiber rather than being tied to cables. By extending multimedia over fiber supplied by others, it enjoys flexibility and degrees of freedom through which it hopes to implement its strategy successfully.

But Microsoft's efforts to develop its online service, MSN, have been less than a rousing success. Launched in 1995, the service has been plagued with difficulties from the beginning. Many customers were upset because the system was difficult to install and the billing system seriously flawed. They found e-mail awkward and the programming style unsuited to their needs. After several strategy shifts and continued efforts to upgrade and improve MSN, Microsoft has peaked at about 2.3 million customers, whereas AOL, after absorbing Compuserve's customers, now serves more than 10 million.[4] MSN has been a humbling experience for Microsoft.

But Microsoft's most critical goal is to develop robust follow-ons to Windows NT that can coordinate and corral thousands of networked microcomputers to satisfy the performance requirements of corporations for their critical central systems. By distributing data over the entire network, Microsoft hopes to reduce the need for central data processing facilities. By increasing its server systems' power and capability, the company also hopes to reduce the power and cost of networked PCs, thereby saving time and money for corporate end users. Microsoft is hedging its bets, however, by planning new products called Windows Terminals that can operate efficiently with centralized systems and compete with low-cost network computers designed by its competitors.

Windows NT follow-ons are being designed from the outset to satisfy the needs of an advanced information highway. By broadening design parameters, advanced NT systems can accommodate both full-function and stripped-down PCs, and other kinds of high-bandwidth network systems from supercomputers to TV set-top boxes. With this strategy, Microsoft is preparing to handle text-based documents, still images, electronic commerce, videotelephony, interactive games, and video on demand. By placing its software at the intersection of these activities, Microsoft intends to be a major participant in the information economy for years to come.

But, just two years after settling with the government over charges it illegally exploited its operating system, Microsoft has again been sued by the Department of Justice. The lawsuit is aimed at the bundling of Microsoft's internet browser with Windows 95, which the government claims should be two separate products to preserve competition. The suit is critical because it aims directly at Microsoft's strategy for preserving dominance by steadily including new features in its software.[5] Claiming it has done nothing illegal, Microsoft intends to fight the suit vigorously—the struggle will likely be protracted and the outcome highly significant for Microsoft and the information industry.

Industry analysts forecast Microsoft's 1998 revenue to be $14 billion, up from $3.75 billion in 1993, with profits of $4.2 billion. By the year 2000, they expect Microsoft's sales to nearly double, approaching $20 billion; some predict growth to continue at 15 to 20 percent per year for another decade or so.

INTRODUCTION

Today, information technology influences the structure and operation of organizations more profoundly than any other technology ever has. Advances in space travel, nuclear energy, medical technology, pharmacology, and chemical fertilizers, and breakthroughs in plant and animal genetics have all been highly important to the world and its people, but none affect organizations in the fundamental way that information technology has.

Information technology shifts power from secretive governments to informed people when CNN broadcasts real-time bombings in Chechnya just as Moscow proclaims bombings are not part of its plan. Power shifts from governments to citizens when intellectual property moves across national boundaries unimpeded by customs agents. As the Post Office seems to ignore advancing technology, power (money) shifts to Federal Express, United Parcel Service, and numerous telecommunications service providers of e-mail and Fax. Executives in organizations throughout the world who disregard information technology do so at their peril and at considerable risk to their reputations, organizations, and employees.

Senior executives in most organizations today expect to use information technology to improve business processes, streamline operations, and embrace electronic commerce by linking their organizations more tightly to customers, suppliers, and business partners. Many want to decentralize operational decision making while retaining centralized control over critical functions. They know that information technology can help them do this. Although they anticipate additional automation and cost reduction in the more routine organizational processes, they clearly have larger expectations for themselves and for their organization as a whole. Their vision includes significant additional support for the activities they conduct personally, but, more importantly, they constantly search for information technology strategies that provide increased leverage to their firms and their employees.

Walter Wriston, former chairman of Citicorp, described an information strategy neatly when he said, "The essence of an information strategy is to turn the burden of burgeoning business data into a bounty of business opportunity. The business organization has to be rebuilt around the goal of managing information productively. The object of the game is to get information to the person or company that needs and can use it in a timely way."[6] His vision of rebuilding organizations, turning information overload into opportunities, and empowering companies or individuals with timely, vital information is central to today's information technology paradigm.

STRATEGIC ISSUES FOR SENIOR EXECUTIVES

For the last several decades, executives authorized expenditures for systems that automate the relatively routine transaction-processing activities of the firm. They invested in systems to help employees make better decisions (DSS) and to help managers operate the business better (MIS). They commissioned networked systems designed to speed information to employees who need it. And, executives also know that resources are being devoted to sophisticated applications such as expert systems, advanced wireline and wireless networks, and many others;

nevertheless, they openly question whether these investments have done anything more than maintain the competitive status quo. Executives in most firms today believe IT investments should produce a strong positive financial return and a competitive edge, not just keep pace with others.

Several important ideas drive executives' thinking on strategic uses of information technology. These are 1) the desire to obtain or maintain competitive advantage for the organization, 2) the need for internal and external interorganizational linking via networking technologies, 3) the objective to maintain decentralized operations with effective central coordination, 4) the requirement to develop flexible and responsive infrastructures for the firm, and 5) the requirement to capitalize on fleeting but critical business information.

As competition drives firms to grow and expand globally via alliances, mergers, and joint ventures, CEOs insist on tight coupling and coordination between operational units regardless of location. They know that network technologies can perform this desirable interorganizational linking. They want the hub or corporate central organization to monitor operations in near real time and perform coordinating activities, while permitting decentralized operational decision making.

Time is of the essence in today's competitive world, and change is a way of life. So the infrastructure, particularly the information infrastructure, must be adaptable and responsive to change. Today's term for this is "agile operations." Information and opportunities are frequently short lived—their value deteriorates rapidly in most cases. Therefore, getting important information to people who need it, when it still has value, is critical. Executives expect that information technology can help achieve these goals. In short, senior executives expect IT to contribute significantly to business results.

Senior executives believe that information technology holds great promise for improved corporate or organizational strength. They expect to employ technology innovatively and to strive for leading positions for the firms they head. Through novel technology applications, executives believe that present and future investments in information technology should result in dominant positions. Because these executives' views are so important, they must direct IT managers' thinking over the long term.

Over the years, but particularly in the recent past, CEOs in many firms observed the productive use of information technology in virtually every aspect of their operations. They have made investments in this technology across the board, significantly affecting nearly every facet of their organizations. Office automation, e-mail, client/server implementations, Internet technology, and links to sales, service, supplier, and customers are almost universal corporate features today. The pervasiveness of information technology in modern organizations demands strategic thinking about its future use.

Finally, the belief that information systems can very significantly impact a firm's strategic direction and its long-term position in the industry is not speculative but validated by many examples from current experience. Senior executives base their desires on specific, concrete precedents and aspirations. For example, they are familiar with the success enjoyed by firms in the airline industry that own important and vital reservation systems. They are aware of significant systems in the brokerage industry and of highly valuable order entry systems. These precedents fuel their craving to see their firms enjoy comparable successes.

For many reasons, senior executives, IT managers, financial officers, and others focus their attention on long-range, strategic implications of information technology and concentrate their energies on attaining or at least maintaining competitive advantage with their IT investments. Therefore, obtaining strategic information systems and using information technology competitively is a high priority issue for business managers in most organizations today.

STRATEGIC INFORMATION SYSTEMS DEFINED

Strategic information systems (SIS) are information systems whose unique function or specific application shapes the organization's competitive strategy and provides competitive advantage for the firm that owns them. SIS may operate in any area of the firm supporting administrative or technical activities. SIS may be visible to customers, such as order entry systems, or may be confidential internal applications like design automation systems. Strategic systems shape the competitive posture and strategy of the firms that own them and perhaps even alter the entire industry.

To the extent that any system gains or maintains competitive advantage for its owner, it forms part of the spectrum of strategic information systems. The attribute of competitive advantage distinguishes a strategic system from all others.

Important, valuable information systems come in all flavors. For example, telecommunications-based transaction processing systems (TPS) are the basis of airline reservation and retail brokerage and banking systems. Decision support systems (DSS), based on confidential algorithms, enable brokerage and investment banking firms to trade stocks and bonds for their own accounts profitably. These proprietary programs permit traders to capture profits from small, fleeting price discrepancies in securities.

Management information systems (MIS) developed by American Hospital Supply Company, now part of Baxter Healthcare, let hospital procurement managers optimize inventory. These systems reduce hospital costs and help tie customers to Baxter. Many manufacturing companies developed programs that enable plant managers to schedule production optimally based on incoming orders and inventory in stock. The term for this type of operation is "agile manufacturing." Customers remain loyal to suppliers who fill orders promptly and accurately, and they become more loyal if their suppliers service or upgrade products quickly, paying keen attention to customer satisfaction. Sometimes this requires sophisticated management information obtained from expert systems.

Today, most complex products cannot be developed or manufactured without massive engineering support systems. Developing and building computers, jet aircraft, and automobiles requires powerful design automation programs. Boeing's 767 aircraft and IBM's System 390 could not have been designed or built without computerized support systems. The strength of these firms today derives in part from their continuous investments in proprietary design and manufacturing systems. Many more examples could be cited to demonstrate that computer applications that provide advantages for their owners are found in all forms and in all parts of modern businesses.[7]

Forces Governing Competition

To understand possible roles that information systems can play in shaping or altering a firm's competitive posture, managers must visualize business competition in its broadest terms. To help visualize competition, Michael Porter developed the model shown in Figure 2.1, which presents a succinct and lucid view of the forces shaping competition.[8] Porter showed that industry consists of firms jockeying for preferred positions while being impacted by the bargaining power of suppliers and customers, and the threats of new entrants and substitute products or services. To prosper and grow long-term, firms must contend with forces governing the competitive business climate.

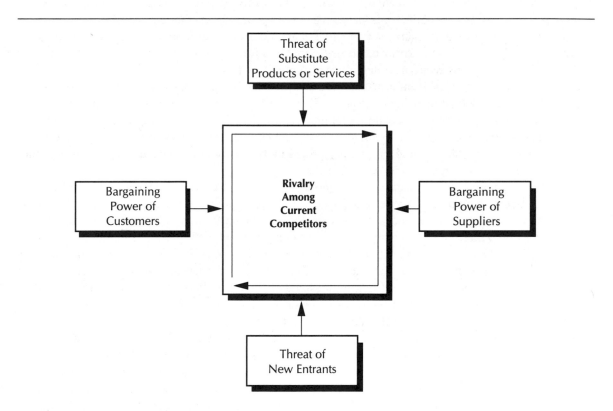

Source: Adapted from Michael E. Porter, *Competitive Advantage: Creating and Sustaining Superior Performance* (New York: Free Press, 1995), 5.

FIGURE 2.1   Forces Governing Competition

This model shows that competitors must take strategic actions to diminish customer or supplier power, lower the possibility of substitute products entering the marketplace, discourage new entrants, or gain a competitive edge within the existing industry. Figure 2.1 is very helpful in thinking about competition in its broadest terms by suggesting areas in which a firm may need to examine its competitive posture. It's a good framework for judging the firm's position and for analyzing various strategies the firm may elect to employ.

STRATEGIC THRUSTS

In 1988, Charles Wiseman presented a detailed addition to the framework of strategy development. He developed the theory of strategic thrusts and provided numerous examples of strategic information systems that embodied them.[9] The essentials of Wiseman's strategic moves are contained in five basic thrusts. I have added "time" as a sixth strategic thrust because reductions in time or increases in responsiveness are critical factors in business competition today. Table 2.1 presents the six basic thrusts.

TABLE 2.1 Strategic Thrusts

1. Differentiation
2. Cost
3. Innovation
4. Growth
5. Alliance
6. Time

Time must be considered an important strategic thrust.[10] Its inclusion is mandated by the increasing importance of telecommunications in strategic systems today and by the growing number of firms that consider themselves to be engaged in time-based competition.

The following paragraphs define the characteristics of these strategic thrusts.

1. *Differentiation.* The firm's products or services are distinguished from competitors' products or services, or a rival's differentiation is reduced. For instance, Automated Teller Machines (ATMs) distinguish the services of some financial institutions from others.

2. *Cost.* Advantage is attained by reducing costs to the firm, the firm's suppliers, or its customers, or by increasing costs to competing firms. Advanced order entry systems or electronic commerce, for example, reduce both the customer's and supplier's business costs.

3. *Innovation.* Introducing changes to the product or process causes fundamental shifts in the way the industry conducts its business. For example, some brokerage firms introduced innovative Web-based systems to provide improved stock trading services for their customers, and soon the entire industry will offer these capabilities.

4. *Growth*. Advantage is secured by expansion, forward or backward integration, or by diversification in products or services. The *Wall Street Journal* and *USA Today* are two examples of daily newspapers that use telecommunication and printing technology to seek and reach broad national markets and to expand their market reach and stimulate growth.

5. *Alliance*. Advantage is attained by reaching agreements, forming joint ventures, or making strategic acquisitions. For example, the Business Vignette in this chapter notes interesting alliances between Microsoft and some content providers. Today, even large firms use agreements and joint ventures as strategic thrusts.

6. *Time*. Competitive advantage is secured by rapid response to changing market conditions or by supplying a more timely flow of products or services. Electronic design automation tools, computer-aided manufacturing systems, and the integration of the CAD/CAM systems and production logistics systems are thrusts that increase manufacturing's response to the marketplace.[11] According to Peter Keen, "innovation via technology requires resources of capital, technology, management, and time. Firms cannot buy time off the shelf. They need a catalyst to integrate the other resources in such a way as to 'buy' them the time they need."[12]

Today, information technology is an important ingredient in the competitive strategies of many firms. It is commonly used to shape strategic thrusts to gain a competitive edge. Executives' desire for strategic information systems stems from their convictions that well-conceived, properly executed systems can provide enormous advantages to their owners.

STRATEGIC SYSTEMS IN ACTION

The next three sections describe classic strategic information systems from the transportation, financial, and distribution industries. Initiated more than 20 years ago and continuously improved and enhanced, these systems still provide benefits to their owners. Some changed considerably during their lifetimes as their owners responded to changing industry conditions or attempted to stay ahead of the competition. In many ways, these systems and the firms that own them are models for thousands of firms worldwide.

Airline Reservation Systems

In the late 1960s, American Airlines developed a rudimentary computerized reservation system called the Semi-Automated Business Research Environment, or Sabre. Sabre capitalized on the advanced third-generation computing equipment available at that time from IBM. American initially invested $350 million in Sabre but did not profit from this investment until 1983. Today, Sabre is a huge reservation system, serving more than 30,000 travel agencies. The system enables agents to provide their customers with airline, hotel, and automobile reservations, and other services. It is one of the world's most widely known and most valuable strategic systems.

The Sabre system combines advanced computerized data processing and telecommunication technology. Reservations are taken by 8000 operators accepting calls at five regional sites and from 130,000 terminals in travel agency locations in more than 50 countries. The Sabre network consists of 45 high-capacity digital lines over which 470,000 airline reservations are booked daily. The system processes more than 5000 transactions per second and has an availability of 99.95 percent.

During the 1970s, Sabre became an important marketing force in the industry and a valuable strategic asset for American Airlines. Travel agencies considering the development of a system of their own posed a competitive threat, so American decided to make Sabre available to them. Because reservation system development is costly, many airlines joined the Sabre system and purchased reservation services from American.

To maintain its dominance, American continually updated and expanded the system and introduced new products and services. Sabre provides a basis for travel agent office automation, serves corporate customers' reservation needs, and makes available a host of travel-related services. American also utilizes the Sabre system internally in its advanced yield management system. Yield management is the process of allocating airplane seats to various fare classes to maximize the profit on each flight. Sabre databases are used to predict demand for each flight and to observe the rate at which seats are booked prior to flight departure. Based on this data, seats are dynamically allocated to full fare or discount fare passengers, thus maximizing seat utilization and flight profitability. In late 1997, Sabre installed an even more powerful yield-management system that is expected to increase revenue by $200 million annually.[13]

The introduction of EAASY Sabre for individual home use and Commercial Sabre for corporate use have been successful growth areas for American. Only schedules are available to these users—customers must make reservations through agents, thus preserving agents' commissions. These new products stem the threat that new low-cost technology, such as personal computers, may provide competition and undermine Sabre's dominant position.

American continually upgraded its system to provide end users with increased flexibility. For example, a feature named "Calltrack" records all incoming calls to Sabre central by agency, caller name, and problem type. This provides American with valuable market and customer relations data. Other innovations such as the Frequent Flier program helped Sabre grow and gave American the competitive advantages of bargaining power and comparative efficiency.

Many companies chose not to compete with Sabre but to lease the service. Air France, KLM, and others are customers of American's reservation system. By 1990, the system generated 5 percent of the gross revenue of the AMR Corporation, American's parent company, and accounted for 15 percent of its profits. By 1995, Sabre earned $380 million—44 percent of AMR's pretax profits.

Hoping to build on Sabre's success, AMR sought opportunities in the computer services business such as processing services for the healthcare or retail industries. But, efforts by AMR Information Services, Inc., to build Confirm, a reservation system for Marriott Corporation, failed, and the parties litigated damages. In 1996, AMR began separating from Sabre by selling 18 percent of its shares to the public for $545 million.

Some companies elected to develop their own reservation systems. Delta Airlines invested more than $120 million in its reservation system, DatasII, and its software update DataStar but, after six years, enrolled only 11 percent of travel agents nationwide. Delta pushed its system by offering back-office capability for travel agents through small networks of PCs and by providing stored video images of cruise ships, rental cars, and other travel products on hard disk. But the company admits to playing catch-up.

Later, in 1989, Delta purchased 40 percent of a firm known as Worldspan Travel Agency Information Services. Northwest owns 33.3 percent and TWA owns 26.6 percent of the firm. It combines Delta's DatasII with PARS, previously owned by Northwest and TWA. The new firm, which combines the fourth and fifth largest reservation systems, hopes to capture 26 percent of the U.S. market.

In the early 1970s United Airlines automated its in-house reservation system and introduced it as a travel agency system, Apollo, in 1976. United spent $250 million to develop Apollo. It gives agents access to listings for all major airlines as well as for hotels, car rentals, and other travel-related services. In 1976, United transferred the system to a separate business unit, the Apollo Services Division.

The division's objectives are to add new, marketable services to the reservation system. One such service is an office automation system for travel agents that provides accounting, reporting, and other managerial features. This service has been enhanced with a new modular Enterprise Agency Management System containing advanced features and functions. United intended to improve the quality of its information processing assets through additional investments.

In 1987, the UAL Corporation decided to spin off a subsidiary centered around its computerized reservation system. This subsidiary became the COVIA Corporation, an independent affiliate of United Airlines. COVIA's mission is to enter new data processing business ventures and to capitalize on opportunities outside the travel industry by using its large worldwide network.

Many legal actions have been taken against United in connection with Apollo. (AMR and Sabre have had many of these difficulties too.) These actions range from charges of monopolizing the reservation system market to charges of bias in displaying and presenting flight information. Many travel agencies have tried to break contracts with United to join other systems. Competing firms have even offered to pay legal fees for agents who elect to break contracts with United in favor of their systems. For the most part, however, United has successfully obtained relief through the courts in these contract disputes. Competitors use every possible means to combat the advantages of United and its system.

The Apollo reservation system was valued at $1 billion in 1988 when UAL Corporation sold 49.9 percent of COVIA for $499 million. USAir, British Airways, and Swissair bought 11.3 percent each, KLM Royal Dutch bought 10 percent, and Alitalia bought 6 percent.

In Swindon, England, a joint business venture of 10 airlines is developing an international reservation system to offer a wide array of services. The joint venture, Galileo, is owned by Aer Lingus, Alitalia, Austrian Airlines, British Airways, KLM, Olympic Airways, Sabena, Swissair, TAP Air Portugal, and the U.S. company, COVIA. Galileo provides services to 35,000 agency locations worldwide. Its services include information and ticketing for 525 airlines worldwide,

48 car rental companies, 20 hotel chains, major theaters, and sporting arenas. One goal of this giant company is to give European travel agents one user-friendly terminal to handle all their customers' needs.

In July 1997, Galileo International sold 32 million shares to the public and raised $784 million for the company and its owners.[14] After the sale, the original founders of Galileo retain 68 percent of the 100 million shares outstanding. Galileo will use the sale's proceeds to invest in other reservation systems including Apollo. In 1996, Galileo International earned $165 million on sales of $1.09 billion.

Europe's largest reservation system is Amadeus Global Travel Distribution SA, jointly owned by Lufthansa, Air France, and Iberia Air Lines. It recently purchased System One, the reservation system of Continental Airlines.[15] The acquisition gives Amadeus entry into the U.S. market and will help System One expand globally. Software development for System One will be shared by Continental, Amadeus, and Electronic Data Systems, which also operates the system.

Beginning with single company systems in the 1960s, computerized reservation systems are now the basis for large, dynamic businesses. Today, public ownership of reservation companies may signal the emergence of a new industry—reservation services.

On another front, however, Southwest Airlines and others use network technology to offer ticketless travel to more than 15,000 passengers daily.[16] When making reservations via phone or over the Internet, customers receive a confirmation code that they exchange for a boarding pass at the gate. This saves the cost of producing the ticket ($15 to $30 according to industry analysts) and reduces ticket counter traffic. About four months of initial development determined the program's feasibility prior to its being phased in throughout Southwest's system. To remain competitive and reduce travel agency fees, many other national and international airlines are adopting Internet-based, ticketless-travel systems.[17]

Stock Brokerage

In 1977, Merrill Lynch introduced its Cash Management Account (CMA) at a New York press conference and began test marketing the product in Atlanta, Denver, and Columbus, OH. The CMA combines a checking account, debit card, and brokerage margin account supported by a computerized cash management system. The system provides customers with current information via phone and detailed printed reports at month end. Customers' net cash balances are invested in one or more money market funds generating interest income for the client. Subscribers' expenditures are applied first against their net cash balance and, when depleted, against their lendable equity in the margin account. This innovative product involved an alliance with BancOne of Columbus, OH, which processes the checking activity for the CMA, thus preserving the separation of banking and brokerage as the law required.

In 1978, Merrill Lynch expanded the CMA to 38 offices in five states. Two years later, the CMA was available in 39 states and serviced more than 186,000 accounts. To expand account growth, Merrill Lynch launched its first specialized version of the CMA designed for estate administrators. The CMA became available in all 50 states in 1981. By then, the number of accounts exceeded the half million mark. The Professional Golfers Association adopted the CMA automatic transfer system that year to manage funds and pay tournament winners. Merrill Lynch continued to invest in additional features to expand services for CMA clients.

The International CMA was launched in 1982, and the Working Capital Management Account, serving the needs of businesses and professional corporations, debuted in 1983. Accessing CMA reserves through cash machines became widespread in 1984. The Capital Builder Account, tailored for the needs of individual investors, became popular in 1985 and 1986. Merrill continued to invest large sums in additional features.

By 1987, its tenth anniversary, the CMA actively served 1.3 million accounts with $150 billion in assets. Merrill Lynch introduced more CMA enhancements including increased ATM access (24 hours a day at more than 22,000 locations) and the new CMA Premier Visa program. The Premier Visa program provides financial benefits and administrative features to customers and an additional $25 annual fee to Merrill Lynch per client.

Ten years after its debut, the CMA was an unqualified success, from which Merrill Lynch derived a major competitive advantage. The minimum balance required to open a CMA is $20,000, but the average balance approaches $100,000 for its 1.3 million clients. The minimum annual fee in 1988 was $65 per client. Commissions on securities transactions, interest charges on debit balances, and service fees on more than $28 billion in Merrill Lynch-managed money market funds substantially augment fee income.

Not until 1984 and later did the competition begin to employ similar information technology. Merrill Lynch carefully protected its position by securing a U.S. patent on the computer program and defending its turf in court. In 1983 Merrill Lynch won a $1 million settlement from Dean Witter, its closest rival at that time, which infringed on Merrill's rights under its patent.

The CMA patent application was filed on July 29, 1980 and granted on August 24, 1982. The patent lists Thomas E. Musmanno as the inventor and is assigned to Merrill Lynch. Displayed on only 11 pages, it contains four drawings (flow charts) and six claims. It represents an uncommon but important form of protection for a valuable strategic information asset.

Additional enhancements continue to be announced for the CMA, implying continued investments by Merrill Lynch. By 1987, industry analysts estimated Merrill Lynch's technology budget to be nearly $1.5 billion, with several hundred million dedicated to software development. Although Merrill's share slowly declines, it holds approximately 50 percent of the market. Today, with the CMA and other services, Merrill Lynch manages about $257 billion in client assets.

Merrill Lynch continues to forge ahead with its CMA program, adding features and functions to help people manage their finances. In addition to brokerage, checking, Visa cards, and money market accounts, Merrill's CMA now features direct deposit capability, bill paying services, automatic investing under dividend reinvestment plans, and statement coordination among various Merrill accounts. Through a series of subaccounts, the CMA statements provide savings and investment information on IRAs, college tuition accounts, retirement accounts, and others. The CMA is growing into a comprehensive, full-service vehicle for managing a family's financial affairs. Special features are available for trust and business accounts, too.

Today, about 1.5 million accounts are active with an average worth of more than $200,000 each. Annual fees are now $100 per year. Still, for high-net-worth individuals, Merrill is proceeding with much more extensive services discussed in the Business Vignette in Chapter 11. For more than two decades, the CMA has exemplified a brilliant combination of information technology, financial services, and strategic development employed for competitive advantage.

But is Merrill Lynch's position secure? Can competitors possibly provide more attractive offerings using information technology?

Competitive offerings from Charles Schwab, Donaldson Lufkin & Jenrette, Prudential, Smith Barney Shearson, Dean Witter, and Fidelity are supplying answers to these questions. Charles Schwab, with its computerized trading system e.schwab, promises significant competition for Merrill Lynch. e.schwab appeals to Merrill's CMA clients by offering a wide range of financial services to investors with PCs. From a PC connected to Schwab's system through the phone network, investors can establish and manage accounts, obtain account status, access market reports or research information, obtain price quotes, and place orders to buy or sell securities. Table 2.2 summarizes some of the system's many features.

TABLE 2.2   e.schwab Features

1. Real-time securities quotations

2. Many types of research reports

3. Dividend reinvestment service

4. Order entry at reduced commissions

5. Online account management tools

6. Check writing ability ($300 minimum)

7. 24-hour/365-day account and trading access

8. 128-bit encryption and other security features

9. Trading through Schwab's Web site (www.Schwab.com)

Schwab's clients need a $5000 opening balance and pay charges of $29.95 for stock trades of up to 1000 shares and $.03 per share thereafter. Unlike some other services, Schwab charges no annual maintenance fee. In 1997, Schwab had nearly one-half of the 1.5 million online brokerage accounts in the U.S. Merrill Lynch expects to begin offering online trading sometime in 1998.[18]

Through widely available network, database, and personal computer technology, e.schwab combines a series of strategic thrusts. It offers a differentiated product, uses innovative information technology, and provides convenient and timely services. Surely these are strong ingredients for success. But sustained marketing capability and growing customer acceptance ultimately determine success.

Distribution Services

Federal Express began its overnight package delivery business in 1973, processing 100 percent of the packages manually. Now, Federal Express leads the U.S. overnight delivery market, yet charges more for its services. Despite significant competition, Federal Express maintains its market share due in large part to Cosmos, a database system that tracks all letters and packages that the company handles. Through Cosmos, the company can tell customers where their package is in 30 minutes or less, thereby easing customers' fears of late delivery or lost parcels.

Federal Express owes much of its growth and competitive advantage to innovation in the systems area.[19] For example, most of its 29,000 U.S. couriers use console-mounted CRTs in their vans to stay online to local dispatching centers through the Courier Dispatching system. It helps drivers track packages and keeps them informed of schedules and locations while allowing the couriers to pick up new orders quickly.

Federal Express uses bar codes to track packages throughout the delivery system. Within two minutes, scanned data is uploaded via telecommunications links into a central database where it is available to answer company or customer questions. Misdirected parcels are virtually nonexistent now and will become even rarer as more technology improvements are implemented. Federal Express sorts packages and prepares bills in a paperless office where terminals display all necessary information. A scan-recognition system sorts packages as they move at 500 feet per minute on conveyer belts. Federal Express processes almost two million packages nightly—advanced technology and solid implementation are necessary to maintain its high service standards.

Federal Express continues to expand its systems and automation, spending about $1 billion annually on technology and research. Bar coding, scanning, and cellular links keep vital information current. When accepting your shipment, the FedEx courier generates and scans the bar code for your package with a hand-held scanner. In the van, the courier inserts the scanner into a Digitally Assisted Dispatch System terminal, which beams the information to Cosmos via satellite. On average, parts of the shipping system including pick-up, sorting, plane loading and delivery, scan packages six times before delivery.

Using the Internet and the World Wide Web, customers can now link to FedEx's systems through FedEx Ship software at their offices. This feature enables customers to order and track shipments and to obtain billing information online. Customers also learn about the company by visiting other FedEx Web site areas, and they communicate with FedEx through thousands of e-mail messages. In addition to FedEx Ship tracking software, the company introduced new systems automation for small business shippers (POWERSHIP), better systems supporting service agents as they deal with customers, and artificial intelligence applications to keep planes and vans on schedule during traffic jams or bad weather. FedEx created a Business Logistics Services (BLS) division to engineer innovative logistics and information management solutions for firms that want to outsource their distribution facilities. Laura Ashley, a London-based retail chain, uses FedEx's BLS to supply its 540 stores in 28 countries, all within 24 to 48 hours of a shipment request.

By continuously developing its innovative information systems, FedEx's business has grown dramatically in the U.S. and abroad. It now serves 95 percent of the U.S. population and 212 foreign countries through 160 airports worldwide with its 560 aircraft. The company established an intra-Asian hub at Subic Bay in the Philippines and plans another in Taiwan to handle growing volumes in Pacific Rim countries. Revenues, cash flow, and net income are projected to grow at rates of 10 percent, 9.5 percent, and 14 percent annually until 2002.[20]

The three examples discussed above relate directly to Porter's model of competition and to the theory of competitive thrusts. These systems and many others rely heavily on telecommunications to expand and grow global businesses and gain efficiency and effectiveness. Some very successful and important systems are composed mostly of telecommunications elements.

Information systems theory and the examples of strategic systems examined in this chapter validate the theory and confirm the practice of strategic systems. Many valuable systems, smaller and less obvious than those discussed, are installed and operating successfully in firms worldwide, although some are proprietary and not discussed publicly. Thousands of firms own and operate information systems that provide them advantages in their competitive environments. Thousands more invest in systems designed to capture similar advantages for their firms.

WHERE ARE THE OPPORTUNITIES?

These examples illustrate some of the numerous opportunities for leveraging information systems investments. Opportunities frequently consist of product or service offerings that have been differentiated from their competitors by applying information technology. All the systems studied exemplify this principle. By using technology innovatively, firms lowered business costs and reduced time barriers. These systems' owners experienced high growth rates and, in some cases, forged important business alliances.

Strategic systems focus not only on customers, but on supplier targets and competitors, too. Sophisticated systems that optimize buying strategies affect organizations that supply raw materials to others. And competitors feel the influence of systems used to design, develop, or manufacture superior products. Firms that use information technology to support or mold strategic thrusts in a competitive environment achieve high potential in many contemporary industries. But important system applications frequently offer other advantages as well.

Some technology applications inspire confidence and promote customer loyalty because they provide superiority through high quality products and reliable services. Information technology applications can improve the cost effectiveness of both the producer and the consumer. Customer loyalty persists even though some significantly enriched services command higher prices. Customers are willing to pay for higher quality. Firms can exploit these leverage points for growth of revenue and profit for the product or service provider.

The Importance of Technology

The introduction of technological advances into business and industry worldwide is a principal driver of competition. Advanced technology shapes the products and services of the future and offers opportunity for innovative organizations to increase their value to the stream of economic activity. The importance of technological advances on international competition becomes evident when considering advances in communication or transportation, witnessing the revolution in chemicals or pharmaceuticals, or observing the importance of information processing.

Information technology is particularly important because it pervades the processes leading to advances in most other endeavors. Because all activities create, transport, disseminate, or use information, advances in information technology have a compound effect on technological advances everywhere. Thus, information technology exerts a remarkable, accelerating effect on global competition.[21]

Advanced technology is important because it alters industry structure, shapes and molds competitive forces within and between companies and industries, and changes forever the behavior patterns of billions of individuals. Technology for its own sake is not important. What is important is its dramatic impact on society as a whole.

The Time Dimension

Time is a valuable, irreplaceable asset and an important source of competitive advantage. Firms must think about time resources as they do about capital, facilities, materials, technology, and management resources. The firm that views time as an important asset will strive to capitalize on it in all business processes.

Driving the notion of time as a competitive factor is the value of responsiveness to customers, markets, and changing market conditions. Responsiveness cuts across all the firm's functions, from product-requirements definition to installation and service. Timeliness involves suppliers and customers and impacts competitors. Agile competitors who capitalize on the time dimension find many valuable opportunities for competitive advantage.

Time factors are highly susceptible to manipulation with information technology, particularly through telecommunication systems. Although examples abound, one high-tech manufacturer's use of just-in-time (JIT) manufacturing illustrates the value of time.

In Irvine, California, the McDonnell Douglas Corporation used the JIT approach to computer-coordinate the flow of parts and raw materials. The computer system required 111 new programs and 97 modifications to installed programs, for a total of 1900 person-hours to program additions and changes. Planning, team meetings, and training required additional effort, and the firm incurred some expenses for tags, labels, and other items. A thorough understanding of how parts, products, and information flowed through the company formed the basis for the JIT approach. Table 2.3 summarizes the impressive benefits McDonnell Douglas reported from using the JIT approach.

TABLE 2.3   JIT Benefits at McDonnell Douglas

Inventory reduction	38%
Work-in-process inventory reduction	40%
Printed circuit board assembly cycle time	80%
Rework reduction	40%
Improvement in inventory turns	100%
Quality-control process yield (now 99% perfect)	80%
Set-up time reductions	50%

Information technology, used in conjunction with parts-logistics control, attacked set-up time. Design process improvements and supplier logistics yielded substantial savings of time and money for this computer products manufacturer. Other computer manufacturers agree. "We've ignored a critical success factor: speed," says John Young, CEO, Hewlett-Packard Corporation. "Our competitors abroad have turned new technologies into new products and processes more rapidly. And they've reaped the commercial rewards of the time-to-market race."[22] And, in speaking about IBM, Louis Gerstner, CEO, says that "one of the unusual things about this industry is that a disproportionate amount of the economic value occurs in the early stages of a product's life. That's when the margins are most significant. So there is real value to speed, to being first—perhaps more than in any other industry."[23]

The Strategic Value of Networks

Telecommunication systems enhance information flow between organizational entities, bridging the gap in space and time. Given information's pervasive role in business, telecommunications offers great potential for competitive advantage through reductions in time and mitigation of distance barriers. Telecommunication products and technology enhance business processes and improve business efficiency. Multinational firms use telecommunication extensively as a necessary condition for conducting business—searching for innovative new applications is a continuous activity for them.

Figure 2.2 illustrates the strategic value of networks by showing how firms use them to improve efficiency and effectiveness, and generate growth by reducing the negative effects of time and distance and by capitalizing on innovation.[24]

	EFFICIENCY	EFFECTIVENESS	GROWTH
TIME	accelerate business activities	improve information flow	obtain early market presence
DISTANCE	reduce geographic barriers	enable integrated control	enter new markets
INNOVATION	enhance current processes	enable new processes	create new products

FIGURE 2.2 Networking's Strategic Value

Reducing time and distance barriers to business processes and improving these processes in other ways improves organizational efficiency. For example, banks update branch records centrally in real time, permit customers to access their accounts electronically from great distances, and enhance customer service by having account details available for use at any branch within the system. Telecommunication improves the efficiency of banking processes and enables banks to serve customers better, too.

Improving information flow enhances effectiveness. It reduces the amount of information in transit (in the mail, for example) and reveals opportunities to improve business excellence. Information flowing rapidly to where it is needed and useful also improves effectiveness by reducing errors and lowering costs. Reducing time and distance factors makes control easier to maintain. New processes can also be introduced, for example, letting customers access their accounts through ATM machines.

In addition to increasing efficiency and effectiveness, telecommunications technology provides leverage for growth and expansion of the firm's business. By reducing time to market, firms can capture an early market presence and obtain advantage over competitors. By extending their reach quickly through Internet applications, for example, firms can tap new markets and new customers, eliminate intermediate processes, and accelerate business growth. Today, the rapid growth of electronic commerce testifies to the importance of these factors.

Using innovative applications to create new products enhances growth in other important ways, too. These products, exemplified by Merrill Lynch's CMA, Charles Schwab's e-schwab, and Federal Express' POWERSHIP, can significantly expand business. Telecommunication applications can also greatly improve a firm's strategic position. In conjunction with other technology applications, they permit restructured forms of business enterprise that support new strategies and respond effectively to competitive threats. IT is tremendously important in today's highly competitive global economy.

In addition to the strategic systems discussed earlier, e-mail, ATM machines, credit card verification, the Internet, the World Wide Web, CNN, WebTV, and PrimeStar provide obvious examples of the impact of telecommunications technology. Additional examples will be discussed throughout this text.

THE STRATEGIST LOOKS INWARD

Many important systems originated from the ideas of individuals or small groups who envisioned a way to capitalize on emerging technology or to streamline some aspect of business. These individual or group insights became catalysts that spawned business opportunity.

Most strategic systems were conceived by analyzing the firm's internal functions. Efforts to automate (informate) the internal activities of the company more fully paid off later in competitive advantage. The Sabre system began this way. Thus, the firm's portfolio of application systems is a logical point to start searching for strategic opportunities.

Many firms own application portfolios holding several thousand programs. Each program supports one or more business processes. Some applications may be candidates for strategic development. The potential thrusts of these applications range from cost reduction to innovative methods for attaining product or process superiority. These internal systems operate throughout the firm from marketing, development, and manufacturing, to sales, service, and administration. Superior market analysis tools, for example, coupled with automated design systems can significantly reduce the time and cost required to respond to changing customer needs and market conditions. Automated manufacturing processes and sophisticated distribution systems speed new products to customers efficiently and at reduced cost.

Information systems to help handle these basic tasks exist in most firms today. How can these systems be augmented or enhanced to improve the firm's posture in the marketplace? What new technology can be employed to improve these processes? What innovative actions will permit the firm to utilize internal resources to maximize its competitive position? These and other questions directed toward current applications form a basis for searching internally for strategic opportunities.

EXTERNAL STRATEGIC THRUSTS

Another useful way to identify potential strategic opportunities considers external factors. These factors include changing industry environment, competitors' recent actions, changing relations among suppliers, and potential business combinations, among others. This view of the firm asks the questions: "What is happening externally that may influence our firm's opportunities to gain competitive advantage?" and "How can we capitalize on external factors by using information technology?"

Answering these questions is different and more complicated than answering introspective questions about the firm. Individuals best suited to this task serve in the company's top positions. They may direct the marketing function responsible for competitive analysis or industry analysis, or they may head product distribution with responsibility for ensuring timely and accurate dissemination of the firm's goods or services. The senior people responsible for guiding the firm's long-term direction have valuable insights. Information technologists who have a vision of what is possible and feasible must influence a senior executive's vision. When these two visions intersect, potential opportunities emerge.

INTEGRATING THE STRATEGIC VISION

A model presented by Wiseman and MacMillan permits the interrelationships of strategic variables to be depicted graphically.[25] Figure 2.3 illustrates these relationships, as well as the internal and external sources and uses of strategic systems. Time, a highly important strategic thrust, complements the five other thrusts in this model.

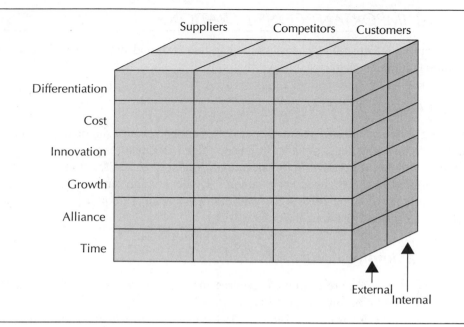

FIGURE 2.3   Integrated Model of Strategic Influences

Most strategic systems cover several areas in Figure 2.3. A system may utilize several of the strategic thrusts itemized on the left side of the diagram, or it may impact more than one of the groups illustrated across the top. For example, cost reductions and time savings obtained internally from the application of innovative technology may lead to growth in both revenue and profits and affect both customers and competitors. Computer-aided design/computer-aided manufacturing systems have this potential. In the computer industry itself, innovative use of advanced technology in sophisticated electronic design automation systems enables firms to spawn new, high quality products in ever shorter development life cycles. Such innovation significantly affects competitors and customers alike.

Systems change character over time as the firm exploits their potential. Some systems start life as an internal thrust and gain external importance to the firm. The Sabre system is one example. Throughout its life cycle, a strategic system may migrate through the model. It may play an important role in several arenas as the firm develops and exploits its potential via redirection and further investment. Because change is a fact of competitive life, managing successful strategic systems involves anticipating, initiating, or reacting to change to obtain or sustain competitive advantage for the firm. Managing strategic systems is a major challenge for senior executives in nearly all organizations today.

The strategic information systems studied in this text, and many others not reviewed, have additional common characteristics worth noting.

Organization and Environment

Strategic systems may alter the market environment and change the field in which competitors engage. Because strategic systems influence the environment as they become effective, they will probably lead to changes in organizational structure. When the firm modifies its structure to accommodate the changed environment, it will probably enhance the system, too. This metamorphosis of competitive status and the firm's response continues unless some form of stability is established among the competitors.

Financial Implications

Strategic systems require continued resource investments to sustain their advantage. Owners of these systems find their competitors always look for ways to build systems or take other actions to negate the owner's advantage. As the examples revealed, staying ahead of the competition is a challenging and never-ending task. For example, nimble, innovative competitors threaten relatively secure positions in the brokerage industry.

Some strategic systems produce their own revenue and become profit centers within the parent corporation. In the most successful examples—COVIA Corporation, Sabre, and Galileo International, the profit center evolved into a major international business entity. Forming the foundation for a publicly owned, international business marks the apogee of success for an information system.

Legal Considerations

Owners of some successful strategic systems eventually engage in legal struggles with their competition. Legal issues involve competitors' appropriate use of the owner's business advantage. Some legal issues relate to protecting the competitive advantage. When faced with severe competition, whether due to superior strategies, better management, or improved technology use, some firms resort to legal actions hoping to nullify, at least partially, or delay their competitor's advantage. In some cases, powerful technology has been used in ways deemed unfair. Over the years, for example, owners of airline reservation systems have been in court almost continuously, litigating issues arising from advantages derived from their systems.

Some Cautions

Several cautions about strategic information systems are in order.[26] Strategic systems usually develop from deliberate attempts to improve or enhance current information systems. They do not begin life as systems separate and distinct from those in the applications portfolio. Most strategic systems do not emerge from specific attempts to meet corporate strategic objectives. Instead, they result from many incremental enhancements and sustained improvements to existing systems. Usually, they do not evolve from radically changed operational systems or from totally new systems.

Successful executives focus on improving corporate performance through constant attention to the many details of their businesses. They search for improvement through new technology, enhancements to systems and operational procedures, and modifications to the organization and its culture. The task of getting ahead and staying ahead of competition is difficult. Organizations must adapt to changes in the business environment and respond to competitive forces. Successful executives know they cannot attain lasting competitive advantage from a few grand strokes.

Although strategic concerns have occupied senior IT executives' attention for many years, they are not fully appreciated if considered in isolation. Important questions frequently accompany or occur in conjunction with strategic issues. "Do the firm and the IT organization have an effective planning process?", "Does the firm have a well-defined information architecture?", and "Do the data resources support the firm's functions and goals in an effective manner?"[27] Missions, goals, and organizational alignment among and between the firm's functional units all bear on strategic concerns. In particular, the vitality of the IT organization and its managers, and its effectiveness in relating to other functions are very important considerations.[28]

The task of improving any one of these areas generally involves improving all of them. The mutuality of these topics demands that managers make progress across a broad front. It is misleading for organizations to believe that competition can be overtaken or beaten by developing and implementing one new strategic system. A significant leap forward using information technology is very unlikely if the organization's planning, alignment, or other critical areas are weak.

SUMMARY

Information technology is absolutely vital to the success of most modern-day firms in the industrialized world. Senior executives have many valid reasons to expect that information technology can provide advantage for their firms. They believe that innovative use of information technology promises improved corporate or organizational leadership. They are prepared to invest in the technology, and they expect their firms to attain a substantial return on this investment. Many realize this investment is not optional but required for the organization's future prosperity. They also realize that recognizing strategic systems in action is far easier than identifying opportunities and capitalizing on them.

Review Questions

1. Why are strategic concerns regarding information systems increasingly important today?

2. What distinguishes strategic information systems from other kinds of systems?

3. Explain Porter's model of forces governing competition.

4. What are the six basic strategic thrusts?

5. What thrusts does American Airlines employ in its reservation system?

6. As a competitive weapon, what distinguishes United's Apollo system from American's Sabre system?

7. In what ways did Merrill Lynch's CMA alter the environment in which brokerage firms operated in 1977?

8. What protection is afforded Merrill Lynch as a result of its patent on the CMA program?

9. Using the information provided in the text and the model in Figure 2.2, trace the Apollo system's evolution.

10. Describe the Cosmos system in terms of Porter's model.

11. What leverage can information technology provide business organizations?

12. What are some common characteristics of strategic systems?

13. What are the advantages and disadvantages of searching internally for sources of strategic systems?

14. What insights can the IT organization bring to the process of searching for strategic systems?

Discussion Questions

1. What is the main thrust of the strategy that Microsoft has been pursuing?

2. What appear to be the strengths of this strategy? Do you think Microsoft's strategy is proactive or reactive, and why?

3. Moore's law states that microprocessor price-performance doubles every 18 months. What does this mean for Microsoft's strategy? What are the implications of Moore's law for IT and other functional managers in most firms today?

4. Research reveals that office automation has been declining in importance as an issue for IT executives. What might this mean in terms of strategic systems?

5. Discuss Porter's model as it applies to the telecommunications industry today. How does the computer industry differ from the telecommunications industry as viewed from the model?

6. The six strategic thrusts presented are not mutually exclusive. Discuss the implications of this fact.

7. Deregulation has encouraged new entrants into the airline industry. Given the enormous advantage of the competitor who owns a reservation system, how can new entrants overcome barriers to entry?

8. Discuss the role that IT managers play within the firm as it seeks to improve its competitive posture. What contributions can they make and in what areas must they take the lead?

9. What additional actions might be taken by Merrill Lynch to capitalize further on its current position?

10. Discuss the importance of telecommunications to Federal Express' strategy.

11. How might firms seeking international competitive advantage rely on information technology? What thrusts might they employ?

Assignments

1. Browse www.Amazon.com and prepare a short report that describes the strategic thrusts this firm uses. List advantages and disadvantages of using Web technology for this firm. Discuss this firm's strategy in light of Porter's model of competition.

2. Using the Wiseman book (referenced in endnote 7) or another reference, select an example of a strategic system not discussed in this chapter, and prepare an analysis of its characteristics. Be prepared to summarize your findings for the class.

3. Study the *Harvard Business Review* article by George Stalk, Jr., listed in the Bibliography, and prepare a report on why he thinks time is an important source of competitive advantage.

4. Access FedEx or Sabre on the World Wide Web at www.Fedex.com or www.Sabre.com and search technical reports or financial data for the latest information on these two important companies. Prepare a short report on your findings for class discussion. Access www.UPS.com and compare UPS with FedEx.

[1] Young, large, and rapidly growing firms such as Intel, with the fifth largest market capitalization in the U.S. (according to the S&P Outlook, July 23, 1997), and Microsoft, with the third largest (larger than the combined market cap of the three largest U.S. automakers), confirm information technology's strategic importance in today's global economy. Their strategies are of great interest to business managers.

[2] Brent Schlender, "What Bill Gates Really Wants," *Fortune*, January 16, 1995, 343.

[3] David Bank, "Microsoft Moves to Rule On-Line Sales," *The Wall Street Journal*, June 5, 1997, B1.

[4] Don Clark, "Feet of Clay," *The Wall Street Journal*, November 5, 1997, A1.

[5] "U.S. Sues Microsoft Over PC Browser," *The Wall Street Journal*, October 21, 1997, A3.

[6] Walter B. Wriston, *The Twilight of Sovereignty* (New York: Charles Scribner's Sons, 1991), 123.

[7] For many other examples see Charles Wiseman, *Strategic Information Systems* (Homewood, IL: Irwin, 1988).

[8] For a complete discussion see Michael E. Porter, *Competitive Advantage* (New York: Free Press, 1985).

[9] See note 7.

[10] "Speed to Market," *Forbes*, May 28, 1990, 350. This special report contains six examples of firms that use or provide time advantages. See also Roy Merrills, "How Northern Telecom Competes on Time," *Harvard Business Review*, July-August, 1989, 108.

[11] Thomas Hoffman, "Visual Tools Key to Chrysler Cost-cutting," *Computerworld*, August 18, 1997, 4. An advanced CAD system costing $2.1 billion cuts eight months and $80 million from the next generation of cars at Chrysler according to reports. This system advances tools built in the late 1980s that shaved previous development times by about one-third. Also see John Killkirk, "Firms Learn That Quick Development Means Big Profits," *USA Today*, November 22, 1989, 10B.

[12] Peter Keen, "Vision and Revision," *CIO*, January-February, 1989, 9.

[13] For a discussion of what yield management means to business and leisure travelers, see Scott McCartney, "Ticket Shock," *The Wall Street Journal*, November 3, 1997, A1. Also see, Dennis J. H. Kraft, Tae H. Oum, and Michael W. Tretheway, "Airline Seat Management," *Logistics and Transportation Review*, June, 1986, 115.

[14] Susan Carey, "Initial Offering of Galileo Shares Raises $784 Million, *The Wall Street Journal*, July 28, 1997, C15.

[15] "Amadeus Acquires Reservations System From Continental Air," *The Wall Street Journal*, April 28, 1995, B5.

[16] Ron Levine, "Southwest Soars With Ticketless Travel," *Lan Times*, April 24, 1995, 31.

[17] See www.Travelcity.com. You can access EAASY Sabre at www.Sabre.com.

[18] According to Forester Research, online investors will manage over $688 billion in more than 14 million accounts by 2002. *Fidelity Focus*, Winter 1997, 7.

[19] Federal Express more than tripled its revenue from 1980 to 1984, more than doubled it from 1984 to 1987, and more than doubled it again from 1987 to 1990. In 1998, analysts expect Federal Express to double its revenue again and earn $405 million profit.

[20] *Value Line*, March 21, 1997, 259.

[21] Michael Porter, "Technology And Competitive Advantage," *Journal Of Business Strategy*, Winter, 1985, 60.

[22] "How Managers Can Succeed Through Speed," *Fortune*, February 13, 1989, 54.

[23] 1993 *IBM Annual Report*, 3.

[24] For another view of this topic, see Michael Hammer and Glenn Mangurian, "The Changing Value of Communications Technology," *Sloan Management Review*, Winter 1987, 65.

[25] Charles Wiseman and Ian C. MacMillan, "Creating Competitive Weapons From Information Systems," *Journal of Business Strategy*, Fall, 1984, 42.

[26] James C. Emery, "Misconceptions About Strategic Information Systems," *MIS Quarterly*, June 1990, vii.

[27] An information architecture describes the interconnections or configuration of information technology components such as networks, hardware, software, and databases.

[28] Sound IT managerial skills are a critical resource for sustaining competitive advantage because, unlike other resources, they are difficult to duplicate. Francisco J. Mata, William L. Fuerst, and Jay B. Barney, "Information Technology and Sustained Competitive Advantage: A Resource-Based Analysis," *MIS Quarterly*, No. 4, 1995.

3

Developing the Organization's IT Strategy

AT&T Adopts a New Strategy

"In the 10-plus years since divestiture, we've converted from a predominantly analog to an all-digital network primed to capitalize on the emerging market for advanced services," stated Chairman Robert E. Allen in 1996.[1] We've taken a business that was exclusively domestic and made major strides in becoming a truly global corporation."

In 1995, AT&T obtained 11 percent of its revenue—$8.75 billion—from operations outside the U.S. and 15 percent—$11.9 billion—from U. S. operations serving international customers. Today, AT&T has calling agreements with 200 telecommunications companies and can provide travelers access to AT&T's network and billing services in more than 100 countries. Its global network carries more than 200 million voice, data, video, and facsimile messages each working day.

AT&T was growing domestically too. Through an exchange of stock valued at $8 billion, AT&T merged with NCR in September 1991. NCR, the world's leading maker of automated teller machines, also actively develops and manufactures powerful digital computers and other semiconductor-based products. AT&T, long a maker of silicon chips and sophisticated digital switches, hoped to achieve synergistic effects between switches and computers. And, to get closer to end customers and stake a major claim to the rapidly growing cellular market, AT&T purchased McCaw Cellular, the largest cellular provider in the U.S., for $11.5 billion in 1994. With these acquisitions, AT&T intended to strengthen its manufacturing arm and broaden its customer base with the latest cellular technology.

Viewing the transformed AT&T, Alex Mandl, then president and chief operating officer, declared, "AT&T *is* the information superhighway."[2]

During the previous 14 years, however, competition had taken its toll on AT&T. In 1984 AT&T held 91 percent of the long-distance market; however, by 1995 AT&T's share had declined by one-third to about 60 percent. Its competitors' share, meanwhile, grew about 11 percent annually, reaching 40 percent of the market in the mid-1990s. WorldCom, an aggressive long-distance company that started operations in 1987, achieved revenues of $3.6 billion by 1995.

Facing increasing competition and declining market share, AT&T decided to take drastic action. On September 20, 1995, nearly 14 years after the court-mandated breakup in 1982, AT&T announced plans to divide itself into three independent, publicly held companies. Over the next 15 months, the single company became three—the new AT&T, Lucent Technologies, and NCR.

Lucent Technologies, second largest of the new trio, manufactures and distributes telecommunications hardware and software systems, high-performance integrated circuits, and optoelectronic components, and designs, manufactures, installs, and maintains business communications systems in more than 90 countries. It is one of the world's premier equipment suppliers. AT&T Global Information Solutions, AT&T's semiconductor and computer arm, changed its name back to NCR before becoming an independent company again. It hopes to lead in high-end computer systems and in transaction processing hardware and software for retailers and financial institutions.

After divestiture, however, the new AT&T remains the world's second largest telecommunications company. (NTT obtains more revenue but its international presence is minimal.) AT&T Wireless Services, the former McCaw Cellular Communications, provides wireless services to more than 100 major cities and is expanding rapidly in the U.S. and abroad. AT&T delivers long-distance calling, voice-messaging, and data servicing to more than 90 million home and business customers in the U.S. Its international services offer long distance to and from the U.S., connecting more than 260 countries for travelers and international businesses. Nearly 150 supercomputer switches, each directing nearly a half-million transactions every hour, flawlessly and reliably guide national and international traffic from source to destination. Its billing systems account for the network's service down to seconds and pennies. AT&T continues to expand and improve its immense network, which includes more than 100,000 miles of underseas fiber-optic cable linking the continents of the globe.

This self-directed trivestiture is a massive strategic move by AT&T to focus more clearly on major industry segments and to become a more agile and aggressive competitor in the new deregulated era. Commenting on the reasons for restructuring, Robert Allen said, "Our decision to go this route reflects our determination to shape and lead the dramatic changes that have already begun in the worldwide market for communications and information services—a market that promises to double in size before we ring in the new century. It was, as well, a determination to act while our position is strong."[3]

The new AT&T's future is critically tied to long distance and to large anticipated revenue increases from local service, wireless services, and several new services like AT&T World Net. AT&T's entry into the Internet service arena is the first and most visible part of its strategy to lock in its long-distance customers and gain entry into online services. Its strategy to form alliances with local access providers and consumer-oriented businesses such as Lotus, Xerox, and Novell is a bold move to capture portions of the local market. To achieve its desired growth, the new AT&T must expand its long-distance, data, and wireless services aggressively and grow online, video, and local services substantially.

In an interview with *Fortune* magazine, AT&T projected revenue growth from $51 billion in 1995 to about $130 billion in the next 10 years. Businesses other than long distance will contribute more than 50 percent of the total, according to the forecast.[4] These projections imply growth rates of about 10 percent across the board, much greater than the firm achieved in the past. With the equipment business now in Lucent Technologies and without NCR, AT&T's performance depends critically on its growth strategy for local, long-distance, and wireless services.

AT&T's ambitious goals will be difficult to reach because building local service is costly and time consuming, and rival long-distance suppliers have their own growth strategies and plans. These innovative firms, accustomed to rapid growth, spell trouble for AT&T in the years ahead.

Following several changes in the executive suite, AT&T's directors selected C. Michael Armstrong, former CEO of Hughes Electronics, as CEO, and Vice Chairman John D. Zeglis as president. Robert Allen, CEO for the past nine years, will resign after a 40-year career at AT&T. The new executive team has its work cut out.

The process of searching for areas in which to develop strategic systems is an important endeavor for the firm. It requires considerable thought and research. However, strategic thinking that goes far beyond strategic systems and covers all the firm's critical areas is even more important. Senior executives of any organization who hope to exploit information technology or other strategic opportunities must apply a systematic, strategic vision to set organizational direction. A strategy is defined as a collection of statements that express or propose a means through which an organization can fulfill its primary purpose or mission. Therefore strategy must focus and coordinate the firm's activity from the top down to accomplish its mission. Thus, a well-developed strategy ensures consistent direction within the firm's units and reduces uncertainty in decision making.[5]

Paul Gaddis states his emphatic position on strategy, "You need to realize clearly that our basic strategic management mode of the past 40 years still prevails. This mode posits that strategy precedes structure precedes systems."[6] This is relevant for IT managers: systems (information systems and management systems) support organizations and organizations support strategic directions, contrary to what some technologists believe.

Successful IT managers in most firms today believe that strategy development is a critical part of their routine business system. For them, strategy development promotes actions leading to sound planning, encourages organizational learning, and helps ensure goal congruence. As Chapter 1 emphasized, developing IT strategies congruent with the firm's strategy is a critical success factor for IT managers. Figure 1.1 shows strategy's important contribution to the organization's business results.

To align IT objectives with corporate objectives, the firm's managers must believe that the IT strategy is a strategy for the entire organization. Goal alignment is not an issue in organizations whose senior management team shares this belief. Intellectual alignment is achieved when written IT and business objectives are internally and externally consistent and cross reference each other. IT and business executives who clearly understand each other's objectives achieve social alignment.[7] Thus, it follows that all senior managers share some responsibility for a sound IT strategy.

Developing a sound strategy begins with a thoughtful understanding of the firm's mission, an analysis of the environment in which the firm and the organization operate, extensive interaction with senior executives and other managers, and a detailed definition of how the firm's business units interact. In addition to understanding current organizational structure, strategists must postulate future structure that may be in place or should be developed. For example, deciding whether to centralize or decentralize may be a part of thoughtful strategizing. Developing organizational dependencies through outsourcing is another important organizational consideration that involves strategic thinking. This chapter develops the processes and techniques essential to successful information technology strategy development and lays the groundwork for strategic planning covered in the next chapter.

When developing strategies, thoughtful managers sense potential business opportunities and recognize possible threats or pitfalls. To the extent possible, successful managers attempt to redirect their efforts to maximize their advantages by capitalizing on opportunities while minimizing risks. The resulting proposed course of action may not capitalize on all available opportunities nor avoid every difficulty; but if optimally conceived, it presents a balanced approach to these conflicting influences. Because situations rarely remain static for long, thoughtful managers must reevaluate their situations periodically. They must reassess their course of action and possibly make adjustments in light of changing conditions, too.

A complete expression of strategy must consider the elements discussed above, among others. Table 3.1 lists ingredients of a strategy statement. The strategy establishes objectives and processes for its maintenance and use.

TABLE 3.1   Elements of a Strategy

1. Mission statement

2. Environmental assessment

3. Statement of objectives

4. Expression of strategy

5. Maintenance processes

6. Performance assessment

The first and most important step in formulating a strategy is to state the organization's mission. The mission of the organization defines its purpose—what it's supposed to accomplish. The IT organization's mission obviously must include elements designed to support the firm's mission. Developing a mission statement for the IT organization may be challenging, particularly on the first attempt. If doubts arise about its validity, obtaining approval for the IT mission statement from the firm's senior executives is wise.

To support the firm's mission, the strategy must describe the IT organization's business and identify its customer base. The customers' needs and the organization's capabilities and resources must be considered. The IT mission should be stated in terms of meeting customers' needs in the context of services the IT organization can provide to the firm overall. Defining the IS markets within the firm itself is essential.[8] In other words, the IT mission statement should be both reactive to, and proactive toward, customers. For example, the mission must include routine transaction processing for the firm but may also include new information technology, such as intranets, for important, new internal applications.

The process of visualizing and understanding opportunities and threats is called "analysis of the environment." Environmental analysis (or environmental scanning) attempts to account for important trends impacting or likely to impact the firm and its functional organizations.

These trends may be political, economic, legal, technological, or organizational. To position their organizations optimally in the future, strategists must understand current environmental trends.

Environmental scanning generates information that enables strategists to accomplish two tasks: 1) to develop the objectives they intend to achieve—i.e., the desirable states toward which action is directed, and, 2) to formulate the course of action, or strategy expression, that guides the achievement of objectives. Strategies rely on assumptions where facts are unavailable, and they account for the nature and degree of risks involved in attaining objectives. All strategies must be somewhat flexible, including options or alternatives. Finally, the strategy statement must contain information necessary for further planning and decision making to occur.

Strategy maintenance consists of reviewing the environment and reassessing the course of action in light of changing events. The environment's volatility determines the amount of strategy maintenance required. For example, businesses in relatively stable environments such as forest-product production may devise long-lived strategies requiring infrequent maintenance activity; however, managers in most firms today find that frequent strategy maintenance is the norm.

A comprehensive strategy, including goals and objectives, provides considerable useful information for the strategist's use in assessing the organization's efficiency and effectiveness. Later, a strategy statement can also help assess organizational performance and evaluate the degree to which the organization achieved its goals and objectives. Thus, a strategy, and subsequent plans built from it, can also be used to measure organizational performance.

Strategy development is the first step in building long- and short-range plans. Figure 3.1 portrays the relationship of these activities to each other and to strategic and operating plans. The next chapter discusses planning activity.

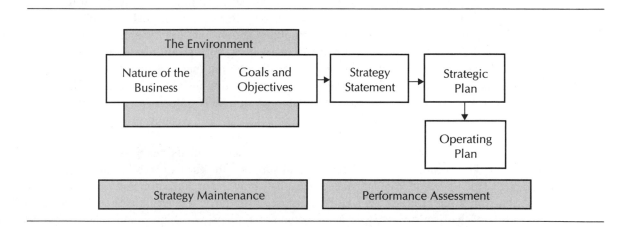

FIGURE 3.1   The Relationship of Strategy and Planning Elements

Essential elements of strategy development and subsequent planning activities are environmental assessment and nature of the business, goals and objectives to be achieved, strategy statement development, strategy maintenance activity, planning activity, and performance assessment. The development of long- and short-range plans flows from the strategy activity. Performance assessment relies on the strategy itself and plans that flow from it. Figure 3.1 suggests the relationship of these items to each other and to maintenance and assessment activities.

Strategies are useful for organizations or groups of people and for individuals. Strategies developed for groups must include formats and procedures designed to enhance communication among the group's members. Good communication not only increases the group's ability to develop strategies but also helps the group use them. To resolve the issues of information architecture, IS's role and contribution, and the use of information systems to integrate across functional lines, strategic thrusts needed to achieve strategy congruence within the firm must be clearly communicated. Succeeding sections of this chapter deal with these aspects of strategy formulation.

RELATIONSHIP OF STRATEGIES TO PLANS

Strategies and plans sometimes seem to run together, but they clearly differ. First, a strategy is a collection of statements that expresses or proposes a means through which an organization can fulfill its primary purpose or mission. On the other hand, a plan is a detailed description of how an organization can accomplish its primary purpose or mission. A strategy forms the basis for a plan but is not a plan in and of itself. The strategy spells out optimum actions required to achieve general objectives but does not provide sufficient detail to carry out the actions. A plan is created from a strategy by adding the many details needed to make it complete enough for the organization to implement. Therefore, planning substantially differs from strategy development. One states optimal actions necessary to attain goals and objectives in the face of uncertainty—the other transforms optimal actions into implementation tasks.

Campbell and Alexander explain it this way: "Strategy is not about plans but about insights. Strategy development is the process of discovering and understanding insights and should not be confused with planning, which is about turning insights into action." To make a further point, they claim that "because executives develop most of their insights while actually doing the real work of running a business, it is important not to separate strategy development from implementation."[9]

The strategy for an organization combines strategies of all its business areas and functional units. The combined detail describes and supports the firm's overall strategic business objectives. This aggregation of statements describes how the organization and all its parts will fulfill the firm's mission and achieve its purposes. Similarly, the firm's strategic plan is an aggregation of the firm's business and functional units' plans, which give detailed descriptions for accomplishing their missions and the firm's mission.

The strategic plan describes detailed actions for the next year or two and less detailed actions beyond two years. The detailed actions for the next year or two are usually called the tactical or near-term plan.

Therefore, business and functional strategies guide and coordinate activities of various business and functional units toward agreed-upon objectives that align with and support the firm's objectives in achieving its mission. Strategy statements, strategic plans, tactical plans, and other short-range or operational plans express this guidance and coordination of actions. Processes for accomplishing these strategizing and planning activities are critically important because synchronized, well-coordinated activities within the firm are essential for achieving the desired business results. As Figure 1.1 shows, the ability to develop strategic directions and plan effective implementations is a vital first step toward successful management.

STRATEGIC MANAGEMENT

The processes of strategy development and strategic planning are essential to strategic management. Over the past several decades, business attitudes about strategy have changed: instead of trying to predict the future, businesses try to create the future. This change in attitude resulted in several phases. Phase 1 consists of basic financial planning, which evolves to forecast-based planning in Phase 2. Phase 3 focuses more on the external environment. It includes externally oriented planning characterized by increased response to competitive pressures and evaluation of strategic alternatives. Phase 4, strategic management, builds and expands on these ideas.[10] Strategic management reorganizes and redeploys resources to attain competitive advantage. It also implements creative, flexible long- and short-range planning systems supporting resource deployment actions. Because information technology has considerable potential for initiating, shaping, and supporting future change, IT managers play a vital role in strategy development for organizations.

> According to George Sawyer, the plan for the management of opportunity deals somewhat with the present but more with the future because in the time required to fulfill any of its plans the business will have moved into that future. The first overview specification is for an attempt to project the evolution of this future, but management is not limited to this projection. Drucker speaks of 'creating the future,' meaning that management has a very real power to design the future and to make it come true.[11]

The following sections discuss several types of strategies and processes needed for their development and maintenance. These sections emphasize IT strategies supporting the parent organization. Chapter 4 will develop in much greater detail the fundamentals of planning for information technology.

In most organizations, functional units such as manufacturing, marketing, or information technology have two types of strategies—functional strategies and stand-alone strategies. Functional strategies describe the unit's broad goals and objectives, whereas stand-alone strategies describe an individual, one-time goal or opportunity. For example, an IT organization's functional strategy describes the actions IT must take to support the entire firm in the long term. It also has a stand-alone strategy to upgrade the firm's central telecommunication network. The IT strategy combines these two. The complete strategy for any function within the firm aggregates its stand-alone strategies with its functional strategy.

The IT functional strategy must contain the basic ingredients shown in Table 3.2. Although these elements apply specifically to the IT function, most also apply to any function in the firm.

TABLE 3.2   Elements of an IT Functional Strategy

Support to business objectives

Technical support

Organizational considerations

Budget and financial matters

Personnel considerations

The firm's strategy consists of overall long-range means or methods to achieve the firm's objectives and its mission combined with more detailed strategies for each business and functional area that supports the firm's strategy. Thus, functional strategies are developed for sales, marketing, manufacturing, engineering, and other functional units. The information technology functional strategy aggregated with strategies from all other areas forms the firm's strategy. Because information technology pervades most firms today, many other functional units' strategies may also include IT activities.

Although the process just described is typical of many firms, some have chosen other approaches, while others ignore strategy development altogether. In all firms, regardless of their inclination toward strategy development, the IT organization must account for, and must work within, the corporate culture. If the firm's senior managers are committed to strategy development, middle-and lower-level managers see the process as valid. They want to invest their time in these activities. If the firm's process is forecast-based, then the IT organization's attempt to shape the firm's future will be difficult, perhaps futile.

The firm's business strategy aggregates product manager strategies and many functional strategies that support the business. Figure 3.2 shows this aggregation. A large firm will have more strategies than Figure 3.2 depicts.

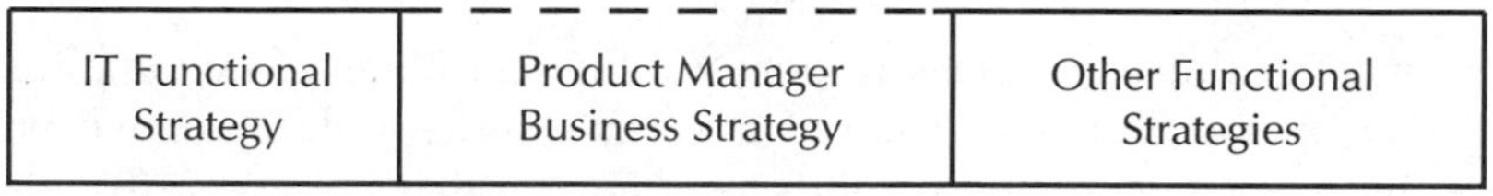

IT Functional Strategy	Product Manager Business Strategy	Other Functional Strategies

FIGURE 3.2 **The Firm's Business Strategy**

Although all units in the firm, including manufacturing, marketing, and information technology, develop functional strategies, product managers responsible for producing revenue develop business strategies that are also part of the firm's strategy. The aggregation of all functional and business strategies comprises the firm's complete strategy. This detailed document outlines the firm's strategic goals and objectives and shows how all the firm's units contribute to attaining its objectives and mission. For the firm to succeed, its overall business strategy must be based on internal goal congruence.

The purpose of functional strategies is to describe how functional units support the firm's business goals and objectives. Functional strategies begin by developing goals and objectives that are congruent with the firm's. IT strategies are most effective when IT goals and the firm's goals are internally consistent. Aligning IT goals with the firm's goals, or achieving goal congruence, continues to be an important issue for many firms and a critical success factor for all IT managers. The management processes described in this chapter aim to achieve this vital alignment.

In a large and complex business, strategy is also complex. Because it contains sensitive, competitive information, it must be held confidential and carefully guarded. Strategies are produced for and used by people in the top echelons of the business who are responsible for directing the firm in the long term. Some firms have a special department responsible for the strategy development process. Individuals in this department also assist senior executives in utilizing the strategy.

Stand-alone Strategies

Occasionally, a functional unit may want or be required to develop a specific strategy to deal with a unique opportunity or threat. Generally, this happens when an embryonic question of key potential significance to the function arises—one not previously considered in detail. In many cases, these specific or stand-alone strategies are required responses to competitive or industry developments. In this sense, stand-alone strategies are *ad hoc* actions to deal with currently emerging opportunities or threats. Once accepted, this strategy is incorporated in the strategic plans of all organizational units that it affects.

A vendor announcement of an important new technology product, for example, creates a situation that might demand a stand-alone strategy for the IT function. If this new product enables achievement of an organizational objective in the short term, the IT function would develop a strategy to capitalize on the new opportunity. If accepted by the firm, planning for implementation begins. Generally, stand-alone strategies are temporary because, as the activity matures, it becomes part of the routine strategies and plans.

Business and Functional Strategies

Business and functional strategies respond to different kinds of opportunities or objectives, but, together, they are the backbone of the firm's strategy development process. Business strategies have revenue and profit objectives for the firm. AT&T's divestiture strategy for maximizing its service revenue is an example of a business strategy. Functional strategies have goals that support the firm's business strategies. If the IT organization does not produce revenue, its strategies are limited to the functional type, i.e., it supports the firm's business objectives. However, if the IT function produces revenue for the firm, and some do, its strategy is also part of the firm's business strategy.

For example, a business strategy for an IT organization that hopes to produce revenue could exploit an application in its portfolio. By offering the program to customers through new telecommunications technology, for instance, the IT organization can generate revenue and profit for the firm. A business strategy based on this proposition would describe how the organization hopes to accomplish these objectives.

In contrast, a functional strategy might address the introduction of new robotic systems supporting the manufacturing plant's cost and quality objectives. This activity would appear in the technical support section of the functional strategy shown in Table 3.2.

Business strategies for the firm direct its functions toward business objectives. Functional strategies, on the other hand, coordinate activities within or between functional units. For example, a strategy within the IT organization to implement an advanced design automation system may be a response to the firm's business strategy for reducing product development time. Usually, the IT functional strategy contains many such supportive activities.

Figure 3.3 shows the relationship among business strategies, functional strategies, and stand-alone strategies for the firm. This illustration emphasizes the IT organization and includes some revenue-producing activities that the business portion of the strategy statement contains.

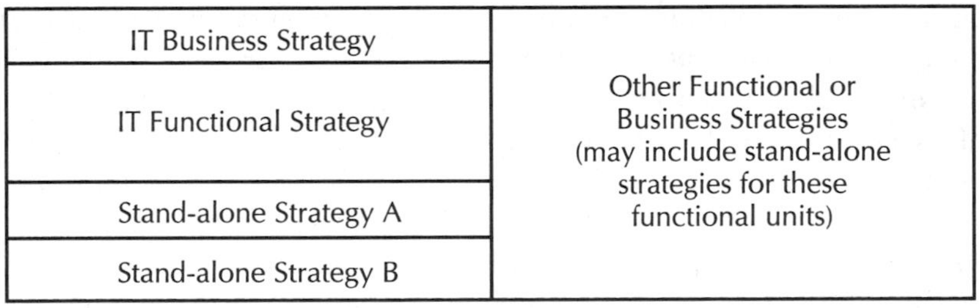

IT Business Strategy	Other Functional or Business Strategies (may include stand-alone strategies for these functional units)
IT Functional Strategy	
Stand-alone Strategy A	
Stand-alone Strategy B	

FIGURE 3.3 Assemblage of Strategies in a Firm

Well-managed IT organizations have an active functional strategy. Because practical limitations on resource allocation must influence strategies, developing and maintaining strategies and their associated plans inevitably becomes an iterative cycle bounded by resources of all kinds.

REQUIREMENTS OF A STRATEGY STATEMENT

A strategy statement is primarily a vehicle for focusing managers' attention on strategic aspects of the firm's business. It is also a means of communicating to those who must review and approve the strategy and to those who use it to guide their actions. Additionally, the document must be available to those responsible for initiating adjustments that account for current input from the environment or business. Those who measure and evaluate the firm's performance or functions also use the document.

By itself, a statement of goals and objectives is insufficient to meet these needs. Information must be added regarding the environment, the basis for selecting the goals and objectives, the assumptions on which they depend, the perceived risks, and available and reasonable options. The firm's long-range plans also need this kind of information because both strategies and plans are snapshots of management's vision that direct the organization's actions.

STRATEGY DOCUMENT OUTLINE

To present the required information coherently, an outline containing the main points of a strategy is usually developed. Table 3.3 shows a sample outline. The subheadings in the strategy do not need to be addressed sequentially, but presentation of the strategy must leave no doubt about whether a statement is an assumption, a course of action, a risk, or an alternative. The strategy must be a logical, coherent entity, clearly indicating the relevance of its parts.

TABLE 3.3   Strategy Outline

1. Nature of the business

2. Environment

3. Goals and objectives

4. Strategy ingredients

The section on the nature of the business must answer questions such as "What business are we concerned with?" and "What are the boundaries of this business?" In many instances answers to these questions are obvious, and this section requires little development. The firm or organization, however, must clearly envision its mission before strategy development proceeds. Failure to understand the nature of the firm's business can be fatal. Thus, questions posed in this section are serious and important.

In this section, describing the IT business carefully is important. What purpose does the IT organization serve in the larger organization, and what type of organization is it? The roles of IT organizations vary considerably. For instance, at Boeing, IT services the main Boeing business, but it also operates a service bureau for outside customers and generates revenue for the firm. In an insurance company, the IT group provides major services for the company and acts as a conduit through which all the firm's transactions flow. A construction company may use data processing for design or project planning purposes, spending only a small fraction of its revenue on IT activities. The IT organization must understand what business it's in and its role within the firm.

The environment section of the strategy document states what is known and assumed about relevant, significant factors and trends surrounding the firm. IT managers must understand those factors that influence the firm's current or future behavior. As a test of relevance, the environment section should include those factors that can potentially influence the attainment of current goals and objectives. Another question to measure relevance is: "Can these factors force a change in our goals?" This question is especially important when later readings on the environment show changes in those factors.

Key environmental assumptions should be reviewed at the conclusion of the environmental section because their credibility and consistency need to be unambiguous. Furthermore, tracking these assumptions enables managers to adjust and update them as part of maintaining the strategy.

What is the IT environment like? What are its current capabilities, and what might they be in the future? IT should know the state of the application portfolio and what new technologies it can apply to the firm's business. Is the IT organization relatively mature and disciplined; or is it underdeveloped, staffed by relatively inexperienced people, and operating under an immature management system? These are key questions to address in the environment section of the strategy outline.

The goals and objectives section must clearly state the ultimate objectives of the strategy. What IT capabilities are we trying to achieve, and what long-term objectives are we going to set for the organization? For instance, is our goal to enhance the application portfolio with several new strategic systems, or are we trying to link our business units with new, modern intranet capability? Perhaps organizational issues such as decentralization or outsourcing also need to be addressed as goals and objectives.

W. R. King developed a methodology for deriving the IT goals and objectives from a firm's strategy.[12] He begins the process by identifying the corporate strategy set and transforming the objectives into IS objectives. For instance, if the corporate strategy calls for reductions in internal investments, then the IT organization might concentrate on inventory reductions through improved inventory management systems. Alternative means for reducing internal investments are developed and then presented to management for decision-making purposes. King's methodology compliments the approach presented in this text.

The IT organization should restrict its search for opportunities to those within the organization's mission. Opportunities must serve the corporate purpose and align with corporate objectives. An important test of a strategy is whether its objectives are attainable and desirable, given the larger organization's goals and objectives and accounting for environmental considerations. A second test is the degree to which the objectives can be used to measure and evaluate unit performance.

Goal setting requires some additional important considerations. Goals should be established within the firm's resources and have a reasonable chance of being attained. Managers must balance easily attainable goals and challenging goals with respect to resources of all kinds. Goals should be explicitly stated and quantified whenever possible. Goal clarity facilitates subsequent planning, measurement, and control activities. A small rather than a large number of goals is preferable. By reducing the chances of ambiguity and conflicts, setting fewer goals also permits managers to direct resources more effectively. Finally, well-planned goals should be time-limited.

Ingredients of a strategy include a course of action and its accompanying and supporting factors. Table 3.4 presents a summary of these items.

TABLE 3.4   Strategy Ingredients

Course of action

Assumptions

Risks

Options

Dependencies

Resource requirements

Financial projections

Alternatives

The strategy statement must describe the organization's course of action in attempting to achieve the strategy's objectives. What steps will the organization take to achieve its goals and objectives? The steps should 1) lead to realizing the objectives, 2) be consistent with the firm's other long-range interests, and 3) be preferred over other possible alternatives.

For example, if a goal of the IT organization is to improve its employees capabilities, will it accomplish this through hiring, training, retraining, termination, or possibly some combination of these actions? The course of action outlines the steps to attain these goals.

On what major assumptions is the strategy based? Assumptions that are inherent in, or exercise significant influence over, the strategy are technical capabilities, functional support activities, and potential competitive reactions. The test of the assumptions is their credibility. To continue our example, one assumption might be that the current employees are trainable; that is, they possess the necessary background knowledge to undertake further training.

Risk is always present. In fact, risk should be a major part of the IT functional strategy. Questions such as "What is the nature of the risks in the strategy?" and "What is their potential impact?" should be answered. In our example, a strategy for hiring skilled employees might run the risk that very few are available to the firm. When some aspects of risk become significant, managers must determine whether optional courses of action offer reasonable insurance against risk.

Because no single course of action has 100 percent probability of success, improving the probability by building options or alternatives into the strategy is usually appropriate. These options should account for specific risks, assumptions, or dependencies that unduly depress the probability of success. Managers must seek options available within this strategy. They must ascertain how long the options are valid and upon what considerations a selection should be made. Also, they should consider whether any options add cost or expense and if so, how much. In the personnel example above, identified options include hiring, training, retraining, and termination.

In most firms, any single strategy will likely depend heavily on other related strategies. For instance, one technical strategy's success may depend on a capability or processes generated by another technical strategy. Thus, "What are the key dependencies of this strategy?", "What is their nature?", and "In what ways are they significant to the strategy?" are some questions that managers must answer.

The strategy must identify resources required to carry out the actions, and it must present financial projections of the revenue, cost, expense, and capital required to implement the strategy. What resources are required to carry out this strategy? Are any unique resources required? Will hardware, software, or staff resources be available in the quantities and on the schedule required? These questions need answers in this part of the strategy statement.

Documentation of alternatives rejected in selecting the strategy and reasons for rejecting them should be retained for future reference.

The steps outlined above represent sound preparation and lead to sound implementation if the subsequent planning process succeeds. Action steps have been identified, and details necessary for implementation and control can now be developed.

STRATEGY DEVELOPMENT PROCESS

The Strategic Time Horizon

Strategy development within a firm usually follows a schedule dictated by the firm's corporate planning director who is responsible for synchronizing strategy development events. This activity is usually an annual affair interspersed with planning, control, and measurement events. Figure 3.4 shows a representative schedule of events for the strategy development process.

FIGURE 3.4   A Schedule for Strategy Development

Measured from the beginning of the current year, the strategic years extend from the beginning of Year 3 to the end of the strategic horizon, as Part (A) of Figure 3.4 shows. In this figure, the strategic period is three years long, but in general, it may extend three to 10 years or more, depending on the business or industry. In this example, strategic activity occurs during the early months of Year 1 and considers the period of five strategic years.

The pattern repeats annually as Part (B) of the figure shows. Each succeeding strategy development activity removes from consideration the first year of the preceding period and adds one year to the future. The period of Year 1 and Year 2 is covered in the firm's tactical or operating plan. The next chapter discusses this activity.

Figure 3.4 suggests the schedule is most convenient for firms using the calendar year for planning and reporting purposes. Organizations such as universities can adapt this process to the academic year, starting in September and ending in August. Other organizations, such as governmental agencies, may adjust this schedule for a fiscal year starting in July and ending in June.

Steps in Strategy Development

Sufficient time must be allotted to develop comprehensive strategy statements. The strategist must thoroughly understand the environment and the area of opportunity or concern. Written statements should be concise and sharply focused. The statement of opportunities or threats should highlight their relevance to the firm's future. If managers believe that remote servers should be consolidated in the secure data center, for example, that statement should clearly indicate why this belief is reasonable. What substantiating evidence or trend information supports this belief? What evidence supports the need to take this action?

The IT strategist should develop a broad, comprehensive understanding of the future environment's influences on the area to be studied. This environmental analysis might include an estimate of future computing costs, anticipated future computing loads, technical advances expected in telecommunications or processing capability, special future business conditions, and other relevant factors. Objectives are set and modified, if necessary, during the iterative strategy development process. Stand-alone strategies may need to identify several possible objectives. Selecting one occurs after testing the options for credibility and attainability. Alternative strategies should span the future environment. They must be clearly expressed so that readers of the strategy can thoroughly understand them.

To exercise a reasonable choice among alternative strategies, selection criteria and the process for using them must be established. The criteria should measure basic, long-range effects on the firm or the department. These effects may be measured in terms of revenue and profit, investment resources required, degree of risk, technological capability exploited, competitive reactions, and other factors important to each individual case.

The best strategy should be selected by using the criteria to measure the expected effect of each option. The selected strategy should then be developed in greater detail. Stand-alone strategies and functional strategies and their supporting data, reasoning, and other forms of evidence must be submitted to the appropriate executives for review and approval. The review and approval process enables and ensures congruent planning. The approval process forces a review of the alignment of IT and corporate goals and keeps executives of all functions informed of important strategic directions. The review and approval process helps integrate strategic actions consistently across functions. It fosters executive or organizational learning, too.

Upon senior executives' approval, IT and other functional strategies will be incorporated into the firm's strategy. When combining strategies, minor changes to some may be required, as may some iteration.

To succeed, the process of strategy development requires considerable effort and thought on the part of the firm's executives. This thoughtful activity focuses the best minds in the organization on the firm's long-term health and welfare. All functions need to be involved, and the IT organization needs to be a full partner in all deliberations. The value of the resulting output is directly proportionate to the level of effort and cooperation among senior executives. Strategy development activities are vital to organizations that intend to remove IT strategic planning from their list of concerns.

THE IT STRATEGY STATEMENT

The IT organization that follows the management process for strategy development will meet certain conditions necessary for success. The process establishes goal congruence, lays the foundation for planning, and carefully reveals the IT organization's role and contribution. Strategy development is an important step in organizational learning, too.

The IT organization must develop and maintain a functional strategy that guides its actions. These actions must support the goals and objectives of the firm and its constituent organizations. The IT function may also have stand-alone strategies that direct projects designed to enhance the objectives of the function and the firm. The senior IT manager, or CIO, is responsible for developing and maintaining these strategies and for ensuring that their synchronization with the firm's overall strategy process.

In most firms, strategy development and planning is one of the top issues confronting IT managers; it is a critical success factor for them. Success in this activity paves the way for success in many other areas because sound and valid planning based on superior strategies is the basis for many other management activities. Failure in this area is a prelude to failure elsewhere. Given the subject's importance to the IT management team, what responsibilities should be considered, and how should they be explored? Table 3.5 outlines some important topics for IT strategies. Minimally, the IT strategy statement should address these topics thoroughly. Other topics specific to the firm should be addressed as well.

TABLE 3.5   IT Strategy Topics

Business aspects

Technical issues

Organizational concerns

Financial matters

Personnel considerations

Business Aspects

The IT organization must maintain a keen awareness of the firm's business goals and objectives and must develop strategies to support them. These business goals may include increased market share, improved customer service, lower production costs, or many other objectives of central importance to the corporation. CEOs expect IT to contribute to the firm's results in conformance with normal business practices. The strategy of the IT organization must reflect and support the firm's business goals.

Goals of not-for-profit organizations, government agencies, and educational institutions, among others, may differ from those noted above. But in all cases and in all forms of enterprise, IT must support the parent organization's goals and objectives.

IT managers, interacting with other senior executives, are key players in developing IT's functional strategy. To the extent that the IT organization is involved in attaining the firm's objectives, IT managers must ensure that the function's actions are in concert with the firm's long-term goals and objectives. An interactive review involving IT managers, their peers, and their superiors must evaluate this involvement. Are strategies congruent? Are the firm's dependencies on the IT organization satisfied? If the strategies are executed as written, will the objectives be met? Answers to these and other questions establish the strategy's validity and help ensure its results.

Technical Issues

IT managers provide leadership in using technology to attain advantage for the firm. The IT strategy is one place for this initiative to occur. The strategy should reveal the practical utilization of advanced technology in support of the firm's goals and objectives. This utilization must be consistent with reasonable risks and available or attainable resources. Additionally, the IT manager must ensure the organization's technical vitality by developing and implementing current or advanced hardware, software, or telecommunication technology.

The IT functional strategy is the vehicle for paving the way toward improved technical health. It gives the organization the formal opportunity to establish objectives designed to maintain and enhance the technical health of the function and the firm. CEOs expect leadership from IT in discovering important technological applications. They expect IT to efficiently use the firm's resources and to work within the organization's norms.

Organizational Concerns

For several reasons organizational considerations are very important to IT strategy. First, introducing information technology tends to have organizational consequences beyond the IT organization itself. These issues are difficult to grasp and, in many cases, hard to resolve. Everyone does not look upon all changes favorably. While many changes are important to senior executives, nearly everyone else resists them for a variety of reasons. The firm needs training and education in the subtleties associated with technology introduction. IT managers need to lead in satisfying these training and educational needs.

In addition, the IT organization's role and its contribution to the firm must not be taken for granted in developing IT strategy. Not everyone appreciates the IT role. Many managers may not fully understand the IT function's contribution, or potential contribution, in supporting new organizational forms and new ways of doing business. Wise IT managers take specific actions to champion their important role. The IT functional strategy reflects these actions. Subsequent chapters discuss this in much more detail.

Financial Matters

When translated into strategic and tactical plans, the IT strategy must satisfy the firm's financial ground rules. In many cases, financial constraints limit the range of opportunities for the IT organization as they do for most other functions. These resource constraints force iteration in the process of developing the firm's business strategy. These constraints also require successive revisions in the IT organization's functional strategy. During the strategy development process, wise IT managers guide their organizations in these matters to conserve energy and optimize results.

Personnel Considerations

No functional strategy is complete without action plans for the task of recruiting, training, and retaining a team of skilled employees. These personnel considerations intimately relate to technical issues, because strong technical people develop solid technical strategies while advanced technical strategies attract strong people. The IT functional strategy must develop these thrusts in coordination with one another.

In many ways, information technology is the productivity engine for the firm. IT managers must improve productivity in their organizations. They must invest capital to increase their own productivity. IT must acquire tools to assist the entire organization in improving productivity. The production sector of the business must reduce its operating costs through superior management aided by information technology. Talented people, skilled managers, and advanced technical resources are some necessary conditions for enabling major productivity improvements and achieving overall firm objectives.

STRATEGY AS A GUIDE TO ACTION

IT strategies are developed with the explicit intent of guiding management to an optimum course of action and appraising management's accomplishments in light of this guidance and direction. In planning and performing their activities, various levels of management take direction from strategies. Implementing a stand-alone strategy to develop and carry out new projects, as well as reshape old ones, may require significant effort. But, referring to strategy gives managers guidance that helps balance and direct the overall functional effort while taking advantage of breaks and unusual opportunities as they arise.

IT managers must exert strategic control—the strategy gives them the basis for doing this. Key dependencies and functional support activities identified in the strategies must be tracked continuously. Analysis of significant departures from expectations can determine whether these variances result from performance weaknesses or present reasons to review and revise the strategy. In particular, unfavorable changes in dependencies should be scrutinized carefully. This strategic control activity preserves the vitality and viability of the strategy between planning periods. It also helps the organization maintain its strategic focus.

Periodically, progress toward achieving the strategy's objectives can be reviewed and appraised. The achievability of the objectives and the degree of stretch they require must be kept in mind. Appraising a moderate degree of accomplishment toward a very difficult objective must differ from appraising a high level of accomplishment toward an easily attained objective.

THE STRATEGY MAINTENANCE PROCESS

Maintenance of a strategy is an essential step in its continued usefulness. Maintenance recognizes that a strategy is constantly subject to change. Frequently, influences and factors beyond the firm's control cause change. A complete strategy statement requires careful documentation of areas where change may occur. This documentation allows actual developments to be tracked against assumptions, dependencies, and risks inherent in the strategy. This is possible only if the strategy has been documented in a way that clearly identifies which factors need to be tracked.

Although tracking identifies deviations that can help maintain and implement the strategy, maintenance of the strategy is neither its implementation nor its use. It's important to realize that strategy maintenance is a vital and essential part of the total strategy process. Tracking each key environmental and strategic assumption, risk, and dependency must be part of the strategy process. Someone in the organization should be specifically assigned the responsibility for tracking each area. In order for the strategy to succeed, a diligent follow-up pattern must be established.

Periodically, or when tracking reveals significant deviation, the entire strategy should be carefully reexamined and updated. The logical process for doing this is to follow the steps for developing a strategy described earlier.

SUMMARY

This chapter describes strategy's critical importance to organizations. It identifies activities of strategy development from the perspective of information technology managers and their organizations. It outlines responsibilities of IT managers in detail and presents processes that apply to the IT organization and the firm. Because IT strategy represents a collaborative effort, these responsibilities include ensuring that line managers throughout the firm actively participate. IT managers and line managers of all the firm's functions must share these responsibilities equally because IT strategy's only purpose is to further the entire organization's business interests.

Development of strategies is a fundamental management responsibility. Success in this endeavor requires significant time and energy. Cooperation between the IT organization and other functions is essential. Because information technology pervades most firms, the strategy process is more difficult and more important for the IT organization than for its peer organizations. In many instances, IT strategy unifies and integrates other functional thrusts, making the senior IT manager's role crucial to the firm's strategy process. Senior managers of all functions touched by information technology share responsibility for the validity of the IT strategy. They must actively participate in its development.

Neatly summarizing the important points of this chapter, Certo and Peter state that "strategic management is a continuous, iterative process aimed at keeping an organization as a whole, appropriately matched to its environment. The process itself involves performing an environmental analysis, establishing organizational direction, formulating organizational strategy, implementing that strategy, and exerting strategic control. In addition, international operations and social responsibility may profoundly affect the organizational strategic management process. It is important that the major business functions within an organization—operations, finance, and marketing—be integrated with the strategic management process."[13]

Review Questions

1. What are the major ingredients of a complete statement of strategy?

2. Besides providing direction to the organization, what other purposes does the strategy serve?

3. How do stand-alone strategies differ from other strategies?

4. Under what conditions can an IT organization produce both business and functional strategies?

5. For what purposes is a functional strategy developed, and how can it be used?

6. Toward what topics is the firm's business strategy pointed? What elements are embodied in a firm's business strategy?

7. How do the functional strategies relate to the firm's business strategy?

8. What is the relationship between strategies and plans?

9. Describe the relationship of a strategic plan to business strategies, functional strategies, and stand-alone strategies.

10. Where do tactical plans fit into this relationship?

11. What are a strategy document's main elements?

12. What role does the environment statement play in the expression of strategy?

13. What are the connections between assumptions and risks? Why is separating these items in the strategy statement useful?

14. Why is the iterative process described in this chapter useful in resolving dependencies, among other things?

15. What main topics should the IT functional strategy address?

16. The strategy development process permits executives to focus on goal incongruities. When and how does this happen?

17. Why do senior managers across the firm's various functions share responsibility for the IT strategy?

Discussion Questions

1. Identify the strategy change that AT&T adopted with its 1995 divestiture plan discussed in the Business Vignette.

2. Discuss environmental forces that caused AT&T to change strategic direction. How does AT&T's new organization help it focus on the new communications environment?

3. Give some examples of IT dependencies likely to be found in the strategy that the iterative process cannot resolve. How can the firm mitigate the effects of external dependencies?

4. Compare and contrast the actions of strategy maintenance and performance appraisal.

5. Identify critical success factors facing the IT manager that the strategy development process outlined in this chapter can address. Establish your own opinion regarding the vital nature of strategic development.

6. An IT organization wants to develop a telecommunications capability to prepare for new services the firm is now considering. What types of strategies will be useful in this connection, and how will they be used?

7. In terms of strategy development, what are the consequences of the pervasive nature of information technology?

8. What are the characteristics of high-quality IT strategy statements? Discuss these characteristics in terms of process and result.

9. What conditions are necessary for the firm's IT organization to achieve success in strategy development?

10. Managing expectations is a key IT issue. Discuss how the strategy development process assists in this task.

Assignments

1. Assume you are the manager of your university's information system organization. Your group maintains student records and the automated registration system. Develop an outline of a stand-alone strategy to incorporate into the system the use of personal computers scattered across the campus.

2. You are considering the purchase of a new, sophisticated personal computer. You will be using your computer for school activities and in your job after graduation. Develop the elements of a strategy to acquire this computer.

3. Search AT&T's Web site, www.ATT.com, to learn about its progress in implementing its strategy. Consider executive speeches, policy considerations, and financial and other news items in your review.

ENDNOTES

[1] Robert E. Allen, *1995 Annual Report* February 11, 1996.

[2] "Tough Newcomer," *The Wall Street Journal*, December 16, 1994, 1. In 1996 Mandl left AT&T to head a small new firm in the telecommunications business.

[3] See note 1.

[4] Andrew Kupfer, "AT&T Ready to Run, Nowhere to Hide," *Fortune*, April 29, 1996, 116.

[5] Henry Mentzberg, "The Strategy Concept II: Another Look at Why Organizations Need Strategies," *California Management Review*, Fall, 1987, 25. According to Mentzberg, a good strategy also helps define the organization and give it meaning.

[6] Paul O. Gaddis, "Strategy Under Attack," *Long Range Planning,* February 1997, 38. See Box 2, p 44.

[7] Blaise Horner Reich and Izak Benbasat, "Measuring the Linkage Between Business and Information Technology Objectives," *MIS Quarterly,* March 1996, 55.

[8] John C. Henderson and John G. Sifonis, "The Value of Strategic IS Planning: Understanding, Consistency, Validity, and IS Markets," *MIS Quarterly*, June, 1988, 187.

[9] Andrew Campbell and Marcus Alexander, "What's Wrong With Strategy," *Harvard Business Review*, November-December, 1997.

[10] Frederick W. Gluck, Stephen P. Kaufman, and A. Steven Walleck, "Strategic Management For Competitive Advantage," *Harvard Business Review*, July-August, 1980, 154. The authors claim that strategic management is largely concerned with creating competitive advantage and shaping the future.

[11] George C. Sawyer, *Designing Strategy* (New York: John Wiley & Sons, 1986), 169.

[12] W. R. King, "Strategic Planning For Management Information Systems," *MIS Quarterly*, March, 1978, 22.

[13] Samuel C. Certo and J. Paul Peter, *Strategic Management* (New York: Random House, 1988), 23.

4

Information Technology Planning

Team-Based Planning Aligns Goals and Objectives

Although corporate planning is a well-established activity, many firms have difficulty building plans that align goals and objectives among and between functions including IT. But not the Indianapolis-based Citizens Gas & Coke utility where a series of management meetings devoted to IT planning are reminiscent of rural New England.[1] Information technology budgets are established similarly to the way a town or village might approve its budget in a series of town meetings. For example, in budget planning sessions, managers approved an electronic mail and calendaring application but deferred a leading-edge CD-ROM-based customer service application until next year.

Driven by the impending deregulation of the utilities industry, a sense of urgency surrounds the corporate strategy process at Citizens Gas. The $350 million firm must maintain its services at the lowest possible cost to remain competitive in the new, emerging environment. Specifically, Citizens Gas is determined to keep its rates 15 percent lower than the average for comparable utilities. To meet this challenging goal, senior executives demand that information systems play a vital role in business activities and that users take ownership of their systems. To help meet these objectives, Citizens Gas established a process that tightly integrates IS strategy with corporate strategy.

To begin its fiscal year on October 1 with a comprehensive plan, the company's 14 business units meet in February, March, and June for three intensive two-day planning sessions. During the sessions, each business unit can sponsor IS projects and each votes on projects proposed by others. Business units propose IS projects that benefit the corporation and require a portion of available resources. After weighing benefits of each other's plans, they apportion the IS budget for the coming year. During these discussions, IS is represented by four managers and each gets one of the 14 votes.

At these planning sessions, business unit representatives outline their visions of the future, present their strategies, and negotiate for a share of the budget. A consensus emerges on common directions and priorities as attendees reach agreement on apportioning the budget. "This is a team-based process focused on customer satisfaction," Don Lindemen, Citizens Gas' president and CEO, reported to *Computerworld*.[2]

At the first meeting, departmental representatives attempt to agree on company direction. Following this discussion, each business unit develops short- and long-term objectives to share in the next meeting in March. Typically, business units collaborate and present joint plans to the group. For example, IS and administration may combine expertise and resources to promote e-mail for the entire corporation. Their proposal contains costs and benefits for Citizens Gas and includes a schedule for completion.

After the March meeting, participants rank each proposal's value on a scale of one to five in preparation for the June meetings during which they decide funding.

Business units use three two-day meetings to establish objectives, strategy, and budgets for the next year.

FEBRUARY	Representatives from the 14 business units present visions, discuss strategy, and argue in support of their particular budget.
MARCH	Business units develop short- and long-term objectives and share them with other units. Negotiation and discussion are intense.
JUNE	Between the March and June meetings, each unit ranks other groups' proposals. Business units decide funding levels at the June meeting.
OCTOBER 1	The agreed-upon budgets and plans take effect as the fiscal year begins.

"People do come in with their own agendas," Bob Steuber, Citizens Gas' Director of systems reported to *Computerworld*.[3] "But then that comes out in front of the whole group, and there's a feeling that you are sub-optimizing the system." However, Steuber continues, "everyone comes out of the meetings understanding why certain decisions were made. We get a common vision out of these meetings," Steuber says. "If the strategies make the cut, there's more dedication to get them done because everyone knows we've all signed off on them."[4]

INTRODUCTION

The preceding chapter concentrated on strategy development and the importance of strategic thinking for IT managers and other senior executives. It presented a process for developing several types of strategies. These strategies express the preferred long-term direction for the organization and are valuable guides for the firm and its functional units. Based on thoughtful analysis of the environment, they account for risks, dependencies, resources, and technical requirements and capabilities. They help coordinate and communicate the goals and objectives of the firm's executives and managers. In addition, these strategy statements are valuable to the firm's managers because they lay the foundation for much succeeding activity.[5]

Strategies establish the broad course of action for the firm and its functions. They outline goals and objectives the organization intends to achieve and provide a general statement of how the firm expects to achieve them. Strategies, however, do not provide many details. They establish the destination and include a general direction for reaching it. Developing the detailed roadmap is called planning. Various plans provide specific information for reaching goals and accomplishing objectives. Formal strategy and planning documents are vehicles for recording and communicating the firm's intended direction.

Planning begins with developing strategic plans from strategy statements and continues through intermediate-range planning to the development of operational plans and controls. Successful IT planning combines four related phases: 1) agreement on the future business vision and technology opportunity, 2) development of business ideas for information technology application, 3) business and information technology planning for applications and architectures, and 4) successful execution of the business and IT plans.6 This planning process is a vital part of the management system for successful IT organizations.

Results of the planning process described in this chapter are an organized framework and a methodology from which IT planning activity can occur.[7] As Chapter 1 noted, sound planning is essential to achieving superior business results and is a critical success factor for IT managers and their organizations. Therefore, plans must be developed in an outstanding manner.

The next sections describe the planning horizon—the time over which various plans are active—and the definitions of strategic plans, tactical plans, and operational plans and controls. The text that follows these sections shows why these plans are needed and how they are used to manage the business.

THE PLANNING HORIZON

This chapter discusses three types of plans: strategic plans, tactical plans, and operational plans. Generally, the time span during which they are valid differentiates these plans. This time span is called the planning horizon. A typical short-range planning horizon is 30 days, for example, and for most firms a long-range planning horizon extends for years. Figure 4.1 presents a typical planning horizon showing the time relationship among various plans.

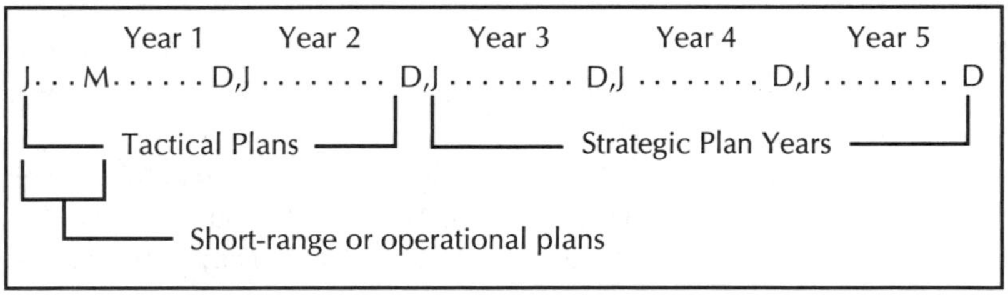

FIGURE 4.1   The Extended Planning Horizon

As Figure 4.1 shows, strategic plans usually cover a period from two to five or more years into the future. They represent long-term implementation of strategic thinking. Strategic plans contain specific details for achieving the organization's mission and for meeting its long-range goals and objectives. They form the basis for short-range and operational plans.

Tactical plans typically address the period from three months or so to two years in the future. They describe actions that begin the implementation of strategic plans. Operational plans consider the near term: from now to three months from now. Tactical plans bridge the period from the operational horizon to the strategic horizon.

Just as managers need stand-alone strategies to cover special business or functional situations, they also need plans to accompany these strategies. In some cases, these plans may necessarily remain separate from the three plan types just discussed. Usually, however, plans associated with stand-alone strategies fold into plans covering the period addressed in the stand-alone strategy. At some high level in the firm, however, resource portions of all these separate plans merge.

Operational planning is broad based and considerably detailed because it relates all important activities happening in the very near term. Operational plans need to be regenerated frequently in response to changes in near-term business conditions. They bridge the gap from the present to the tactical time period and contain detailed information that is assimilated and implemented at fairly low levels in the organization. Operational plans give structure to the firm's everyday activities for first-line managers and their teams. They provide the basis for taking short-term action and the reference points for measuring short-term results. Short-term variances, derived by comparing planned versus actual performance, lead to corrective actions that keep the organization on track toward meeting its goals and objectives.

Tactical or intermediate-term plans are less detailed and are effective during a longer period than operational plans. Typically, these plans cover the current year plus the next year. They provide overriding direction to operational planning. In the ideal situation absent significant variances, the implementation of succeeding operational plans leads to the tactical plan's successful implementation. Because activities seldom go as expected, managers must contend with variances in results of both operational and tactical plans.

Strategic plans extend from the end of the tactical period to the end of the extended planning horizon. Just as tactical plans guide operational plans, strategic plans provide direction for tactical plans. Well-executed tactical plans pave the way for accomplishing strategic objectives. Thus, implementing the planning process results in a succession of overlapping plans moving forward through time.

Just as the three time frames form a continuum over the extended planning horizon, the three types of plans must form a unified picture of the firm's future actions. The plans cover different time periods and contain different amounts of detail. They portray a coordinated description of how the firm intends to carry out its strategies from now into the foreseeable future. The strategic plan supports the firm's strategy. The tactical plan provides shorter-range support to the strategic plan. Operational plans give day-to-day or near-term meaning to the tactical plan. Without unanticipated changes in the environment during this entire time period, and with flawless development and implementation of its plans, the firm will attain its strategic objectives. Because these conditions are seldom realized in today's rapid-paced business climate, the planners' true challenge lies in developing strategies and valid plans that account for anticipated environmental changes, yet contain flexibility for revision in response to unexpected events.

STRATEGIC PLANNING

Strategic planning converts strategies into plans. It takes information found in the strategy statement, adds detailed actions, and allocates resources to attain stated goals. Resources consist of people, money, facilities, and technical capabilities, blended together and focused on objectives. The plan must resolve assumptions, risks, and dependencies found in the strategy. The plan converts assumptions in the strategy into reality and resolves dependencies. Risks noted in the strategy must be mitigated to the fullest extent possible by specific actions designed to contain and offset them. And, most importantly, the planning process must generate action plans to accomplish goals and objectives.

The strategic plan contains a detailed resource allocation schedule arranged to attain strategic goals and objectives. An organization's strategic plan is also a financial statement of projected revenues, expenses, costs, investments, and profits. The strategic IT plan also includes financial information that is usually limited to IT costs and expenses. The plan is detailed enough to be tracked over its lifetime by comparing planned to actual performance. When the strategic plan has been translated into tactical and operational plans, tracking takes place in the near term. Because the strategic plan translates strategic goals and objectives into strategic actions two to five or more years in the future, it forms the foundation for building the intermediate-range plan.

As an example, consider an IT group whose long-term goal is to improve market response times by intense automation of product design, development, and manufacturing. The strategy includes installing CAD/CAM systems for its design and development engineers and for its plant. The systems are to be effective in 48 months. The strategy statement contains assumptions for using CAD/CAM and outlines dependencies and risks. Resistance to new systems, for example, might be a risk. The strategic plan adds details of schedule, actions, and resources: space requirements for equipment, training for users, capital resources, and operating expenses. The plan states that space will be leased and that training will overcome any resistance to the new systems. As the tactical plan is developed, more detail is added, particularly for the near term.

TACTICAL PLANS

Tactical or intermediate-range plans generally cover the current year in detail and the following year in less detail. Because managers commit to implementing tactical plans, their plans are used to measure their performance. Thus tactical plans are sometimes called measurement plans. Tactical plans form the basis for assessing the firm's performance too; they also provide objective information for measuring the firm's progress toward long-range goals. In addition, senior executives use them to measure unit performance. Tactical plans contain more detail and bear more directly on near-term activities than strategic plans. Thus they guide very short-range activities and link near-term actions to long-range goals.

To continue the CAD/CAM example, many activities will be planned for the first two years. Some of these are selecting hardware and software, installing hardware and software systems, training users and IT people, converting from previous processes to new procedures, and establishing measurement systems. Questions must be answered, such as, "From whom will the space be leased?", "Where will it be located?", and "What type of training is required to overcome resistance to new technology?". Managers must schedule these activities, assign human and material resources, and develop detailed budgets to complete the plans. Implementing managers must commit to the plan and will be evaluated on the degree of plan attainment. The tactical plan forms the basis for day-to-day CAD/CAM system implementation activity.

OPERATIONAL PLANS AND CONTROLS

Operational plans add detail and direction to the firm's very near-term activities on a day-to-day or week-to-week basis. First-line managers or nonmanagerial employees use them as they carry out their responsibilities. Operational plans usually require more detailed analysis than longer-range plans. They must contain more detailed controls as well. Variance analysis between planned and actual results shows managers whether daily or weekly activities are on schedule and proceeding according to plan.

The CAD/CAM installation contains many near-term tasks. For example, these include identification of hardware and software vendors, setting vendor criteria, developing space alternatives, and performing systems analysis of current procedures. In the meantime, planning for future short-term goals, budget management, and cost tracking continues.

PLANNING SCHEDULES

In most firms, planning is a regularly scheduled event, closely related to the firm's fiscal calendar. For most organizations, the fiscal year and the calendar year are identical and planning activity is seasonal. Therefore, common practice is to develop tactical plans so they can be approved just before the tactical period begins. For most firms operating on the calendar year, this means that the tactical plan for the next two years is developed and approved in the few months before the new year begins. The organization begins the new calendar year under control of the new tactical plan for that year and the following year.

In the Business Vignette, Citizens Gas' fiscal year starts in October, so planning begins in February for the next fiscal year. February, March, and June meetings lead to agreement on funding levels for each business unit. Detailed budgets are developed and in place on October 1 when the new fiscal year starts.

In a model planning calendar, developing strategies and strategic plans occurs shortly after the new year begins and is completed and approved around mid-year. When completed, the plans cover the period from 30 months in the future to the end of the extended planning horizon, perhaps five or more years hence. Figure 4.2 illustrates the second iteration of planning in relation to current plans. The new tactical plan incorporates the first year of the previous strategic plan, and the new strategic plan adds an additional future year.

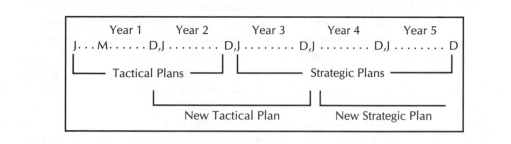

FIGURE 4.2   The Second Planning Iteration

Organizations with well-established planning processes do not start planning at the beginning with each iteration. As Figure 4.2 illustrates, the new tactical plan revises the second year of the previous plan and adds to it the first year of the previous strategic plan. The new strategic plan is a revision of the last years of the previous plan with a new year added. Revisions reflect changes due to new or revised strategies, changing business conditions, or changes in the environment. Failure to achieve prior goals and objectives or alterations related to risk management or dependency management may also lead to revisions.

The entire planning process is cyclical and repetitive, with revised longer-range plans becoming more near-term with each passing cycle. Usually, operational plans are not as tightly scheduled, and, if the tactical plans' implementation proceeds smoothly, these plans may very well be of an *ad hoc* nature. The need for short-range plans depends on the circumstances and on the function's and firm's activities.

The CAD/CAM example discussed earlier could proceed similarly. Details of hardware, software, space, training, and many other items develop over time as succeeding plan iterations and implementations occur. If all goes well, systems become operational as planned, and the firm reaps the benefits of shortened development and manufacturing cycles and increased responsiveness to changing market requirements. Ideally, the firm achieves its strategic objectives with the new systems.

A PLANNING MODEL FOR IT MANAGERS

So far the discussion has related mostly to common planning elements—elements applicable to most functions in most firms. In reality, the planning horizon varies between firms, and the planning schedule may change considerably, depending on the industry and the firm's desires. For example, electric utility companies usually have a long planning horizon, on the order of 10 years or more, and firms in the forest products industry in the Pacific Northwest plan for 50 or more years in the future. Firms in highly volatile or highly competitive industries may benefit most from shorter-range thinking. Firms on the edge of bankruptcy think only of survival. A year may be a very long time for a troubled firm.

The information technology function and other functions within the firm must adapt to internal and external forces and constraints in searching for a management system to govern their planning. The firm's planning system establishes many parameters to which the IT organization and others must conform. For example, the planning department establishes the planning calendar and planning horizon, schedules review procedures, and coordinates approval processes. Usually, the planning department also prescribes plan formats and content outlines.[8]

One important ingredient the IT organization brings to the firm is a perspective of future technology developments. Technology trend information is important because it permits IT to present credible technical detail pertinent to the extended planning horizon. Subsequent chapters present methods through which IT can secure this important trend information.

Although planners can specify some strategic plan elements in detail, many other items represent the planner's best judgement. Superior long-range planning elegantly blends solid facts, imaginative intuition, and seasoned judgement. Available facts, however, are always preferred in the planning process.

Because information technology is widely dispersed in most firms, chief information officers' and other IT managers' responsibilities extend far beyond the boundaries of their line organizations. Organizational factors strongly influence planning activity—the CIO position is advantageous in this regard. Factors such as the perceived importance and status of the IT manager, the corporate culture, and overall management style aid in planning. IT managers' strategies and plans must reflect their own organizations' actions while including the entire firm's information processing activities. Their view must encompass both their staff and line responsibilities. Their vision must include both technological and organizational factors. It must provide technical direction or at least generate practical technical alternatives. And it must consider the organization's maturity and sophistication. Intelligent IT executives do not attempt to transform organizations from relative technological immaturity to maturity in one planning cycle.

Obviously, the firm's senior executives must include the entire organization's environment in their planning activity. This is especially important for information technology-intensive firms because the technology frequently leads to organizational and other changes. The corporate planning environment must anticipate and prepare to respond to IT-induced changes of all kinds.

The planning environment at Citizens Gas does this exceptionally well through a series of meetings that virtually eliminate surprises in the final result. The Business Vignette described how the company uses well-coordinated planning meetings to set expectations, obtain consensus, and establish unity among business functions. Upon completion of the meetings, units have congruent goals, objectives, and plans. IT and the functions it supports obtain agreement from the CEO and other senior executives on the direction for the next year. And, in March, when everyone's plans have been in place for six months, the group begins the planning process again. Citizens Gas is an outstanding example of good corporate governance and planning.

Since planning considers constraints, ambiguities, facts, and uncertainties, what planning procedures might best serve the IT function? What kinds of activities need to be planned? What tools and information assist in the task of IT planning? A complete IT plan developed from a rigorous process answers these questions. An effective plan is a detailed and thoughtful document describing actions for implementing strategy over the planning period. The plan must address both line and staff responsibilities of IT—it consists of the ingredients outlined in Table 4.1.

TABLE 4.1   IT Management Planning Model

Application considerations

System operations

Resource plans

Administrative actions

Technology planning

Table 4.1 presents items critical to achieving the business results noted in Figure 1.1, and they are among the critical success factors for IT managers, too. Thus, for several important reasons, successful planning is a high priority item for managers.

Application Considerations

The applications portfolio consists of the complete set of application programs the firm uses to conduct its automated business functions. As will become clear later, managing applications assets is a very difficult task, and planning for this activity must be conducted skillfully. The plan for dealing with the application portfolio must include the selection of projects to be implemented during the plan period and the scheduling, control, and evaluation of these projects during implementation. Resources required for development, enhancement, maintenance, and implementation must also be planned.

Project selection describes which application programs merit maintenance or enhancement activity and what this activity accomplishes. Managers must know the activity's schedule and its resource requirements. Applications supporting new or enhanced business processes, not currently in the portfolio, must be defined and a plan for acquiring these programs must be developed.

Identifying systems for enhancement or acquisition to support new strategic directions or new or altered business activities is sometimes difficult. Programs may be purchased or developed locally or by subcontractors. In all cases, the plan describes what will happen, when it will happen, how it will occur, and how much it will cost. The plan must justify the action, describe why the action is being taken, and identify who is responsible for taking the action.

Managers must know the answers to "who, what, when, where, why, and how" for all planning activity. If they consider these fundamental items in planing the portfolio, they have the basis for controlling plan implementation and for monitoring deviations from it. It bears repeating that effective plans must contain controls.

Major difficulties often surround the planning direction for the applications portfolio. These difficulties usually arise from competition for resources among various work items. Their resolution usually requires involvement of the organization's senior executives.[9] IT managers must ensure that these issues are resolved—Chapter 8 presents a rigorous method for doing so. The results of the method described in Chapter 8 provide the ingredients for the applications portfolio plan.

System Operations

System operations consist of processes for running the firm's applications. Usually this means that data is collected or developed, applications are executed using current and possibly historical data, and results are stored and/or distributed as planned. The processes may be continuous online systems, networked client/server systems, periodic, scheduled systems, or, more likely, some combination of these. In any case, results generated by the applications are used by the firm's members to carry out their responsibilities, stored for use by secondary applications, or distributed to customers or suppliers. The organization depends on these outputs for successful and efficient operation.

IT managers must plan system operations to satisfy customers within the firm and outside it. IT must have a plan to satisfy customers' expectations about the IT service they receive. Planning service levels is very important in system operations.[10] Essential disciplines for meeting service expectations consist of problem management, change management, recovery management, capacity planning, and network planning and management. Problem, change, and recovery management are critical for ensuring the attainment of service levels through operational management processes. Capacity planning and network planning ensure service levels through resource planning. Table 4.2 lists the planning elements of system operations.

TABLE 4.2   System Operations Planning Elements

Service level planning

Problem management

Change management

Recovery management

Capacity planning

Network planning

Managing system operations is critical to IT managers' success. System operations management is a critical success factor they must handle properly in both the short and long term. Part IV discusses in detail basic considerations involving the elements of the system operations plan. This discussion includes the details of service-level planning and the operational disciplines of problem, change, and recovery management. These items' importance to IT planning will be explained later. When used properly, they ensure that managers attain planned and committed service levels.

Service-level attainment depends on resources adequate to process the applications. IT and functional user organizations must possess sufficient computing and network capacity to handle the application workload satisfactorily. Physical resources include computer-processing units such as servers and personal workstations and all peripheral equipment such as auxiliary storage and input/output devices. In most organizations today where client/server and intranet computing is well developed, individual workstations and associated equipment are important components of the overall system's capacity. Chapter 14 intensively discusses performance analysis and capacity planning for computing systems serving the applications.

Network planning and management are the subjects of Chapter 15. Envisioning the future, senior managers establish policy and consider architecture in network planning, thus setting the stage for network design and implementation.[11] Like the system's computer components, performance analysis and capacity planning are required to meet network service levels. The disciplines of problem, change, and recovery management are also important elements of the IT plan.

Resource Planning

The next part of a complete IT plan is the resource section. IT resource items consist of money, equipment, people, and space. Technology is also a resource, but technology planning will be discussed separately. Table 4.3 lists elements commonly found in the plan's resource section.

TABLE 4.3   Resource Elements

Equipment plans

Space plans

People plans

Financial plans

Administrative actions

The IT plan describes the critical dependency on available resources throughout the planning horizon. A summary of the hardware and equipment required to operate the firm's information processing activities appears in the plan's resource section. This section includes the space requirements for housing equipment and special facilities for running equipment. If the firm's system operation is growing, space requirements can be substantial. Special facilities such as heating, ventilation, air conditioning, electrical power, cabling conduits and cabinets, and perhaps raised floors must also be considered. In many installations, uninterruptible power sources ensure reliable service but also add significant cost.

In addition, the plan must include network equipment and services such as switches, routers, and perhaps dedicated leased lines. The organization's financial obligations consist of purchase and lease costs, and expenses for third-party services. The IT's complete telecommunication system financial plan integrates equipment and use charges with software and applications costs.

People Plans

The next essential ingredient of the IT resource plan is people—by any measure, the most crucial element of any plan. People enable the firm and the IT organization to meet their goals and objectives. Without skilled people, abundant equipment, space, and money are relatively useless. Although people management is a necessary condition for success, it is not sufficient by itself. Because effective management of people resources is a critical success factor for IT managers, it is essential that this portion of the plan be particularly well conceived and developed.

IT managers and administrators can provide much of the people-related information. Staffing, training, and retraining plans go hand-in-hand with the people management activities of performance planning and assessment. Because these matters are so critical to good management, plans for the organization must reflect their current and anticipated actions.

The personnel plan identifies requirements for people according to skill level and considers their development and deployment over the extended plan horizon. It must address attrition, hiring, training, and retraining. The IT plan identifies sources of new employees and formulates recruiting actions. For people currently in the organization, managers propose individual development plans. Managers also maintain plans for people in skill groups such as systems analysts, programmers, support specialists, telecommunication experts, and managers. The staffing plan must be consistent with the plans to accomplish technical and organizational objectives. Working with this detail, IT planners can calculate the total cost of the people resource by summing their salaries, benefits, training costs, and so on for the financial plan.

Financial Plans

The financial plan for the IT organization summarizes its equipment, space, people, and miscellaneous costs. The firm's code of accounts segregates and identifies specific costs and expenses. The timing of all expenditures reflects the rate at which the organization acquires or expends resources consistent with work activities and event completion. When the plan is approved, the budget for the planning years is constructed by funding the various accounts to the level the plan requests. In almost all cases, the planning process is iterative and intense discussion and negotiation determine the final budgeted amount. The plan's early phase, typically its first year, represents a commitment. Executives evaluate IT managers by comparing actual performance to planned or expected performance.

Administrative Actions

Throughout the planning cycle many administrative actions assist in the planning process and in implementing the approved plan. Before beginning, planners develop plan assumptions and provide plan ground rules. For planning purposes, the firm might state that it expects to increase revenue by 10 percent per year. This is an example of a planning assumption. Holding the number of programmers and analysts constant throughout the plan period is an example of a plan ground rule. An established list of assumptions and ground rules ensures that the resulting plan remains within the bounds of reason. Without a set of reasonable assumptions and ground rules, the plan process may iterate endlessly.

When planning, IT managers must obtain the cooperation of managers in adjacent functions and coordinate the emerging plan with the many units that it affects. This happens in a variety of ways; nevertheless, planners must ensure complete and unambiguous communication. It's important to brainstorm to get a variety of ideas, for example, and to develop win-win situations among participants.[12] Plan-review meetings should be held with managers of all functional areas in the firm. For example, the plan must be reviewed with marketing managers to ensure properly planned marketing requirements for information technology support. Such communication meetings help ensure plan congruence and assist in organizational learning.[13] This process clarifies the role and contribution of IT to its users. A well-coordinated plan review process like the one Citizens Gas employed mitigates many concerns noted in Chapter 1.

To improve IT planning effectiveness, many firms establish steering committees consisting of representatives from its functional areas. These committees' purpose is to guide the IT planning and investment process. Active steering committees increase the mutual understanding of users and IT personnel and tend to increase executive involvement in and understanding of IT matters.[14] Steering committees, consisting of senior executives from all parts of the business, provide valuable assistance in ensuring plan synchronization between IT and the rest of the firm. This approach is especially valuable for firms with widely dispersed, decentralized IT operations.

Finally, managers and administrators are responsible for implementing the plan and for tracking activity within the organization to actions specified in the plan. Because control is one of management's fundamental responsibilities, IT managers must establish measurement and tracking mechanisms designed to ensure that the organization's activities are on target. Control functions must address the plan elements individually.

For example, actions affecting the application portfolio are routinely subject to project management control and monitoring. Controlling the arrival and departure of equipment and facilities and the related space is essential. System services must be measured and controlled according to the service-level agreements negotiated with the users. The firm's financial organization usually tracks consumption and utilization of financial resources. When actual expenditures differ from planned expenditures, the IT manager must explain account variances to the firm's controller and others. Schedules, budgets, events, and accomplishments are the principle control elements in IT plans.

Technology Planning

The IT organization has a continuing responsibility to monitor advances in technology and to keep the firm informed of progress in the field. Senior professionals throughout the IT organization must keep pace with state-of-the-art developments in their disciplines. Programming managers must track programming technology. System support personnel and telecommunications experts must maintain a current awareness of technology and must improve their expertise. These experts assess technology for the firm's user organizations through the senior IT executive. Table 4.4 lists some of the many technology areas that should be evaluated during the planning process.

TABLE 4.4   Technology Areas

New processor developments

Advances in storage devices

Telecommunications systems

Operating systems

Communications software

Programming tools

Vendor application software

Systems management tools

Although new technology may have many benefits, its introduction leads to some predictable problems. Costs are usually underestimated and benefits frequently overestimated. To help avoid these difficulties, senior IT managers should fund a small advanced technology group whose purpose is to make assessments and provide models for using emerging technology within the firm. The ad tech group researches emerging technologies and evaluates new concepts. It establishes feasibility and begins the development and introduction of the new concepts. Later, the group transfers feasible concepts to development and user departments for implementation and production. These activities assist in technology introduction by reducing risks and providing useful insights for planning purposes.

OTHER APPROACHES TO PLANNING

Nolan's stages of growth and Rockart's critical success factors, two ideas important to information technology managers, are particularly relevant to IT planning. Although both are incomplete IT planning processes in themselves, they add value to the methodology outlined in this chapter and are useful in testing the credibility of results derived by other means.

Stages of Growth

Nolan's theories provide predictive insight to organizations within the firm and to the firm itself. To use the stage theory, managers must assess the current posture of the organization or the firm. Then managers can base plans on an extrapolation of observed trends. For example, if the firm, or some part of it, is in the control stage, it would be wise to anticipate and plan for the integration phase followed by the data administration stage. If one function is in the data administration phase and others are in the integration phase, understanding the reasons for this disparity is advisable. Perhaps the firm should concentrate on bringing the laggards up to the level of the more advanced function. For competitive reasons, understanding the position of other firms in the industry may also be useful.

Before IT managers take action, however, they must communicate their assessments to the functional heads involved. Conditions may be as they are for very good reasons. In any case, decisions must be made by all interested parties, including the steering committee, not just the IT organization acting alone. Analyses of the stage theory and its use in planning should be integrated into the more formal process outlined earlier in this chapter.

Critical Success Factors

Critical success factors are excellent tools to use in conjunction with the formal planning process. They lend structure to, and improve, planning by focusing on important managerial issues.[15] They can be used to audit results of the planning process to ensure that the plan contains necessary conditions for success.

For instance, consider critical success factors 2a and 2d (strategic and competitive issues), and 4b (operational items) identified in Chapter 1. If the IT plan is not completely synchronized with the firm's plan, the resulting implementation will not totally support the firm's objectives. This unsatisfactory condition must be corrected. If the planning process sets unrealistic expectations or does not ensure realistic long-term expectations, the IT organization will suffer in later years. A third example from the operations side of the organization is worth considering: if the IT plan does not contain enough resources to deliver timely services, the IT organization will certainly experience short-term difficulties. Results of the formal planning process can be evaluated by reviewing the plan against the critical success factors for the IT organization and the firm. This evaluation can also help level expectations within the firm.

Chapter 1 presented critical success factors founded on prominent industry issues and on other factors important to the firm. Critical success factors are organized into long-range issues, intermediate-range concerns, short-range items, and business issues to coincide with the planning time frames discussed in this chapter. Careful IT planners always ensure that the plan includes all necessary conditions for success.

Knowing the critical success factors of the senior IT executive's superior is vitally important. Even more important is knowing the firm's critical success factors. Plans prepared by the IT organization must account for these factors so the IT organization can meet all necessary conditions for success.

Business Systems Planning

Another well-known approach to IT planning is the methodology of business systems planning (BSP) developed by IBM. BSP concentrates on the firm's data resources and strives to develop an information architecture supporting a coordinated view of the firm's major systems' data needs. The BSP process identifies the firm's key activities and the systems and data that support these activities. Data is arranged in classes, and an architecture is developed to relate data classes to the firm's activities and its information systems. In essence, BSP strives to model the firm's business through its information resources. With BSP, planning emphasis changes from portfolio applications to supporting databases.

The BSP process is very detailed and time consuming.[16] It requires a bottoms-up effort and tacitly assumes that a data architecture can be developed in one step. The process works best in centralized environments where the data, applications, and structure for processing are physically near one another. It is a useful approach in departments that operate departmental computing systems. BSP gains importance as data warehousing and data mining become more popular in modern organizations.

Current trends toward decentralization, broad deployment of information technology through internetworking, and the increasing importance and value of information technology greatly complicate the IT planning task while greatly elevating its importance. For these reasons, a variety of planning methods must be considered.[17]

THE INTEGRATED APPROACH

This chapter presents the important types of plans, plan ingredients, and some important planning approaches that IT organizations currently use. Each of the many planning tools discussed in the literature offers advantages under certain circumstances and limitations or disadvantages under other conditions.[18] Most methods are weak when information technology is critical to the firm, when the technology is widely dispersed, and when the organization is decentralized.

Cornelius Sullivan studied 37 major American companies to understand their planning systems' effectiveness of in relationship to factors indigenous to the firm.[19] Sullivan found two factors correlate with planning effectiveness: infusion, or the degree to which information technology has penetrated the firm's operation, and diffusion, the extent to which information technology is disseminated throughout the firm. Firms that considered themselves effective IT planners were tabulated in a matrix according to the type of planning system in use. Figure 4.3 presents Sullivan's very revealing conclusions in matrix form.

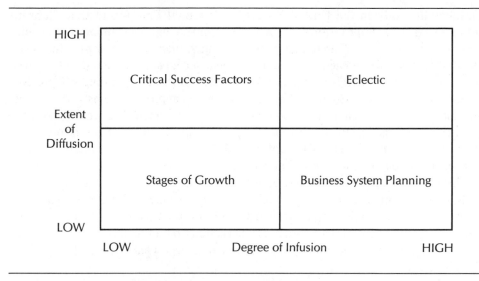

FIGURE 4.3   IT Planning Environments

Sullivan's research reveals that the stages-of-growth methodology works effectively in firms in which technology is concentrated and does not have a high impact. Most firms today do not find the stages-of-growth methodology effective as their primary planning mechanism. Critical success factors (CSF) are more important to firms in which technology is dispersed and its impact is moderate. For many firms, the CSF methodology is a valuable adjunct to the formal planning mechanisms discussed earlier. BSP is more effective for firms in which information technology is centralized and has a high impact on the company.

Many of today's firms, however, are in the upper-right, Eclectic quadrant of Figure 4.3 or moving toward that environment. For these firms, the planning process is more complex because their IT operations are very complex. Stages-of-growth concepts are of limited value, critical success factors continue to be important, and concentration on information architecture is valuable. As a planning mechanism, BSP also has some utility. What planning methodology best serves the firm in the Eclectic quadrant? How can IT executives plan effectively, given the fast-paced transformations common among firms today?[20]

Firms in the Eclectic quadrant, or moving toward it, face some predictable planning issues. Major changes take place in applications portfolio acquisition with increases in purchased applications, joint development activity and alliances, and client/server and intranet application development. Today, organizations are being redesigned around information systems; thus IT managers and advocates are becoming architects of the changing environment. Organizations being reengineered around information systems to serve customers better are depending on IT to support the new structures. Widely distributed information systems elements linked with new telecommunication technology, increasing reliance on networked systems, and growing internetworking technology greatly complicates IT planning in modern organizations.

In an important study in the United Kingdom, Michael Earl found that planning approaches fall into five general categories.[21] The Technological approach attempts to develop an "information systems-oriented" architecture around which planning develops. This seems to be a variant of the BSP method and offers some of the same advantages and disadvantages. The Administrative method focuses on bottom-up resource allocation building off previous plans. This method deemphasizes strategy. The Method-Driven approach begins with previous IT plans and may consider consultants' recommendations. This method's final outcome is likely to emphasize external factors. In the Business-Led approach, the firm assumes that business plans eventually lead to IT plans. Unfortunately, Earl found that business plans are generally not specific enough for detailed IT planning.

Earl concluded that the fifth method, the Organizational approach, is consistently superior to the other four. In this method, planning consists of frequent interchanges between the IT function and client functions to develop projects that lead to shared outcomes. (This technique resembles that used at Citizens Gas.) Functional interactions on an as-needed basis supplement other long-range planning activities. This approach selects what is needed, when it is needed: it is indeed eclectic.

Successful IT managers must understand business trends in their firms and must plan to support their firm's business activity, sometimes in the face of considerable uncertainty. Understanding management and organizational issues associated with the firm's application portfolio and managing the portfolio effectively are critical goals. Supporting and managing user-owned and user-driven computing and its telecommunications infrastructure are also critical activities. And establishing relationships with executives and client organizations to maintain flexibility and facilitate change efficiently is mandatory. Subsequent chapters address all these topics. In part, planning in this environment is always eclectic.

MANAGEMENT FEEDBACK MECHANISMS

This chapter emphasized the need for control processes or management feedback mechanisms. Control consists of knowing who, what, when, where, why, and how for all the organization's essential activities. Managers who know these answers operate under control. Those who don't operate out of control. An out-of-control organization causes very serious difficulties for its managers and for the firm. Control is vital not only to an IT organization's success, but to the entire firm because information technology is so important to so many parts of the business.

Sound strategy development and planning activity is the basis for sound control mechanisms. In particular, the IT plan must provide data to answer control questions. Control processes must be designed to compare the organization's actual performance to the expected performance detailed in the plan. Chapter 3 discussed the notion of performance assessment in connection with the strategy development process. Vehicles for gauging and assessing actual performance are the plans describing how the strategy is implemented. In short, plans are the basis for assessing progress toward achieving strategic goals and objectives.

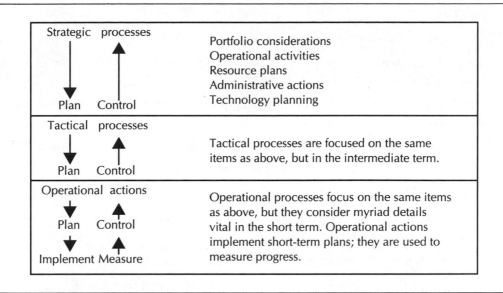

FIGURE 4.4   Management Feedback Mechanisms

Figure 4.4 shows how these ideas fit together. The left side of the figure shows how strategic processes lead to tactical processes which, in turn, are expressed in operational actions. Measurements at the lowest level lead to control assessments. These control assessments focus attention on tactical processes which, in turn, bring focus to strategic processes. Thus, the entire process is recursive. Strategies and plans are validated or altered with each planning cycle based on measurements made at several levels. Measuring progress by frequently comparing actual position against planned position using control items allows managers to keep the organization on track toward meeting its long-term objectives.

SUMMARY

Planning for the IT organization and for the use of information technology within the firm is complex and best accomplished through systematic processes. Planning techniques rely on separating the extended planning horizon into strategic, or long-term; tactical, or intermediate-term; and operational, or short-term components. Because the planning cycle is an annual event, each cycle adds another year to the strategic period as each plan advances one year. This systematic process is synchronized through a planning calendar followed by the entire firm.

Regardless of how they are developed, plans are only as good as their implementation and control mechanisms. Strategic and tactical processes and operational actions are related to measurement and control actions through management feedback mechanisms. Control is a fundamental management responsibility—especially important to people managing volatile, rapidly changing activities. Measurements are also very important because they are the foundation of performance appraisal. Information technology planning is a necessary condition and a critical success factor for IT managers.

Review Questions

1. How do the strategy and the strategic plan differ?

2. What do we mean by the extended planning horizon? Describe the planning horizons discussed in this chapter.

3. What is the relationship of strategic plans to tactical plans? How are operational plans related to tactical plans?

4. For the strategic plan, what resources must be blended together and planned?

5. What financial items does the firm's strategic plan include?

6. Describe the purpose of tactical plans.

7. Who are the main users of operational plans? How does the firm's controller use operational and tactical plans?

8. Differentiate between the planning calendar and the planning cycle.

9. What are the elements of the IT organization's planning model?

10. What considerations surround applications portfolio planning?

11. What questions need to be answered to achieve proper control?

12. What are the elements of system operations planning?

13. Which elements ensure that service levels are attained through management processes and which through resource planning?

14. What elements comprise the resource portion of the IT plan?

15. How is the financial portion of the IT plan constructed?

16. What does the term "plan ground rules" mean? How do plan assumptions differ from plan ground rules?

17. What role does technology play in planning for the IT organization?

18. When does the business systems planning methodology become important in IT planning?

19. What does eclectic planning mean, and for what firms is it important?

20. What is the connection between operational plans and management control?

Discussion Questions

1. What important elements of the planning process at Citizens Gas are described in the Business Vignette? How many of the critical success factors presented in Chapter 1 are present in the Citizens Gas planning process?

2. Chapter 1 presented a list of critical success factors for IT managers. Which of these relates to IT planning?

3. Using an intranet installation as an example, discuss the continuum of time frames inherent in IT planning and describe the levels of detail in the plans throughout these time frames.

4. Discuss the planning cycle and the planning horizon for a university operating on the academic year starting in September and ending in August.

5. Name several planning assumptions and ground rules that might apply to the university in Question 4.

6. What is the role of the firm's planning department in the plan development process?

7. What additions might be included in the planning model for an IT organization that provides services to customers outside the firm?

8. Discuss the relationship between the list of critical success factors and planning for system operations.

9. The need to include auxiliary items such as space, cabling, supplies, and maintenance often complicates equipment plans. Discuss how an IT manager might determine that an equipment plan is complete.

10. What is the role of IT managers in the process of people planning? Discuss why this activity should be continuous, not just at plan time.

11. Why is technology assessment vital to the IT organization? Describe how you think this assessment can be brought into the planning process.

12. Discuss some ways in which you think senior IT management should exercise control.

Assignments

1. Study BSP or one of the other planning techniques mentioned in this chapter. What are its advantages and disadvantages? How would it fit with the process discussed in this chapter?

2. A firm wants to replace its present servers with more powerful, lower cost servers. What kinds of items must be planned to accomplish this objective?

3. Read the article by Mary Louise Hatten and Kenneth J. Hattan in *Long Range Planning*, Vol. 30, No. 2, page 254, titled "Information Systems Strategy; Long Overdue—And Still Not Here," and prepare a short analysis for your class.

ENDNOTES

[1] Leslie Goff, "I Walk Aligned," *Computerworld*, August 8, 1994, 68.

[2] See note 1 above.

[3] See note 1 above.

[4] For a rigorous description of team-based planning, see Gregory Mentzas, "Implementing an IS Strategy—A Team Approach," *Long Range Planning*, February 1997, 84. This article outlines the kinds of teams needed and how they must relate to solve the problems he sees with casual planning activities.

[5] In spite of some criticism, most executives value strategic planning. See Paul O. Gaddis, "Strategy Under Attack," *Long Range Planning*, February 1997, 38.

[6] Marilyn M. Parker and Robert J. Benson, "Enterprisewide Information Management: State-of-the-Art Strategic Planning," *Journal of Information Systems Management*, Summer, 1989, 14.

[7] Focusing on planning processes and finding solutions to problems is considered very important according to William Gilmore and John Camillus. See "Do Your Planning Processes Meet the Reality Test?," *Long Range Planning*, December 1996, 869.

[8] This discussion assumes that the organization has some kind of formal planning process. The absence of a formal planning process greatly increases the risk of plan incongruence and may lead to business failures.

[9] Martin Buss, "How To Rank Computer Projects," *Harvard Business Review*, January-February, 1983, 118. This article discusses the issue of who should make decisions on applying resources to the portfolio.

[10] Edward A. Van Schaik, *A Management System For The Information Business: Organizational Analysis* (Englewood Cliffs, NJ: Prentice-Hall Inc., 1985). In particular see Chapter 7, "Service-Level Management", pages 155-225. Also see John P. Singleton, Ephraim R. McLean, and Edward N. Altman, "Measuring Information Systems Performance: Experience With the Management by Results System at Security Pacific Bank," *MIS Quarterly*, June, 1988, 325.

[11] Peter G. W. Keen, "Business Innovation Through Telecommunications," *Competing In Time: Using Telecommunications For Competitive Advantage* (New York: Ballinger Publishing Co., 1988), 5.

[12] See note 7. Subdividing plans to deal with complexity and prototyping planned ideas when plans are implemented improve success in IT planning according to the authors.

[13] John C. Henderson and John G. Sifonis, "The Value of Strategic IS Planning: Understanding, Consistency, Validity, and IS Markets," *MIS Quarterly*, June, 1988, 187. See also Terry Byrd, V. Sambamurthy, and Robert Zmud, "An Examination of IT Planning in a Large, Diversified Public Organization," *Decision Sciences*, January-February, 1995, 49.

[14] D. H. Drury, "An Evaluation of Data Processing Steering Committees," *MIS Quarterly*, December, 1984, 257. Drury found that approximately half of the firms polled had a committee that had existed an average of five years. The typical committee had six members and was most effective when it held regularly scheduled meetings and when agenda items originated outside the IT department.

[15] Andrew C. Boynton and Robert W. Zmud, "An Assessment of Critical Success Factors," *Sloan Management Review*, Summer, 1984, 17.

[16] For other perspectives on data planning, see Dale L. Goodhue, Laurie J. Kirsch, Judith A. Quillard, and Michael D. Wybe, "Strategic Data Planning: Lessons From The Field," *MIS Quarterly*, March 1992, 11.

[17] A. L. Lederer and V. Sethi analyzed four IT planning methodologies in "Meeting The Challenges of Information Systems Planning," *Long Range Planning*, April 1992, 69. The authors find many problems with each but conclude that, regardless of method, top management involvement is a must.

[18] Albert L. Lederer and Vijay Sethi, "The Implementation of Strategic Information Systems Planning Methodologies," *MIS Quarterly*, September, 1988, 445.

[19] Cornelius H. Sullivan, "Systems Planning in the Information Age," *Sloan Management Review*, Winter, 1985, 4.

[20] Marilyn M. Parker discusses planning for organizations in transition in *Strategic Transformation and Information Technology*, (Upper Saddle River, NJ: Prentice Hall), 1996. In particular see Chapter 13, "Creating A Transformational Planning Context".

[21] Michael J. Earl, "Experiences in Strategic Informations Systems Planning," *MIS Quarterly*, March 1993, 1.

PART
Two

Technology and Industry Trends

Advances in computing and telecommunications hardware and software provide many opportunities for today's managers and their organizations. Rapid advances in client/server and Internet-based systems enable organizations to integrate business activities and leverage their assets more effectively. Government actions and telecommunication industry developments are accelerating trends toward global business and electronic commerce. New technologies are changing managers' roles and employees' work and permanently altering the way organizations conduct their affairs. Part Two includes Hardware and Software Trends, Modern Telecommunication Systems, and Legislative and Industry Trends.

Successful information technology managers remain alert to technology and industry trends. They use trend information to prepare their people and organizations for the future.

5 Hardware and Software Trends

Founded in 1976 by Steve Jobs and Mike Markkula, Apple Computer was the darling youngster of the PC business. Its product line advanced from the Apple II, announced in 1978, to the Macintosh in 1984, to the Macintosh Plus in 1986, and on to the Macintosh SE and Macintosh II in 1987. Its solid base of talented and innovative employees attracted others with similar, valuable skills and work ethic. They developed many creative and strategically important products; however, some were carelessly exorcised in the Apple management jungle. Examples include the Mac OS for Intel microprocessors, an advanced video graphics chip called QuickScan, and a project to transform a TV set into a personal computer, which became the basis for WebTV Networks Inc.

Nevertheless, offering many advanced, innovative features like graphical interfaces, the mouse, and stereo sound, the Macintosh led the field by years. Thousands of devotees found its ease of use and excellent what-you-see-is-what-you-get publishing capabilities far superior to its competitors' command-driven characteristics. Despite major failures in product, marketing, and pricing strategies that jeopardized Apple's future, legions of fans bonded to Apple and to its Macintosh with a fervor rarely seen in business.

By 1993, however, competition and management blunders took their toll on Apple. Competitors imitated formerly unique Macintosh features. By failing to license its product to others, its market share dwindled, application programs materialized slowly, and its financial performance declined dramatically. Gross profit margins dropped from 43 to 26 percent in 12 months, and 2,100 of 13,000 permanent employees were laid off. Could competition do anything Apple could do and do it just as well? Would Apple continue to develop easy-to-use, innovative PCs for its loyal followers? The answers were critical to Apple's survival. Many feared that the visionary spirit of Steve Jobs, Mike Markkula, Steve Wozniak, and John Sculley had disappeared.

The over-hyped Newton device showed poorly in the marketplace, adversely contributing to the company's declining image. Apple touted this small, hand-held personal assistant as revolutionary technology, boasting of its ability to read handwriting and to bring its bearer instant communication. In fact, reading handwriting was much less successful than anticipated and Newton was poorly received. Eventually, it seriously drained cash from the company.

From the beginning, Apple's strategy was to be a niche player. Its systems were incompatible with those built from Intel parts, like the IBM PC and all its clones. Apple relied on Motorola microprocessors; hence its software and applications were also incompatible. In addition, Apple refused to license its operating system to other manufacturers, preserving its uniqueness while sacrificing its market share. To increase market share, Apple needed to lead the field with innovations.

Feeling pressured and hoping to improve its position through alliances, Apple signed joint development agreements with IBM leading to two software projects and a hardware project to develop the PowerPC, a microprocessor designed by IBM and manufactured by Motorola and IBM. Apple also designed a new version of Newton that stored handwritten notes without interpreting them and licensed the Newton architecture to several Japanese firms—a bold new step for Apple. Internally, Apple beefed up management controls on advanced technology projects, shortened development cycles, and searched for technology to reinforce its stock in trade—user friendliness. After years of internal debate, it began licensing its system to clone developers, too.

But these actions were too little and too late—Apple's business continued to slide. The software projects were not fruitful. Apple's partners lost interest in them and in the continued development of the PowerPC. With profits declining, Gilbert Amelio was elected Chairmen and CEO. His mandate was to stop the revenue and profit erosion and reverse Apple's fortunes. His task proved harder than many imagined.

Today, after several failed recovery plans, revenue continues to decline as losses mount. Employment at Apple is declining, research expenses have been reduced several times, and projects not central to the Macintosh system have been abandoned. New notebook computers are selling well, but uncertainty surrounding Rhapsody, a new operating system to replace the aging Mac OS, continues to plague the company. Outside application developers hesitate to write new software for Apple's systems, further compounding problems by eroding customer confidence. Apple acquired NeXT Software and retained Steve Jobs to help revitalize the company; however, in the nine months following October 1996, nine officers left Apple and the executive suite was in turmoil.

Several times during the past decade, buyers expressed an interest in Apple, but, for various reasons including unrealistic prices, sales did not materialize. Still, many believe Apple's only chance for survival is to find a strong, well-positioned buyer.

Following Amelio's departure and an unsuccessful campaign to attract a chief executive, Steve Jobs is now running the company again. He reversed the decision to license clones, bought Apple's largest clone maker, and obtained an infusion of funds from his old nemesis, Microsoft. Late 1997, Apple launched a new G3 series computer, placed manufacturing on a build-to-order basis, and improved distribution by letting customers buy Macintosh computers through retail or reseller channels, through direct sales in education, and directly from Apple.[1]

"The real question for Apple," claims Jim Carlton, a follower of the company since its inception, "is whether it has a future at all. The measure of how far it has fallen is reflected in the fact that its fate lies not in its own hands but in market forces over which it no longer has control." He continues, "But the company's plunge has taken place so rapidly that it is tantamount to a snowball picking up speed and size as it hurtles down the mountain. Can anyone stop it or slow it down? Maybe. But the odds are not good."[2]

Recent sales and earnings figures highlight causes of concern. Apple's net earnings peaked at $530 million in 1992, dropped to $231 million in 1994, and showed a loss of $742 million in 1996. Further losses are projected for 1997. Although debt free for many years, Apple's balance sheet now carries about $950 million of long-term obligations.[3]

INTRODUCTION

Advancing technology and its importance to managers and organizations is a theme of this text. The development of new forms of communication technology, powerful digital computers, and their integration into revolutionary new computer/communications systems defines the age in which we live—the information age, the age of *telematique*.

Over the past four decades, continuous technological innovation has fueled unprecedented advances in digital computing and communication capability.[4] These innovations occurred in the fields of semiconductor technology, recording technology, communications systems, and, to a lesser extent, in programming. Together, these four technical activities form the foundation of the information technology revolution. Separately, each endeavor is subject to limitations imposed by the laws of physics, human or organizational limits, system or interrelational barriers, fabrication limits, and limitations reflecting economic reality. During the last forty-plus years, practitioners in these fields have marshaled intellectual, financial, and production resources to push these technologies' practical limitations toward their theoretical limits.

Developers have been extraordinarily successful in their endeavors. Progress is measured in orders of magnitude (factors of 10). Hardware capability doubles in performance and costs decrease by half every few years, following Moore's law.[5] Component sizes decline proportionately, contributing significantly to performance increases and cost reductions. Only programming—the production of computer instructions—proceeds at a more conventional pace.

Moreover, with the exception of programming, most observers predict that this rapid innovative pace will continue for the next five years or longer. Although we can be less certain about the practical utilization of these technologies within our social and business structures, we can expect dramatic and highly meaningful results for ourselves and the organizations in which we operate.

But are we truly justified in extrapolating current trends much beyond the present? Can it be that we deceive ourselves about the technological future? Are we about to reach a technological dead end?

In reality, our anxiety about the future should concern not whether the torrent of innovation will continue—it will—but how we can harness innovations to better the lives of individuals and society as a whole. Hence, technological progress and its effect on our lives is critically important to business managers.

Although observers conclude that innovations will continue, what evidence is there that a technological brick wall isn't lurking just ahead? How can one be reasonably certain that technological innovation will continue as so many predict? Answers to these questions can be found by examining three areas of information technology: 1) logic and memory semiconductor technology, 2) electronic and optical storage systems, and 3) communication bandwidth deployment.

This chapter and the next explore these important topics and make careful projections concerning their future form and direction. In particular, the text relates them to our organizations' activities and extracts meaning from them for ourselves as managers, for our employees and associates, and for our organizations. Successful managers always remain alert to trend information; exceptional managers use leading technological indicators to prepare themselves and their firms for the future.

The invention of the transistor in 1947 set in motion a stream of innovations and inventions that spawned today's semiconductor industry.[6] The industry grows rapidly by developing many important new products. In addition, advanced semiconductor technology and numerous innovative products fuel the growth of the data processing and telecommunications industries. And, according to some, "the microchip has become the dynamo of a new economy marked by low unemployment, negligible inflation, and a rationally exuberant stock market."[7] Since 1948, revenues of the U.S. information technology industry have grown to $866 billion, making it the nation's largest. According to estimates, worldwide spending for information technology will exceed $1,400 billion in the year 2000.[8]

Semiconductor technology is significant to managers because it's the foundation for the dynamic, highly important information industry. Semiconductor chip technology makes today's microprocessors possible and permits system designers to pack ever more computational performance into systems of all sizes, at steadily decreasing unit costs. Advanced chip designs fuel the growth of the information handling and transport industries and have created the information age as we know it. Managers must have knowledge of the trends in this industry to perform their jobs effectively.

Trends in Semiconductor Technology

The exceptional growth rate of the electronics industry depends on the ever-decreasing size of the integrated circuit itself and on continued increases in the density of devices on the carrier. Industry growth rates directly relate to circuit density increases or growth in the number of devices per unit area of carrier. As circuit density increases on the chip, semiconductor device reliability improves with time, and unit cost rapidly declines. Table 5.1 summarizes these trends, which are favorable for the ultimate consumer and the semiconductor industry.

TABLE 5.1   Semiconductor Technology Trends

Diminishing device size

Increasing density of devises on chips

Faster switching speeds

Expanded function per chip

Increased reliability

Rapidly declining unit prices

As transistor device sizes become ever smaller and a chip holds more devices, device switching speeds and chip functionality increase. In the 30-year period from 1960 to 1990, chip density grew by seven orders of magnitude, from three to about 30 million circuits per chip. Projections to the year 2020 range from slightly less than 1 billion to more than 10 billion circuits per chip, depending on assumptions regarding processes and materials used to build these superchips. Today, experimental laboratory memory chips contain one billion bits, and experimental microprocessors can perform more than one billion instructions per second.[9]

Figure 5.1 shows the growth of circuit density versus time from the late 1950s to the present, and projections to the year 2020.

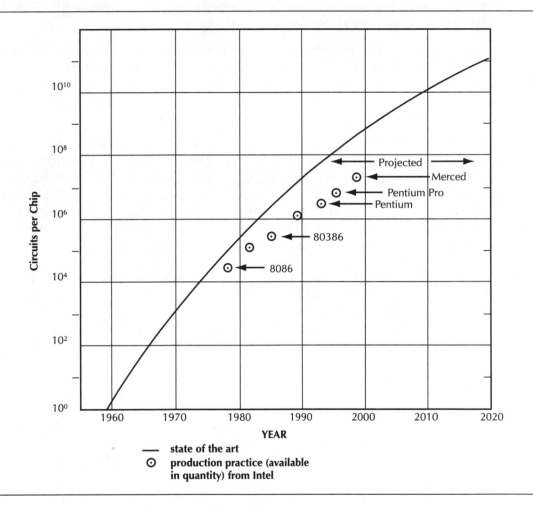

FIGURE 5.1   Semiconductor Chip Density vs. Time

Shrinking the size of integrated circuits and packing more of them on a chip reduces overall cost, increases chip functionality, minimizes device switching time (the time needed for the circuit to go from one binary state to the other), and reduces on-chip communication delays. Thus, ever-decreasing circuit dimensions reduce unit cost, increase unit performance, and drive continued semiconductor industry growth.

Present manufacturing technology produces circuit lines on a silicon chip that are fractions of a micron wide. (A micron is one ten-thousandth of a centimeter, or roughly one one-hundredth the width of a human hair.) Advanced photolithography and other techniques can further reduce the line widths, in some cases down to 0.1 micron. Reducing the line width from 0.5 micron to 0.25 micron quadruples the number of circuits per unit area and reduces signal travel distances, thus improving chip performance. Therefore, industry growth rates, which depend on increasing performance and diminishing unit costs, are inversely related to the fundamental circuit's size.

Robert Keyes presents an exciting insight when he states:

> I am writing this article on a computer that contains some 10 million transistors, an astounding number of manufactured items for one person to own. Yet they cost less than the hard disk, the keyboard, the display, and the cabinet. Ten million staples, in contrast, would cost about as much as the entire computer. Transistors have become this cheap because during the past 40 years engineers have learned to etch ever more of them on a single wafer of silicon.[10]

Today, 40 years after the commercial introduction of the single transistor chip, the number of transistors that can be fabricated on one chip has grown to more than 100 million—a factor of 10 to the 8th power, an average growth rate of 100 times each decade. While performance increased more than 100,000 times, chip cost remained relatively unchanged. Transistors on a microprocessor chip can be purchased for less than 10 cents per 1000, and microprocessor speed is available for about one dollar per million calculations.[11] Semiconductor memory chips, much less expensive to develop and manufacture than microprocessors, sell for about 0.5 cents per 1000 bytes.

Although computer chips are cheap as measured against performance, their very small size makes them about twice as valuable by weight as gold. Ancient alchemists sought to turn lead into gold—today's manufacturers do them one better by turning sand into computer chips. The semiconductor industry has been extraordinarily successful because it invests huge amounts of money in research and optimizes all important factors in commercializing research discoveries in chip production and use.

As silicon devices shrink and approach their minimum practical dimensions, scientists search for new technologies and materials to maintain the downscaling process.[12] One such material, gallium arsenide, possesses these physical properties. Gallium arsenide potentially reduces device size by another factor of 100, increases switching speeds up to 200 billion times per second, and reduces cost per unit of function by orders of magnitude. Industry, government, and academic laboratories around the world are researching this technology and bringing it into production practice.

Intel's Pentium series and Digital's Alpha microprocessor chips are good examples of the rapid pace of semiconductor development. First shipped in 1993, Pentium is the follow-on to the popular 386 and 486 chips. When introduced, Digital's Alpha chip was faster than Intel's but was not nearly as successful in the marketplace. Table 5.2 displays the advances in some of these chips and chips under development.[13]

Part Two: Technology and Industry Trends

TABLE 5.2 Performance Advances in Microprocessor Chips

Product	Clock rate in MHz	Availability
8008	.06	April 1972
8086	.3	June 1978
80386	5.0	October 1985
80486	20	April 1989
Pentium	100	March 1993
Pentium Pro	300	March 1995
Pentium II	500	March 1997
Alpha 21164	400-533	July 1997
PowerPC	500	July 1998
New Alpha	1000	December 1998
80886	1000	2000 (estimated)
801286	100,000	2011 (estimated)

Many changes in production technology, microprocessor architecture, and functionality occur with each new microprocessor version. Today's chips contain wide data flows, high-speed caches, and hundreds of other advanced features not found in early microprocessors. Today, some chips contain an advanced controller function to accommodate a second processor on the motherboard that enables symmetric multiprocessing. New versions of operating systems will support this advanced feature. As commercially-available chip densities approach 10 million transistors per chip, line widths must decline to one-tenth of a micron or so.[14]

Although Intel is by far the largest producer of microprocessor chips (nearly 90 percent of the market), it faces competition from several sources. In March 1997, Digital announced three new processors in its Alpha series with internal speeds ranging from 400 to 533 mHz.[15] With these products, DEC takes the lead for microprocessor speed, beating Intel's Pentium series, IBM/Motorola/Apple's PowerPC, and SPARC from SUN Microsystems. Figure 5.2 displays the growth in processing speed over time of several popular microprocessor chips.

Although DEC's new chips are faster than Intel's and competitively priced, too, Intel captures the lion's share of the market because of abundant support from Microsoft systems and software. Although Digital's design may be better, the Wintel marketing combination is difficult to beat. In 1996, for example, Intel shipped more than 84 million 80×86 style chips vs. its competitors' combined total of 6.5 million. Digital shipped 16,000 Alpha chips in 1996.[16]

Semiconductor laboratories around the world continue to improve silicon chip design and manufacture. For example, by reducing line width to 0.1 micron, experimental chips attain switching speeds of about 1/100th of a nanosecond. Such achievements help close the gap between silicon and gallium arsenide. In 1997, IBM and others announced process improvements that increase silicon wafer size from eight inches to 12 inches in diameter.

Two and one-half times as many chips can be cut from the larger wafers at a cost increase of 1.7 times, thus further reducing chip costs. With extensions to existing technology for photolithography and other processes, silicon will continue to be the primary material for chip fabrication in the next decade.

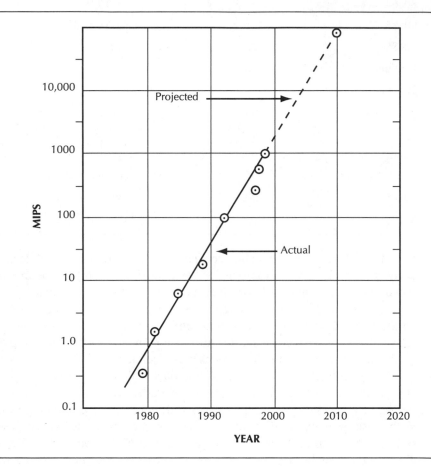

FIGURE 5.2   Microprocessor Chip Speed vs. Time

How fast can microprocessors be made to operate? What is their ultimate speed? Although no absolute answers to these questions exist, experts have made some predictions. Michael Dertouzos, Professor at the MIT Laboratory for Computer Science, answers the questions as follows: "Moore's Law will run out within a decade, limiting single-processor computers to a speed of 2-10 ghz (gigahertz). From there on, multiple processors will be ganged together. That will give us another factor of 1,000 in speed. All that, of course, does not say what these machines will do, just how fast their wheels will spin."[17]

The transistor's invention and its continued development over more than 50 years has been and will continue to be enormously important in many ways. Chip densities will increase by orders of magnitude through advances in physics, metallurgy, and manufacturing tools and processes. We can count on smaller, faster, lower cost, and more reliable memory and logic circuitry. IT systems, however, require more than semiconductor logic and memory to function. Large capacity storage devices are a vitally important adjunct to powerful computer chips. Fortunately, scientific advances abound in the recording technology field as well.

RECORDING TECHNOLOGY

Scientists and technicians are making rapid progress in logic and memory circuit development; moreover, their progress is equally dramatic in recording technology. This is fortunate, actually mandatory, for digital computing because thousands of processors ganged together and connected to gigabit-per-second fiber transmission networks demand terrabytes of storage to function properly. Advanced, high-speed processors have enormous appetites for instructions and data. Thus, computer system performance depends very heavily on huge, high-performance auxiliary memory for instruction and data storage. Approximately equal progress in both fields is essential for systems development. What is the outlook for mass storage?

Magnetic Recording

Magnetic recording devices allow permanent storage of massive amounts of data, measured in gigabytes and terrabytes, at lower cost per byte than semiconductor devices. With today's high performance hard drives, data access is under 10 milliseconds and data transfer rates are on the order of 40 megabytes per second. For example, the Western Digital WD4360WC direct access storage device holds 4.63 gigabytes of data with a seek time (the time needed for the read/write head to move to the desired data track) of eight milliseconds. The disk spins at 7200 rpm so the rotational delay (the average time the disk takes to spin to the desired data) is 4.17 milliseconds. The manufacturer warrants the device for five years. At a retail price of about $1000, the cost per megabyte of storage is 22 cents, about one-twentieth of that of semiconductor memory. Lower performance drives are available at costs under five cents per megabyte. U.S. manufacturers produce over 100 million hard drives annually, exceeding CPU production.

Magnetic recording technology is impressive. On a rigid disk, data is read or written by a read/write head flying over the surface on a cushion of air about 10 millionths of an inch thick. The disk under the head travels at a speed approaching 100 mph. Manufacturing tools and processes required to construct rigid disks in the factory rival those found in semiconductor plants. Tolerances on surface finish approach 1 millionth of an inch, and the purity of materials in the thin-film heads is similar to that of the silicon used in chip production. Like semiconductor plants, disk manufacturing plants are highly capital intensive.

Advances in head design, recording materials, and production practices ensure continued growth in capacity and performance of both rigid disks and magnetic tape devices. In late 1997, IBM claimed that its research yielded data storage at a density of more than 11 billion bits per square inch—enough to hold 725,000 pages of double-spaced, typed pages. This technology will reach the market in 2001. The industry estimates that it will ship 465,000 terabytes of storage in 1998 at an average price of about three cents per megabyte. A terabyte is 1,000,000,000,000 bytes. At five cents per megabyte in 1997, this book's contents can be stored electronically for less than a dime—much less expensively than on paper, even today.[18]

And limits to storage are nowhere in sight. Theoretical calculations and experiments with holography indicate that 10 terabytes of information can be stored in the volume of 10 stacked credit cards. Research physicist Lambertus Hesselink expects transfer speeds in the range of a gigabit per second within five years.[19] The question is not whether we can store all the data we have, but rather, what we will do with all the data we've stored.

As with silicon chip technology, advances in magnetic recording technology will migrate from the higher performance, higher capacity devices to the lower performance, more widely available devices. In addition, improvements in disk subsystem architecture will greatly increase storage system reliability.

Fault-Tolerant Storage Subsystems

The common practice of interconnecting large numbers of workstations into local area network (LAN) configurations increases the need for reliable operations of critical network components. Failure of the server (the computer that coordinates network operations and acts as a data repository) or its storage devices may halt the entire LAN operation, disrupting operations and perhaps causing lost revenue. Critical systems need to be fault tolerant. Fault-tolerant devices and networks are configured to reduce the impact of failure by introducing redundancy and back-up mechanisms.

To reduce exposure to failure in storage subsystems, disk devices are organized to maintain some form of data redundancy. This concept is called RAID—Redundant Array of Inexpensive Disks. RAID is a set of standards (RAID 0 through RAID 6 plus some others) describing various types of redundancy that can be built into disk subsystems. For example, RAID 1 employs complete redundancy by writing each data set to two different disks. RAID 5 stores data on alternate drives in sector-sized blocks with parity interleaved. RAID 6 is the same as RAID 5 but includes redundant disk controllers and power supplies. RAID 10 combines RAID 1 and RAID 5.

Redundancy greatly improves a storage subsystem's mean time between data loss (MTBDL). For example, replacing a four-disk non-RAID subsystem with a five-disk RAID 5 subsystem increases MTBDL from about 38,000 hours to 48,875,000 hours (essentially never a data loss), an improvement of about 1300 times.

The operating system considers the disk array as a single unit, but components in some systems can be replaced during system operation. For example, a diskette showing signs of failure can be replaced without terminating system operation. This is called a hot replaceable or a hot pluggable system.

The notion of redundancy in storage systems can be extended to other system components such as power supplies, cabling, and controllers. Redundancy improves system reliability at an increased cost but may reduce slightly instantaneous system response. Technicians and managers must understand the trade-offs when making system decisions that include redundancy.

CD-ROM Storage

Magnetic recording will remain highly important for a long time, but optical devices prevail for many applications involving multimedia and operating system storage. The ability to store and combine video, audio, and text in many useful patterns has countless applications. Users have just begun to tap the potential of multimedia applications.

The compact disc, known mostly for its ability to record sound to near perfection, is also a powerful medium for storing computerized data. One disc, about five inches in diameter, can hold more than two Gbytes—about 90 sets of encyclopedias. If this isn't enough for an application, devices resembling jukeboxes are available for accessing many CDs from a single computer. Today, most personal computers are equipped with read-only CD drives; some have writable optical disc drives with capacities up to 2.6 Gbytes. Today's writable optical discs are of the write-once-read-many (WORM) variety, but soon write-many-read-many (WMRM) devices will become available. Most organizations find CD-ROM storage attractive—commercial applications are predicted to grow explosively in the future.

Numerous large data bases, such as the 1990 census data, are available on CDs for computer users. Newspaper and magazine references, abstracts of academic research, legal and medical files, very large data bases of financial information, and countless other material are all available at prices that decline with improvements in technology and economies of scale in production and sales. Now widely available, this storehouse of information augments the printed media; it greatly improves the process of information retrieval for many purposes.

COMPUTER ARCHITECTURE

Computers constructed from building blocks of logic and memory, online and offline storage devices, and a variety of input and output devices span a wide performance spectrum. The spectrum is typically, but somewhat arbitrarily, divided into micro-computers, mini-computers, mainframes, and supercomputers, although these divisions are becoming increasingly blurred. The low end of the spectrum typically includes embedded micros and, on the high end, supercomputers constructed of many individual processors. Embedded micros are small computing devices found in machine tools, automotive or aircraft control systems, and household appliances such as microwave ovens and bread machines.

Before the introduction of minicomputers in 1966, mainframes were our only CPUs. Personal or portable computers originated shortly after 1970. At about this time, small computing devices (frequently single-chip devices) were embedded into various equipment for control purposes. With CPUs intermediate in performance between mainframes and personal computers, minicomputers are a vital component of the computing spectrum. They serve small- and medium-size businesses as central processors or as servers on departmental or enterprise networks.

The Microcomputer Revolution

Although systems of all sizes have grown rapidly in function, performance increase of personal computers since their introduction in the early 1970s has been the most rapid of all. Soon PCs will be available that can process nearly 1000 MIPS. Even though its architecture differs vastly, today's personal computer puts the power of yesterday's mainframe on millions of users' desks. This phenomenon has important social and business implications because it profoundly and irrevocably changes communication methods and the role of computing in modern society.

Within the decade, 1000 MIPS desktop computers containing gigabytes of storage will connect through high-speed data links to larger computers, giving users access to many more gigabytes of information. As high performance networks are deployed and large numbers of powerful processors are connected to them, experts believe the total capability of networked systems will grow as the square of the number of processors.[20] In other words, doubling the number of processor nodes quadruples the network system's capability. Just as the telephone network's value increased as the number of telephones connected to it increased, so too will the power of network-centric computing increase as the number of processors connected to the network increases.

Ten years hence, office workers will have enormous computing and communicating power at their fingertips. They, their managers, and their firms will be challenged to use this capability advantageously. The field is ripe for inventors, innovators, and visionaries.

Supercomputers

Firms in several countries manufacture supercomputers, the highest performance systems of all, to serve the needs of scientific laboratories, government agencies, and large commercial customers. Many institutions in the U.S. and more than 20 other countries are conducting research on these advanced systems.[21] The technology and architecture employed in these machines is leading-edge, and they are precursors to tomorrow's more widely available machines.

Uniprocessors with capabilities exceeding several hundred MIPS are common today, but many system designers turn to multiprocessors to enhance performance. Multiprocessors are not a new phenomenon—they have existed since the 1960s—but present-day processor technology differs substantially from that of three decades ago.

Early multiprocessors for commercial applications usually consisted of a front-end processor serving as a job scheduler for one or more back-end processors. Some applications divided the workload by type between the processors; one processor handled communications and another performed batch production, for example. Although today's commercial multiprocessors contain 10 or more individual processors, the trend of advanced technology is toward systems with as many as tens of thousands of interconnected processors.[22]

Because individual processors are cheap, adaptable, and powerful, hundreds or thousands can be combined in multiprocessor configurations ideally suited to advanced information-age applications. Intel's efforts to build a one-trillion operation per second multiprocessing computer illustrates this trend.

Because these large machines are typically used for scientific calculations, their performance is measured in floating-point operations per second (flops). Floating point is a term that describes the computer's ability to track the decimal point over many orders of magnitude while performing arithmetic calculations. One million flops is one mega-flop, or Mflop, and one billion flops is one giga-flop, or Gflop.

Supercomputers containing many parallel processors are capable of one trillion flops, or one Tflop. But even more powerful machines are planned—some with as many as 32 million processors. Designed for high-speed image processing, these machines execute the same instruction simultaneously in all processors. This architecture is called single instruction, multiple data.

Computer manufacturers and government agencies are discussing the development of ultra-fast computers with projected processing speed of one petaFLOP, or one million billion floating-point operations per second. Such performance can be achieved only by interconnecting large numbers of very high performance processors in a massively parallel arrangement. Presently, computer scientists are considering machine designs that have one million or more processor nodes, each capable of executing instructions at the rate of 100 Mflops. Calculations indicate that a computer of this capability would have more computing power than all networked computers in the U.S. today.

Numerous major engineering and programming problems must be solved before these computers' power can be realized. Operation of one million interconnected 100 Mflop processors implies massive internal data flows. Developing high-bandwidth internal communication channels among and between processors and to the huge main memory poses tremendous engineering challenges. Developing operating systems to manage these massively parallel machines and constructing application software that takes advantage of these massive computer resources are also formidable challenges. But, machines of this type are ideal for scientific research, such as weather forecasting and nuclear reactor modeling, and for commercial applications, too.

The usefulness of supercomputers and parallel processing, however, extends far beyond scientific studies. Wal-Mart uses them to obtain vital information about its vast store of sales and purchasing data.[23] It uploads 20 million point-of-sale transactions from its 3000 stores daily, adding information to its 24-terabyte data warehouse. Several thousand complex queries are issued daily against the database, two-thirds of which are stored for use by parallel-processing machines. Wal-Mart began moving to parallel processing in 1989 and credits its sophisticated information system with helping it master the details of sales and inventory management.

Advances in supercomputers show no signs of slowing and the technology and architecture they embody will migrate to a broader range of machines in the future. Some analysts forecast that Gflop capability will be available in desktop computers in the near future. The technology of computing advances rapidly and relentlessly from one end of the computing spectrum to the other.

As we have seen, performance, capability, and function of computer systems continue to increase rapidly through advances in semiconductor and recording technology and through new internal architectures using powerful new microprocessor chips. Novel multiprocessor architectures that combine thousands or more uniprocessors promise performance increases of many more orders of magnitude. Because many of these advances result from economic as well as technical factors, new, more powerful computers will be both available and widely affordable, i.e., cost per unit of function will continue to decline significantly.

Given our confidence in these trends, what alternatives are available to deploy this capability within the enterprise? What systems considerations or architectural innovations are being employed to translate continued hardware improvements into improvements in organizational effectiveness?

Client/Servers

The abundance of low-cost, high-performance personal workstations, convenient and reliable means for interconnecting them, and the organizational need to use technology to empower workers and achieve productivity improvements is driving a strong trend toward moving processing power from centralized mainframe systems to local workstations. Organizations are moving from distributed data processing models (departmental islands of computing using mid-sized computers) to end-user computing environments (individual islands linked together and sometimes linked to departmental or centralized systems). During this evolutionary period, the computing workplace has been restructured as more function moved to the workplace and as interconnectivity increased. This transformation developed into a very popular architecture called client/server.

The first application of local networks to data processing consisted of workstation terminals (keyboard and display) connected to mainframe processors on which large data stores and application programs resided. Terminal operators entered or updated data in mainframe storage and reviewed the results of application processing on their displays. The application program contained the format of the data presentation on the display terminal. In most cases, computer operators working to a production schedule initiated mainframe application processing. In other cases, such as program development, operators entered source language statements at terminals, initiating language compilers from the terminal, and displayed the compiled results from compiler-created data sets.

As their capabilities advanced, personal workstations were used to customize presentations of mainframe application output and, for modest applications, to process the applications themselves. Voluminous data stores, such as archival records and sensitive information such as payroll records, remained securely protected on mainframe or departmental facilities. These larger CPUs provided file space and print facilities for numerous users at linked personal workstations. Used in this manner, the larger CPUs became known as file servers or print servers. Data and program storage, application processing, and control functions became rebalanced as application architecture migrated from traditional host-supported terminals through file and print server modes to client/server models of operation. Figure 5.3 shows the evolution of client/server architecture.

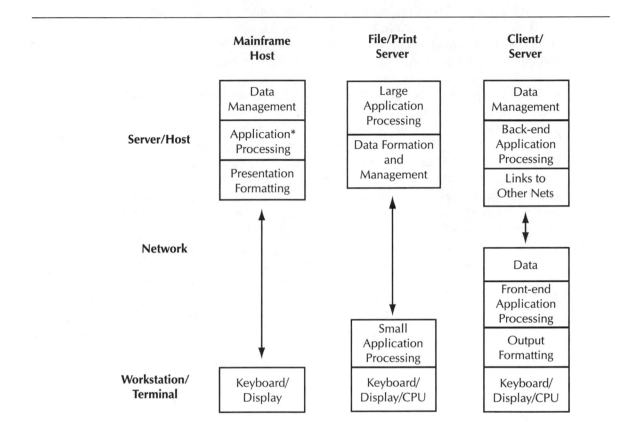

	Mainframe Host	**File/Print Server**	**Client/ Server**
Server/Host	Data Management Application* Processing Presentation Formatting	Large Application Processing Data Formation and Management	Data Management Back-end Application Processing Links to Other Nets
Network	↕	↑	↕
			Data Front-end Application Processing Output Formatting
		Small Application Processing	
Workstation/ Terminal	Keyboard/ Display	Keyboard/ Display/CPU	Keyboard/ Display/CPU

*NOTE: Control (initiation) of application processing may reside with the workstation operator or the host operator.

FIGURE 5.3   Evolution of Client/Server Architecture

In client/server mode, application data is entered into workstation storage and preliminary data processing occurs there, frequently in combination with data from the server. Similar operations occur simultaneously at many workstations throughout the business. Periodically, results of workstation processing are combined, and the server completes final or summary processing.

For example, in an order processing operation, individual operators at workstations receive phone orders and enter them into order-entry applications. These workstation applications (front-end processing) prepare the screen; obtain customer data such as previous orders, credit limits, or product information from the server; perform validity checks; and record the order after verification is complete. Periodically, orders from all order-entry stations are collated for processing (back-end processing) on the host or server. The host application prepares stock picking and packing information for the warehouse, integrates shipping data for the distribution center, and performs many other administrative and logistical tasks.

To complete the example above, order clerks enter information from phone or mail orders; some customers enter information directly via inter-organizational system links and electronic data interchange. Extending the network to customer offices and linking them electronically to the ordering system is a powerful way to reduce costs, enhance customer service, and gain competitive advantage. In a similar manner, suppliers can access inventory records and update inventory when it's needed—just in time.

The order-entry example is typical of many made possible by client/server architectures. Financial operations such as payroll and accounts payable, engineering tasks such as product design or bill-of-material development, and many office and administrative tasks are performed more effectively in this mode. In fact, client/server facilities transform many traditional business processes. They empower employees with increased responsibilities and sophisticated data processing capabilities to carry them out. The process reduces information float and paper records and is more cost effective. Businesses, customers, and suppliers benefit from higher-quality, lower-cost operations.

Figure 5.4 displays a simple client/server system operating on a local area network. In many organizations the server depicted in Figure 5.4 performs more functions than those shown here. For example, the server may manage a bulk printer for all workstation users or connect to a facsimile line to transmit workstation-initiated messages. The server may also act as a bridge or router/gateway connecting this LAN to other LANs or MANs (metropolitan area nets) that the organization uses. It may also connect to a large enterprise mainframe, sharing processing and data storage functions hierarchically. The possibilities are almost limitless for innovative network architects and business process engineers.

Because of the many important advantages to downsizing mainframe applications and converting them and their associated business processes to distributed processing modes of operation, businesses make substantial investments in client/server implementation, applications, and operations. Client/server operations have revolutionized the implementation and management of information systems, with far-reaching ramifications for traditional business organizations. IS professionals and operations people are profoundly affected by the wave of networked processing.

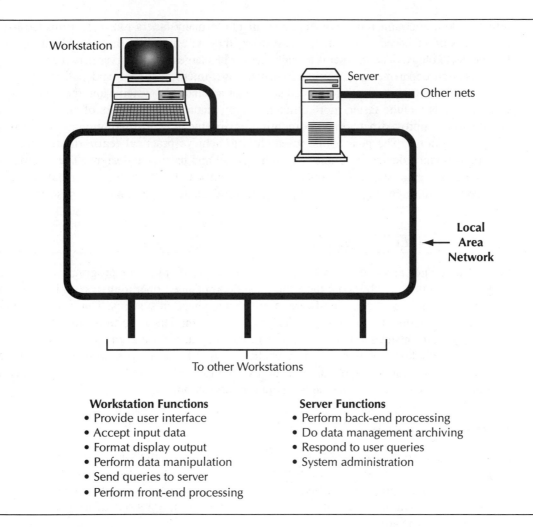

Workstation

Server

Other nets

Local
Area
Network

To other Workstations

Workstation Functions
- Provide user interface
- Accept input data
- Format display output
- Perform data manipulation
- Send queries to server
- Perform front-end processing

Server Functions
- Perform back-end processing
- Do data management archiving
- Respond to user queries
- System administration

FIGURE 5.4   Client/Server Architecture

Intranets and Extranets

Client/servers represent a major step in computing's evolution from stand-alone mainframes to fully integrated but widely dispersed facilities. They mark the beginning of a more general environment of fully integrated network computing.

Today, many firms use Internet technology to augment other forms of communication within the firm and to communicate beyond the firm, too. Departmental or headquarters Web servers support intranets for important internal employee information. When accessible to persons outside the organization, these nets are called extranets. Today, organizations add to the explosive growth of the Internet by using Web technology to foster their interests internally and to make information and commercial applications available to many others too.[24] As firms exploit the advantages of advanced communications systems, they will find many uses for client/server and Internet technology. (The Internet is actually the world's largest client/server network.).

Research departments at computer firms, government agencies, and public and private research labs in this country and abroad search for advances in systems and their applications. Current trends include faster, more efficient servers, increasing numbers of nodes in network architecture, improved networked computing environments, and revolutionary software and operating systems. The goal of the research is to achieve practical realization of inherently high performance devices through new architectural and hardware designs. The Internet, still in its developmental stage, is destined to evolve rapidly and achieve greater social importance through breakthroughs in programming, architecture, and telecommunication systems.

PROGRAMMING TECHNOLOGY

Parallel computer architecture raises very significant challenges for programmers, language designers, and people who write the enormously complicated operating system programs that provide the interface between hardware, users, and application systems. The task of designing language translators and making parallel-processing compilers is formidable indeed, but the control programs needed to manage large parallel computers pose even more demanding challenges. Nevertheless, solutions to these challenges will be developed so that users and their applications can capitalize on the enormous potential of parallel-hardware architectures. Obviously, this requires a tremendous amount of innovation.

Operating Systems

An operating system, such as Microsoft Windows NT or Apple's Rhapsody, is a computer program that interfaces between the computer hardware and its users. Its purpose is to provide a friendly environment in which users may execute programs. Thus, an operating system's primary goal is to make the computer system convenient to use. Its secondary goal is to permit the computer hardware's efficient use.[25]

The first uses of computers involved an individual with a problem who wrote a program to solve it, then personally ran the program on the computer. This method of operation was called "hands-on." The programmer/user managed input/output devices by including device drivers, usually a small deck of punched cards, along with a deck of punched cards for the application program. Schedule management for the computer consisted of a sign-up sheet, and set-up time played an important role in this type of operation. Some programs ran longer than anticipated, inconveniencing the person next in line, while others failed to run at all, causing idle time in the schedule.

As hardware and applications gained complexity, operating systems needed improvements to make the total system efficient and convenient. Through which stages did operating systems advance? What can we expect from future operating systems? Table 5.3 lists advances that operating systems achieved.

TABLE 5.3 Operating Systems Advances

Programmer/operator

System monitors

Uniprogramming operating systems

Multiprogramming operating systems

Multiprocessing operating systems

System monitors and permanent computer operators solved some of the problems discussed earlier. The computer operator batched similar jobs, reducing set-up times and improving efficiency by managing peripheral devices. The system monitor contained all the device drivers, compilers, assemblers, and the job control system. Small, peripheral computers performed card-to-tape and tape-to-printer tasks, thus reducing input/output time and more efficiently using the main processor.

Later, multiprogramming several application programs in main memory simultaneously improved CPU efficiency. Multiprogramming operating systems schedule execution of programs according to predetermined algorithms designed to minimize idle system resources. These operating systems handle memory management, task scheduling, main memory and data protection mechanisms, timer control, and other functions. Multiprogramming operating systems are extremely complex; their development requires large amounts of time and resources. Today, multiprogramming (multitasking) operating systems are available for most personal computers.

Multiprocessor hardware configurations have existed for two decades or longer, and operating systems to manage them have evolved with the hardware. These operating systems handle all the functions mentioned above; in addition, they manage several interconnected multiprogramming CPUs. These systems' complexity qualifies them as one of the most complex human undertakings.[26] Complexities of operating systems supporting large parallel processors will probably increase by orders of magnitude in the future.

In addition to the functions already noted, contemporary operating systems contain extensive communications and telecommunications software that permit application programs to utilize networks with ease. They also provide application programmers with data-management systems that handle and organize the large amounts of data associated with many modern data-warehousing applications. Most organizations have specialists who install and maintain their operating systems. Called system programmers, these skilled individuals assist application programmers and others in using these operating systems' many features.

Today's desktop computers typically use one of the three most popular operating systems. The Microsoft DOS/Windows operating system, used on IBM systems and on IBM compatibles, has the lion's share of the PC market. Today, most users run Windows 95 or Windows NT, although some still use DOS. The next revision of the popular Windows 95 is expected to be Windows 98. Apple's Mac OS, used only on Apple Macintosh computers, has about five percent of the market. Apple expects its new operating system, Rhapsody, to gain popularity and help increase sales for Apple. Developed as a competitor to Windows and DOS, IBM's OS/2 seems unable to gain ground against Microsoft. OS/2 is a more expensive system, requiring more computer power. It is better suited to larger systems in sophisticated networked applications.

Network Operating Systems

Other operating systems have been developed to manage services on large computer networks. Windows NT, Unix, and NextStep are in this category. These operating systems are very important given the trend toward network computing, so we can expect considerable development activity by computer suppliers.

In many instances, client/server operations provide opportunities for firms to improve productivity significantly. But the client/server architecture itself still offers many opportunities for increased efficiency. Generally, networked systems' inefficiencies fall into four categories:

1. Available but inaccessible system resources

2. Processing or storage redundancy

3. Weak or ineffective system controls

4. Excessive need for manual operator intervention.

The client/server architecture splits the application functions in two—one part executes on the server and the other on the personal workstation. Client workstations talk to servers only. In the next step toward the network computing model, processors are considered peers and each element can talk to the others. Instruction execution, file storage, and database management resources are interchangeable among system elements. The goal of network computing generalizes the client/server model to give users easy access to all system resources required for their applications. Meeting this easily identifiable goal is extremely difficult—in practice totally achieving it may be difficult.

By reducing or eliminating inefficiencies in client/server configurations, well-developed network operating systems improve the performance of networked systems. Application interfaces are more consistent, easier to use, and permit more flexible work flows and improved organizational effectiveness.

Sophisticated operating systems for large computer systems evolved over many years, but the evolutionary pace for small computer systems has been more rapid. Technologies of large systems quickly migrate to small systems, and new technologies for interconnected personal workstations are developing rapidly, too. Alert IT managers carefully observe these trends and attempt to profit from them.

Software associated with future systems will be extremely complex and differ substantially from software commonly used today. System designers, and perhaps system programmers, must change their approach to data-processing problems if they are to realize the full potential of hardware advances. Their new vision must include not only advanced hardware and telecommunications systems but must embrace new methods of operation such as client/server and networked computing models. This new vision will make solving increasingly sophisticated problems possible—problems thought intractable in today's sequential world. Users of computing technology will benefit greatly from expanded vision and inventiveness.

Application Programming

In the past, compiler designers focused their attention on the efficient execution of sequential operations. Conventional languages specified tasks to be performed one at a time in a single thread mode. They brought together program instructions and data for discrete sequential operations. Achieving faster operation required increases in the speed of purely sequential activities. Parallel computers, however, demand languages designed to attack problems in a multi-thread or simultaneous fashion. Thus, new approaches to computer languages must be taken to capture the benefits of parallel hardware.

Due to the large inventory of sequential programs and the ease of working sequentially, first attempts at parallel programming involved converting sequential operations to simultaneous operations. For example, programmers examined programs containing loops for parallel operations with some success, particularly in scientific vector or matrix calculations. Computer programs have been developed that search for parallelism in routinely sequential instructions and then generate code that uses the implicit parallelism they discover. Other methods based on data flows have been marginally successful.

In the single instruction, multiple data machines described earlier, main memory is divided into increments, one for each processor (perhaps tens of thousands of increments). Data is distributed over the memory increments, one data element per increment. Then, each processor performs the same instruction in synchronized fashion, transforming the data in all memory increments simultaneously. This parallel approach to problem solving works well on certain classes of problems such as image processing. The computer requires a supervisor processor controlled by a supervisor program to coordinate activities of all other processors.

Parallel-processing compilers are being developed for commercial applications. For example, IBM Clustered FORTRAN executes jobs across 12 processors at speeds up to 10 times faster than a single processor can. Firms developing parallel processors and some firms using them are actively researching parallel processing. Significant advances are expected.

Rapidly declining costs and high-speed hardware permits implementation of computer programs in new ways, too. Some frequently used software is now implemented directly in hardware with great improvements in speed. This is called Application Specific Integrated Circuits (ASICs). Data communications, robotics, and even traffic lights and household appliances provide extensive examples of programs implemented in hardware. As ASICs develops further, choosing a computer also means choosing some hardware embedded applications. This has great potential significance for the computer industry.[27]

COMMUNICATIONS TECHNOLOGY

The worldwide telecommunications industry has capitalized on the same technology as the computer industry, for all the same reasons. Digitization of analog communication signals permits the industry to apply enormously productive digital technology effectively. New devices, new media such as satellites and fiber-optics, and changes in industry structure and environment, such as the court-mandated breakup of AT&T, the 1996 Telecommunications Law, and privatization of many national phone companies, all operate in concert to expand communications capability phenomenally. These and many more factors bring the industry tremendous growth and benefit users by delivering greatly enhanced communications capability.

For example, the bandwidth explosion in wireline networks created by fiber-optic systems and the increase in our ability to transmit information through cable, fiber, and satellites yields bandwidth growth rates of about 100 percent per year. Doubling capacity every year means that network bandwidth increases by a factor of more than a thousand in a decade. If we ever reach the physical limits of fiber-optic technology, all the phones in Europe can be connected to phones in North America simultaneously—over a single fiber.[28] More about these exciting developments in the next chapter.

For reasons cited earlier and many more, we can count on remarkable technological advances far into the future. Any concerns that advancing electronic technology will hit a brick wall, a technological dead end, are unfounded. They will undoubtably continue—but can their introduction into society be managed fairly? "In the rush for rich pickings," says Colin Blackman, editor of *Telecommunications Policy*, "I fear that governments and the telecommunication industry are in danger of creating a global information-rich elite while condemning the rest of the planet to the information slums."[29] Information disparity is only one of several important questions surrounding technology introduction.

IMPLICATIONS FOR MANAGERS

Because of very favorable economics, technology improvements have greatly increased the information processing capability available to individuals and organizations. Growing numbers of large systems, intermediate-range systems, and microcomputers, all with significantly improved performance compared with their predecessor systems, express this capability. Tomorrow's managers will command orders of magnitude more computing capability than today's and will face the continuous challenge to achieve the full potential of this rapidly evolving capability.

Huge data stores containing all kinds of detailed information will accompany the increased power of computer hardware. The firm will generate information in the data warehouse, mostly from within, but increasing amounts will originate from public sources available to wide audiences. Access to these huge data stores will be rapid, easy, and inexpensive. The potential for enormous benefits from sophisticated data analysis will increase substantially.

Enormously complicated, communication-based operating systems will make advanced hardware with great power and complexity functional. Most systems will interconnect through extensive, fast, and easy-to-use telecommunications networks that will enable rapid communication throughout the organization and to suppliers and customers.

Nearly all employees will use electronically interconnected devices for rapid communication and nearly simultaneous parallel problem solving. Employees of many firms today already engage in parallel processing with intranets and client/server systems. On a global scale, firms are linking to their customers and suppliers through Electronic Data Interchange, the Internet, extended client/servers, and other interorganizational systems. Telecommunication will continue to reduce or remove time and distance barriers and increase the efficiency of international alliances and global partnerships. Alternative forms of business organization will flourish in this environment.

These changes are very significant for managers. Within the next 10 years, information technology for business and other activities will exceed today's capability by orders of magnitude. More important are the ways we will apply technology, but, here our vision is much less clear. The editors of *Scientific American* said, "Imagine the citizens of the 18th century trying to envision the shape of the future that would include electrical power, telecommunications, jet transportation, and biotechnology. We who are alive at the end of the 20th century are having the same trouble discerning the impact of an evolutionary force that is reshaping our world: the fusion of computer and communication technology."[30] Tomorrow's employees and managers will face the opportunities and the challenges of advancing technology in ways we cannot possibly envision today.

Potential benefits of information technology will not come without some eventual pitfalls. Managers must not only harness this growing wave of potential but must manage inevitable personal and organizational problems riding in its wake. Recalling Drucker's comments that opened the first chapter, we realize that one challenge is to assimilate technology while facing organizational transitions, managerial reductions, and personnel dislocations.

Facing many rigorous challenges, skillful general managers must develop strategies to introduce emerging, rewarding new technology. Their strategies must capture the technology's advantages for their organizations while mitigating the myriad difficulties that arise.

IMPLICATIONS FOR ORGANIZATIONS

The first chapter asserted that major organizational changes will accompany the conversion to knowledge-based organizations. Four decades ago, some scholars thought introducing digital computers and automating thousands of routine tasks would generate widespread unemployment. Instead, many newly created jobs and upgraded old ones resulted in more job opportunities today than individuals qualified to fill them. Managers and employees who constantly maintain relevant skills are always in high demand.

As firms streamline operations to become more competitive in our knowledge-based society, they continue to create new and better jobs for novice and experienced employees who acquire and maintain critical skills. In today's rapid-paced business environment, our organizational lives will not remain static for long. Managers can anticipate evolutionary, if not revolutionary, changes in organizational structure and in the content of managerial roles. Adapting to ever-shifting currents of change via structural adjustments will challenge future managers.

Other organizational changes are underway, too. Firms create alliances and form joint ventures to exploit technology, while transforming themselves in many other ways. For example, Volvo's integrated network tightly couples its subsidiaries throughout Europe and North America to its headquarters in Sweden for improved management control. Corning's wheel and spoke management structure centralizes control information while decentralizing operational decision-making. Both radically depart from the norm of 20 years ago. Subsequent chapters of this text discuss these and other manifestations of information technology.

Rapid advances in technology impact organizations in yet another way. Alliances form among major corporations to speed the development and deployment of new technology by sharing financial and intellectual resources. For example, Oracle and Time Warner's CNN are collaborating to improve CNN's World Wide Web news service so consumers can personalize information they want to receive. Because personalization is a major selling point for Web news providers, competition has increased—Microsoft joined NBC, and the Walt Disney company launched a Web news site. Many telephone and cable companies have engaged in joint ventures so they can share financial, intellectual, and physical resources in the competitive race to bring the public advanced services.

SUMMARY

The rapid technological advances experienced for nearly four decades will continue into the foreseeable future. To continue, however, many technical problems in metallurgy, mechanics, systems architecture, programming systems architecture, and other scientific or engineering fields need solutions. Rapidly advancing information processing technology will challenge our ability to utilize it beneficially, and, like today, technological solutions will wait for problems. Management and enterprise-wide issues will continue to restrain our ability to capitalize fully on rapidly emerging technology.

Technological advances and new products will create huge opportunities for managers in all lines of work. Many managers will be able to exploit these opportunities to benefit their employees, their firms, and themselves. Employing sophisticated technology is not an optional activity. In our increasingly competitive global economy, failure to stay current technologically will be a high-risk option.

Review Questions

1. What are the technological foundations for advances in digital computers?

2. Which of these technologies has advanced at a moderate pace, and why?

3. What are today's observable trends in semiconductor technology development?

4. Why are semiconductor manufacturers striving to shrink the size of integrated circuits on chips, and how much smaller can they make them?

5. Why are advances in logic and memory technology and simultaneous advances in recording technology essential for progress in systems development?

6. Why are rigid-disk manufacturing plants similar in sophistication to semiconductor plants?

7. Why are optical recording devices for use in digital computing systems increasingly popular?

8.   Why are RAID devices becoming popular, and on what principle does their success rely?

9.   What are the important trends in supercomputers?

10. What challenges do parallel processors pose to operating system developers?

11. In what ways does the operation of today's PCs resemble the operation of mainframe computers 20 years ago? What are the implications for IT management?

12. How is a client/server system organized, and how does it operate?

13. What is the relationship between communications technology and other parts of information technology?

14. Summarize the dominant trends in computer technology today.

15. What is the managerial significance of technology trends? How will these trends affect organizations?

Discussion Questions

1.   Considering the facts in the Business Vignette and using Figure 2.1, analyze Apple's position today. Which forces governing competition have been most critical to Apple?

2.   Discuss the business skills referred to in the statement, "Alpha is a superior design but its success ultimately depends on DEC's business skills."

3.   The Intel 80386 cost about $100 million to develop. It's estimated that the follow-on product, the 80486, took about four years and $300 million to develop. What risks were involved in this development, and how might these risks be quantified?

4.   What is the annual compound growth rate in unit performance per dollar for a technology whose unit performance doubles and cost halves over a four-year period?

5.   Michael L. Dertouzos answers a question when discussing ultimate microprocessor speeds but raises a more important one. What is this question? Discuss the consequences of some possible answers.

6.   Discuss the concepts of RAID storage and their application to system components other than storage subsystems.

7.   Discuss the advantages and disadvantages of CD-ROMs in comparison to hard drives, floppies, and magnetic tape.

8.   Discuss how client/server operations might be extended to a firm's customers and suppliers in a catalog sales retail operation. What are the possibilities and limitations of this service, and what might be its strategic significance?

9. Discuss why technology advances increase management risk. What actions might managers take to mitigate risk factors?

10. What factors encourage the rise in alliances, consortia, and joint ventures? What are the international implications behind these movements?

11. Discuss the elements of risk and reward for corporations engaged in high-technology development.

Assignments

1. Research Apple Computer's Web site at www.Apple.com, and write a short report on Apple's latest financial filings or on Apple's latest product announcement. How does this information relate to the material in the Business Vignette?

2. Determine how much money U.S. corporations spend on research and development activities in the areas of semiconductor technology, recording technology, telecommunications technology, and programming. Compare these amounts with the R&D activity in the pharmaceutical industry. What conclusions or inferences can you draw from the data you collected?

3. Obtain the descriptive details of one of the new operating systems mentioned in this chapter, and summarize them for the class. In particular, describe where this system would be used and why the developers expect it to succeed.

4. Many alliances and joint ventures are in process among IT companies today. Select one and discuss its goals, objectives, and prospects. Prepare a summary of your findings for the class.

ENDNOTES

[1] Apple press release, Nov. 10, 1997.

[2] Jim Carlton, *Apple, The Inside Story of Intrigue, Egomania, and Business Blunders* (New York: Times Books, 1997), 429.

[3] Revenue is projected to be $7.08 billion in 1997, down from $11.1 billion in 1995. *The Value Line*, October 24, 1997, 1083.

[4] During the 1940s and 1950s, analog computers under development were thought to have great promise, but the invention of the transistor and the development of silicon chips allowed digital computers to eclipse their analog counterparts totally.

[5] Gordon Moore is a founder of Intel Corporation and was its CEO until 1987. Since the first computer chip was fabricated, transistor densities on chips have increased and switching times have declined at rates such that computing power per dollar doubled about every 18 months. This phenomenon, known as Moore's law, is predicted to continue for the next decade or so.

[6] John Bardeen, Walter Brattain, and William Shockley demonstrated the first transistor at Bell Laboratories on December 23, 1947. They received a Nobel prize for their invention in 1956.

[7] Walter Isaacson, "MAN OF THE YEAR...driven by the passion of Intel's Andrew Grove," *Time*, December 29, 1997/January 5, 1998, 48.

[8] Steve Lohr, "Information Technology Field is Rated Largest U.S. Industry," *The New York Times*, November 19, 1997. Year 2000 estimates by Montgomery Research.

[9] Gary Stix, "Toward 'Point One'," *Scientific American*, February 1995, 90.

[10] Robert W. Keyes, "The Future of the Transistor," *Scientific American*, June 1993, 70.

[11] The 200 mhz Pentium Pro, first shipped in 1995, contains 5.5 million transistors, and one could be purchased in 1997 for $600. See *Computer Shopper*, June 1997, 713.

[12] Laboratory researchers are experimenting with dimensions near 30 Angstroms or about 0.003 microns where severe quantum effects make further downsizing very difficult. See note 9.

[13] *The Wall Street Journal Reports—Technology*, June 16, 1997, R4.

[14] The Intel chip, code named Merced and due to be available in 1999, is expected to contain about 12 million transistors.

[15] April Jacobs, "NEC Joins High-End Pentium Pro Battle," *Computerworld*, March 24, 1997, 41.

[16] Jaikumar Vijayan, "Intel Chugs on; Rivals Still Lag," *Computerworld*, May 26, 1997, 41.

[17] Michael L. Dertouzos was the guest speaker discussing computing in the 21st century on the *Business Week* Online Talk and Conference, August 1997.

[18] In November 1997, IBM announced a 16.8-gigabyte drive for PCs that can hold up to eight hours of full-motion video on its 3.5-inch disk. Its list price is $895.

[19] Mark Halper, "A World of Servers," *Forbes ASAP*, June 3, 1996, 122. See page 126 for data.

[20] This is called Metcalfe's Law. In May 1973, while working at Xerox's Palo Alto Research Center, Robert Metcalf invented Ethernet, a transmission protocol for local area networks.

[21] Ralph Duncan, "Parallel Computer Construction Outside the United States," *Advances in Computers*, Vol. 44 (San Diego: Academic Press, 1997), 180. This activity casts doubt on the U.S. policy to keep advanced computers from entering certain countries.

[22] IBM's chess-playing machine, the RS/6000, uses 256 processors to analyze 200 million chess board positions per second.

[23] Charles Babcock, "Parallel Processing Mines Retail Data," *Computerworld*, September 26, 1994, 6.

[24] According to Jim Barksdale, CEO of Netscape, 86 million people are connected to the Web. "Nightly Business Report", Sept. 3, 1997. More than 125 million will use e-mail by 2001.

[25] James L. Peterson and Abraham Silberschatz, *Operating Systems Concepts*, 2nd ed. (Reading, MA: Addison-Wesley Publishing Company, 1985).

[26] For details on IBM's S/390 Parallel System Complex and its operating system, see *IBM Systems Journal*, Vol. 36, No. 2.

[27] Thanks to an anonymous reviewer for suggesting that this topic be included.

[28] Emmanuel Desurvire, "Lightwave Communications: The Fifth Generation," *Scientific American*, Special Issue, Vol. 6, No. 1, 1995, 54.

[29] Colin Blackman, "To Have and Have Not," *Telecommunications Policy*, January/February, 1994, 3.

[30] "The Computer in the 21st Century." See note 28, p. 4.

6

Modern Telecommunications Systems

"A web of glass (fiber) spans the globe. Through it, brief sparks of light incessantly fly, linking machines chip to chip and people face to face. Ours is the age of information, in which machines have joined humans in the exchange and creation of knowledge."[1]

INTRODUCTION

This is the age of *telematique*, the integration of computers and telecommunication systems. Supported by software such as e-mail, file transfer programs, and browsers, telecommunication systems link many of the world's computers and change their essential character from computational devices to communication instruments. With hundreds of millions of individuals expected to be online in the next decade, networked systems and software are changing the computers' function, increasing their utility, and rapidly transforming them to commodity products. In the age of *telematique*, the term personal "computer" is becoming a misnomer.

The importance of software-enabled networks will continue to grow, so managers need to understand network basics as well as their critical business applications. Once they understand the basics, managers know that networking is a highly technical discipline. Like hardware engineers and system programmers, network specialists are technically sophisticated. Just as financial managers must know the basics of the tax code to manage tax accountants, IT managers must understand the basics of telecommunications to communicate with and manage network specialists effectively. This chapter provides these basics.

Telecommunications is the science and technology of communication by electronic transmission of impulses through telegraphy, cable, telephony, radio, or television, with or without physical media. Tele stems from the Greek word for distant; communicate has Latin origins meaning to impart. Telecommunications impart information over a distance.

This chapter deals with electronic communication methods and devices tied together or linked into systems or networks. In telecommunications systems, both the type of interconnecting media (wires, cable, etc.) and the form or architecture of the interconnections are important. The interconnecting media and its architecture form the network. The transmission medium—wire, cable, or the atmosphere in terrestrial microwave—carries signals through the system. The devices the network connects define the network type as, for example, telephone network, television network, or computer network.

THE TELEPHONE SYSTEM

The telephone system is a switched, interactive, narrow band, public network. A switched network is one form of network architecture. Network architectures (the method of interconnecting the primary elements) can be of many types. Traffic patterns, message types, frequency of use, and other factors dictate their form. Telephone traffic is usually sporadic and consists of two-way or interactive voice signals. In a telephone system, any one phone may need to communicate with any other phone at any given time, but not all phones will be used simultaneously. These communication characteristics help determine the network architecture.

The part of the system nearest the user is organized in a star pattern. Up to 10,000 access lines originate at the local exchange, fan out in a star-like pattern, and terminate at the residence or business phone. Figure 6.1 illustrates this architecture.

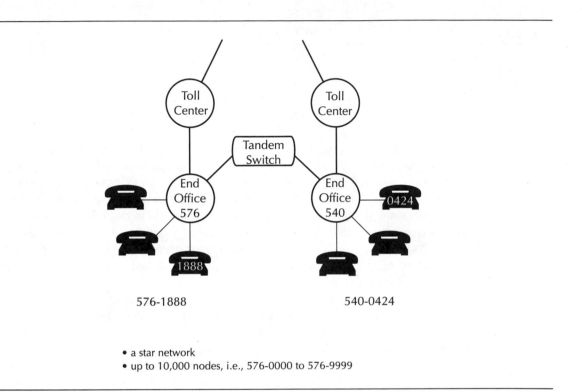

576-1888   540-0424

- a star network
- up to 10,000 nodes, i.e., 576-0000 to 576-9999

FIGURE 6.1   End Office Network Configuration

The devices at the customer end of each access line are usually ordinary telephones but may be fax machines, computers, or other terminal devices. Collectively, these devices are called customer premise equipment. The last four digits of the phone number identify the access lines, which connect the customer premise equipment to the local exchange. The first three digits of the seven digit number (576 and 540 in Figure 6.1) identify the local exchange to the network. Access lines are called the local loop or the twisted pair, referring to the pairs of copper wires that usually connect to the end office.

One purpose of the local exchange is to connect any one phone to any other phone on demand. To accomplish this, the exchange contains a switch that makes the desired connections between any two of up to 10,000 access lines. Thus, the exchange forms a switched star network. It is very effective for traffic patterns in which most access lines are idle at any given time. At this level in the phone network, on average about 1,500 of the 10,000 lines are busy at peak traffic times.[2]

The local exchange connects to other exchanges in the immediate area through tandem switches and connects to toll centers, too. Figure 6.2 shows this arrangement or architecture. Connections between the higher-level centers (toll, primary, and sectional centers) are organized like a star network configuration but include additional direct toll lines between many adjacent centers. Links connecting the higher-level centers to each other and to the end offices are high-capacity trunk lines capable of handling hundreds or thousands of individual calls simultaneously. In the U.S. this network (with several important modifications) links about 140,000,000 individual access lines.

The switched phone network begins operating when the customer raises the instrument's handset, initiating a connection through the local loop to the exchange. The dial tone signals the caller that connection to the exchange is complete and the switch awaits the destination phone number or address. The phone number sent to the exchange in the form of tones (pulse counts in older equipment) drives the switching equipment. Notice in Figure 6.2 that several switches may be involved in long-distance calls.

A long-distance call or toll call, indicated by a leading 1, signals the caller's exchange to route the call to a toll center that can connect to the destination exchange. This may involve a primary or sectional center, depending on the communicants' respective locations. Because of network loading, link failure, or degradation in toll line capacity, may change routing between successive calls to the same number. After the signal reaches the destination exchange, its switch connects with the desired access line. Introduced in 1951, direct distance dialing (calling outside the local area without operator assistance) is now available in about 100 foreign countries. The word "dial" carries over from older pulse code equipment as a permanent word in our vocabulary. Conceptually simple, the switched telephone network is operationally complex and relies on computers and high-technology switching equipment for effective operation.

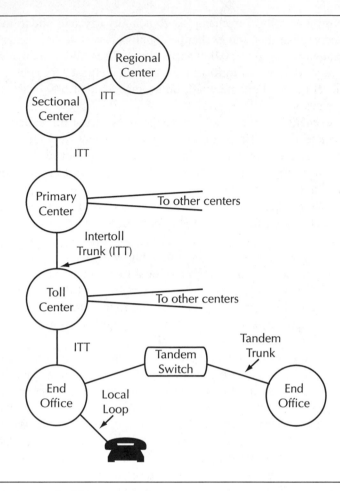

FIGURE 6.2 Bell System Switching Network

Telephone Network Characteristics

In contrast with cable TV and broadcast radio or TV, where communication flows in one direction only, the telephone system is bi-directional. Because the network provides bi-directional, on-demand communication, it is called interactive. Various governmental agencies regulate phone companies as common carriers or public utilities and mandate that they make services widely available to the general public. Because of its great national importance, the government regulates the privately owned phone network as a public system.

In addition to being switched, interactive, and public, the phone system is narrowband. Narrowband is a term that describes the network's information carrying capacity or performance. To fully understand networks, we need to examine the signals they carry and their capacity to transmit information.

Telephone Signals

Networks can carry two types of signals: analog and digital. Analog signals vary in amplitude (signal strength) or frequency (signal pitch or tone).[3] Voice and music signals are examples of analog signals: They vary continuously in amplitude (loudness) and frequency (pitch). For voice communication, the telephone converts voice sounds (sound vibrations) into faithful electrical reproductions for transmission and converts received electrical signals into voice sounds again. From end to end, the telephone network's design accomplishes this conversion, transmission, and reconversion. Figure 6.3 shows how an analog signal varies in amplitude and frequency.

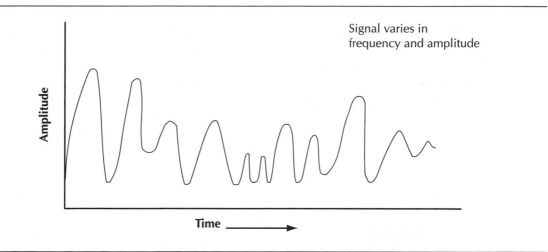

FIGURE 6.3   Analog Signal

On the other hand, digital signals are discrete or discontinuous signals. A signal that exists in only two states, plus and minus for example, is a digital signal. Signals transmitted in one of only two states are called binary signals. Most naturally occurring events are analog, however; all digital events in information technology are binary. Figure 6.4 illustrates a digital signal.

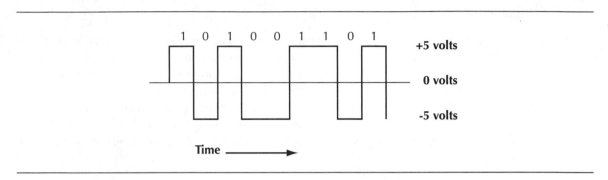

FIGURE 6.4   Digital Signal

Telephone circuits were originally designed to accommodate the range of voice frequencies found in normal speaking tones. Although the human ear can detect frequencies from 20 hertz to 14,000 hertz or so (hertz equals cycles per second, i.e., 1000 hertz equals 1000 cycles per second), effective normal voice communication takes place in the range from 300 hertz to 3400 hertz, or so. Voice telecommunications systems customarily allot 4000 hertz to one voice transmission path or channel but limit signal transmission to the range 300 hertz to 3400 hertz. This provides some space (termed guardbands) on each side of the voice channel to prevent signal overlap with adjacent channels and to leave room for the phone system to transmit the signaling information needed to operate the network.

Multiplexing

Guardbands and adjacent channels are essential to the important concept of multiplexing. Multiplexing is the subdivision of a transmission channel into two or more separate channels. One common form of multiplexing divides the frequency range of the channel into narrow bands so that separate signals can be transmitted in each band independently. For example, let's assume two incoming voice signals, A and B, are each limited to a frequency range of 300 hertz to 3400 hertz. Suppose signal B is electronically added to a 4000 hertz tone so the resulting signal now varies between 4300 hertz and 7400 hertz. The frequency-shifted signal B combined with signal A, are both sent down the line. Figure 6.5 shows this example. At the receiving end, signals in each band are separated, and 4000 hertz is electronically subtracted from the higher frequency signal. These operations result in the reproduction of the two original voice signals.

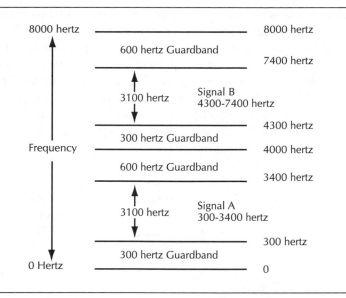

FIGURE 6.5 Two-channel Voice Grade Line

Notice that guardbands provide a signal-free space of 900 hertz to prevent signal overlap or interference between the multiplexed signals A and B. The phone system uses this free space for transmitting call management (setup, maintenance, and termination) and network management information. Using guardbands for this purpose is called inband signalling.

The process in this example can be repeated many times by combining incoming voice signals with tones at multiples of 4000 hertz and stacking them on the line for simultaneous transmission. At the receiving end, signals are demultiplexed into individual voice signals, each limited to 4000 hertz. Technical factors such as type and length of line, and line amplifier factors like linearity and distortion limit the number of subdivisions or stacks in this multiplexing method. This method, called frequency-division multiplexing, is useful for transmitting analog signals.

Network Capacity and Performance

The 4000 hertz bandwidth of the normal voice grade channel compares to a bandwidth of about 500,000,000 hertz, or 500 megahertz, for coaxial cables used in cable TV, and to about 10,000 megahertz for microwave transmission. Coaxial cable and microwave transmissions are called broadband. The phone network is called narrowband because of the limited bandwidth of the access line from the customer's equipment to the local exchange. However, the microwave or fiber-optic trunk lines that carry hundreds of multiplexed conversations between toll switching centers are high bandwidth telephone network components.

Bandwidth measures the information carrying capacity of transmission lines or links; high-bandwidth links can transport more information per time unit than low-bandwidth links. In 1933, Harry Nyquist developed the theory of information transmission. Nyquist showed that the maximum number of discrete bits or digits of information that a line of X bandwidth could transmit is 2X. This means that a voice grade line having a bandwidth of 4000 hertz can transmit 8000 bits per second. Inversely, if one wants to reproduce a signal varying at the rate of 4000 hertz by periodic sampling, one needs to sample the signal 2X times or at least 8000 times per second. This is known as Nyquist's sampling theorem. These concepts are critical in converting analog signals to digital signals.

In 1948, Claude Shannon of Bell Labs refined Nyquist's theory to account for the average amount of inherent noise found on the transmission line.[4] According to Shannon's theory, for a typical application with a signal to noise ratio of 20, a 4000 hertz bandwidth line is theoretically capable of transmitting 12,178 bps. These results differ from Nyquist's calculation, but, in actual practice, both these theoretical calculations somewhat overstate the transmission capacity of a communication link.

Digitizing Voice Signals

Applying the enormous power of digital computers to telephone systems requires converting information from analog to digital format for processing and transmission. Upon reception, the signals must be restored to analog format for human use. Converting signals from analog to digital requires sampling techniques and relies on Nyquist's sampling formula.

Analog to digital conversion is a two-step process involving pulse amplitude modulation and pulse code modulation. Figure 6.6 illustrates these processes.

The top portion of Figure 6.6 shows an analog signal varying in amplitude over time, quantized into eight discrete levels or pulse amplitudes. The bottom portion of Figure 6.6 shows the quantized signal being sampled at discrete time intervals. The sample heights are converted into binary digits; this case of eight quantized levels requires three bits to identify the signal amplitudes. The process of converting samples to binary digits is called pulse code modulation.

Increasing the number of quantizing levels improves the representation of the analog signal and increases its fidelity. With 128 levels, for example, it takes seven binary positions to represent the sampled signal. Adding one control bit converts the sampled signal to eight bits or, in computer language, one byte. Further, using Nyquist's theorem, a 4000 hertz analog signal must be sampled 8000 times per second to be faithfully reproduced. This means that a digitized voice signal limited to a bandwidth of 4000 hertz converts into a bit stream of 8000 samples of eight bits each, or 64,000 bits per second.

This process is the basis for digitizing analog signals in the voice telecommunication system. Digitized signals are crucial to communication systems. With these signals, powerful digital computers can greatly enhance parts of the system such as switching systems.

1 The signal is first shaped so it occupies a discrete set of values.

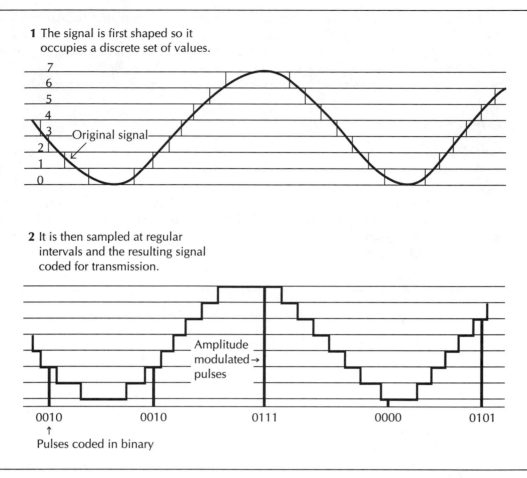

7
6
5
4
3 —Original signal—
2
1
0

2 It is then sampled at regular intervals and the resulting signal coded for transmission.

Amplitude modulated→ pulses

| 0010 | 0010 | 0111 | 0000 | 0101 |

↑
Pulses coded in binary

FIGURE 6.6   Pulse Amplitude and Pulse Code Modulation

Telephone Switching Systems

One valuable characteristic of the telephone system is its ability to connect any two phones within its worldwide network. To properly connect, local exchanges and toll centers must be able to read destination addresses or phone numbers. Complex, high-technology switches in various network centers perform this task.

The Bell system network first installed electronic switches in 1965. They represented the first step in integrating digital computer technology into the phone system. Like digital computers, a program stored in the switch's memory controls electronic switches. The program analyzes the address number, searches for a physical path to the address, and claims that path for the call's duration. From all possible lines running between the caller's exchange and the destination exchange, the switch seizes one for its exclusive use. This process of finding and claiming one of many lines is called space-division multiplexing.

Replacing older mechanical equipment with integrated computer technology greatly improved network efficiency, flexibility, and reliability. Digitizing voice signals and processing both signaling information and the signals themselves in binary format using digital signal processors offered many additional advantages. Entirely digital modern switching and transmission systems employ another form of multiplexing called time-division multiplexing.

Time-Division Multiplexing

Time-division multiplexing is the process of successively allocating time segments on a transmission channel to different users. In the phone system, this means digitized information from many low-speed inputs is assembled in sequence for transmission over one high-speed digital line. One popular implementation is to interleave digitized signals from 24 voice grade lines for transmission over one line running at 1.544 megabits per second. This line is called a T-1 line, and the assemblage of information is called the T-1 frame. Figure 6.7 illustrates the construction of the T-1 frame.

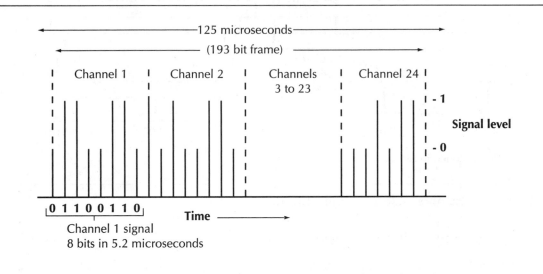

•Channels 2–24 each contain 8 bits. Therefore, 24 digitized signals are time multiplexed, each signal is sampled every 125 µsec, or 8000 times each second.

•Each frame contains 8 x 24 = 192 bits plus 1 frame alignment bit for a total of 193 bits per 125 µsec. Therefore, the data transfer rate is 193 x 8000 = 1.544 megabits per second.

FIGURE 6.7   Time-division Multiplexing and the T-1 Frame

Notice that the T-1 frame sequences 24 channels, each containing eight bits of information plus one control bit per frame. The eight bits represent one sample of the original analog signal. Bytes of information from each of 24 channels are interleaved at about five microsecond intervals (one microsecond equals one millionth of a second), and the sequence repeats every 125 microseconds. Thus, each analog signal is sampled once every 125 microseconds or 8000 times per second. The data transfer rate is 193 bits per frame multiplied by 8000 frames per second, or 1.544 mbps. This is called the T-1 data rate. The phone network uses multiples of T-1 lines.

In digital systems, the digital signal processor manages the bit stream that includes network and signal information, acts on the data determining where the signal information should be sent, and routes the call interleaved with many others over a high-speed line. The switch is, in fact, a programmable digital computer. As such, it can implement functions such as call forwarding (looking up a number and finding the alias), call waiting (detecting a busy line and signalling the called party), or automatic number identification (sending the number of the calling party to the called party). It also stores information such as the address called and call length for billing. Digital technology makes these and many other features possible.

The phone operations discussed here are premised on the ideas that the switching system establishes the call routing and physical connections needed to complete the communication path, and that the call's route remains constant for its duration. This convention is called circuit switching. Circuit switching works quite well for voice communication because conversations take place at a rather constant rate and utilize the circuits nearly full time (usually one party or the other is talking). With data communication, however, transmission rates can vary widely and communication can be sporadic or in bursts. Sometimes the circuit is fully loaded; at others, it is idle for prolonged periods. Another type of switching, called packet switching, helps improve efficiency in data transmission.

Packet switching breaks the data file to be transmitted into short packets, usually less than 1000 bytes, and prefixes them with addressing and control information. Each packet is transmitted in turn, received by a forward node, stored briefly, then forwarded to the next node until it reaches its destination. Eventually, the destination device receives all packets for a given message file and assembles them in sequence; then the process terminates. This type of operation more efficiently uses the network for bursts of data but requires more sophisticated switches and networks. Packets are not required to take the same route through the net because instantaneous network traffic partly determines the route. Packet switching is important to advanced data transmission techniques.

Digitization of voice signals, multiplexing, programmable switches, and broad bandwidth lines greatly improved the phone network's capability. Broad bandwidth lines have been state-of-the-art between exchanges for some time; however, today phone companies use new technology to increase the capacity of the narrow-band line from the customer's equipment to the exchange.

If time-division multiplexing, circuit switching, and packet switching seem to require that a lot be accomplished very quickly, recall that a modest size, general-purpose digital computer can process 100 million instructions per second. In the five microseconds taken to place one byte of information on the line, it can perform 500 add, subtract, or compare operations.

Computers specially designed for signal processing and switching operate even more efficiently than general-purpose computers. Processing binary signals, whether voice, data, or video, exemplifies the integration of computer and communications technologies.

Network Transmission Media

Transmission links can use either conducting media or radiated media. Copper wires, like those connecting the phone set to the wall outlet and to the local exchange, are conducting media. Conducting media imply a direct physical connection between network components for signal transmission. On the other hand, broadcast radio transmission exemplifies radiated media: electromagnetic energy passes through the atmosphere or through space without a physical connection. Modern networks use both media types.

Today, most phone connections consist of four insulated wires twisted around each other, within a jacket forming a cable. Twisted pairs grouped together into larger cables (a cable 3.35 inches in diameter carries 1200 copper pairs) are found today near exchanges where lines fan in from offices or residences. To reduce the volume of wires and increase transmission capability, many network applications benefit from coaxial cable transmission links.

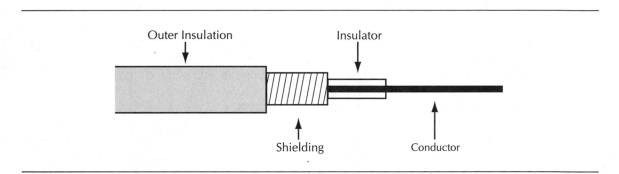

FIGURE 6.8   Coaxial Cable Construction

Coaxial cable has many telecommunications applications, particularly for distances of a few miles or so. Figure 6.8 shows the construction of a single conductor coaxial cable. Its diameter may range from one-fourth inch to one inch. Its outer shield protects against electrical interference and physical damage. Permanent taps are easily placed in the line, thus aiding physical rearrangement of terminal equipment. Coaxial cable permits data-transmission rates on the order of millions of bits per second and is relatively immune to noise or interference. These characteristics make coaxial cable suitable for short-range computer networks, telephone toll lines, or cable TV lines. Large diameter coaxial cables augmented with signal amplifiers can be used for longer distances in phone or TV networks.

Ideal for many applications, coaxial cables are very common in networks. For applications requiring very high bandwidth and low attenuation over long distances, however, cables made of optical fibers are more economical.

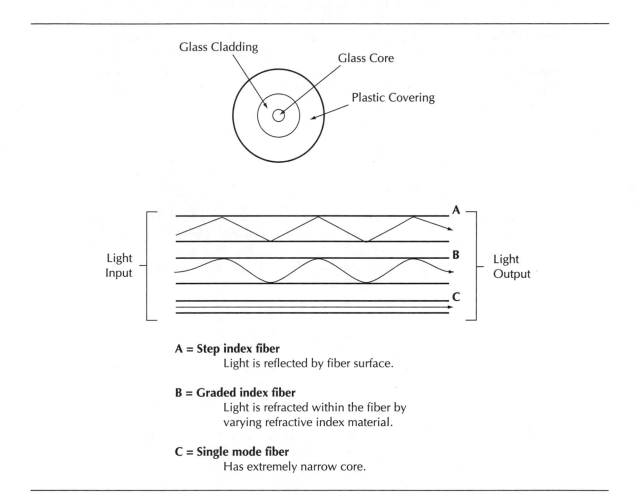

A = Step index fiber
　　Light is reflected by fiber surface.

B = Graded index fiber
　　Light is refracted within the fiber by varying refractive index material.

C = Single mode fiber
　　Has extremely narrow core.

FIGURE 6.9   Construction of Optical-fiber Cable

Optical fibers consist of a glass or plastic fiber surrounded by glass or plastic cladding or covering. The core fiber's construction determines the type of transmission involved. Figure 6.9 shows an end view of a fiber-optic cable and three transmission methods within the core. In the step index type, the fiber contains light signals reflected off the walls as they progress down the cable. This was the earliest type of fiber used. The graded index fiber is constructed of core material whose refractive index varies from the center of the core to the outside. Refraction of the light signal keeps it within the core. Single-mode transmission uses a very thin core of glass (about 1/100th of a millimeter or less in diameter) to contains the signal. Still under development, this technology promises to be the most effective of the three types.

Fiber-optic cables offer many advantages over earlier cable types. They are suitable for long distances because they have low signal attenuation and are relatively cheap. Optical fibers are high bandwidth components: They are capable of transmitting billions of bits per second.[5] Because light is the signal source, optical fibers are free of electrical interference. Fiber-optic cables are difficult to tap and very secure. However, taps and termination devices are relatively expensive, so short fiber-optic links are not economical for some applications. Their many advantages and still unrealized potential make optical fibers increasingly popular in voice, data, and video networks today.

ADVANCED NETWORK SYSTEMS

Digital Subscriber Lines

To increase capacity from the local exchange to the customer's location without replacing the twisted pair, phone companies are adopting a new technology called Digital Subscriber Lines, or DSL. DSL technology integrates digitized voice and data for transmission over the copper circuitry currently installed in the phone system. DSL eliminates the analog portion of today's networks and makes the network digital from end to end. Integrated Services Digital Network (ISDN), the first of these services, is becoming widely available.

First adopted in Europe, ISDN increases the local loop's capacity about sixfold and brings digital network capability directly to the customer's location for use by phones and computers. It offers many new features and capabilities to phone network users.

ISDN features two classes of service, basic and primary. Figure 6.10 depicts these services. The basic service, called 2B+D, consists of two 64k bps B channels and one 16k bps D channel. The B channels transmit either digitized voice or data, while the D channel controls signals and routes information.

The primary service, named 23B+D, consists of twenty-three 64k bps B channels and one 64k bps D channel. The primary service supports 1.544 Mbps bidirectional transmission. The B channels can be used for a mixture of voice and/or data communication needs. ISDN is useful for interconnecting data processing facilities within the firm, for connecting sites within the firm for voice communication, or for integrating voice, facsimile, and data transmission. ISDN is the first step in the march toward advanced digital communication capability for residences and businesses.

The next step may be Asymmetric Digital Subscriber Lines (ASDL), now being tested by the Regional Bell Operating Companies and others. ADSL runs on normal phone lines equipped with special high-speed, microprocessor-based modems at *each* end of the copper lines that connect subscriber equipment to the central office switches. The technology is called "asymmetric" because the transmission rate to the customer's equipment is greater than the return rate by a factor of about 10. For moving data to the user, ADSL is reported to be about 20 times faster than the 28.8 kbps modems generally used.

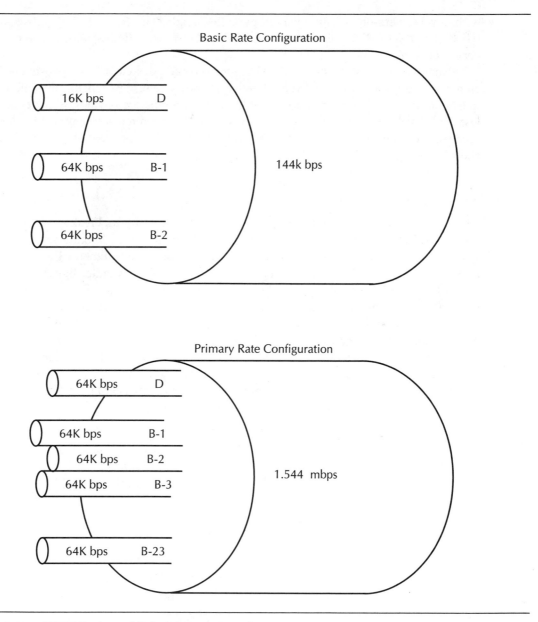

FIGURE 6.10   ISDN Basic and Primary Rate Interfaces

One advantage of ADSL is its use of existing copper links, avoiding expensive upgrades. Another is that the modem at the central office connects directly to the digital switch without modification, again avoiding expense. ADSL's line length is limited to about 3.5 miles, enough to reach 80 percent of the phones in the Bell regions. Developers are experimenting with several versions of the technology called High Data Rate Digital Subscriber Lines (HDSL) and Very High Data Rate Digital Subscriber Lines (VDSL), which have much higher bandwidths than ADSL.[6]

Because adoption of DSL in the U.S. must proceed despite a huge investment in analog circuitry, and because the relatively effective phone system must remain operable during the transition, the financial barriers to a speedy implementation are obvious. Some countries with less well-developed phone systems or with systems that need replacement may be motivated to accelerate the implementation of ISDN. In parts of Europe, national strategies support early adoption of digital technology.

T Services

Responding to the new competitive environment spawned by its breakup, AT&T made T-1 service available to the public in September, 1983. Implemented internally within the phone system for many years to connect exchanges, it was not made commercially available earlier. A T-1 link can be constructed of copper wires enhanced to handle digital signals at rates exceeding 1.5m bps, but it is generally made from other media. T-1 links can handle up to 24 simultaneous phone conversations at approximately four times the cost of an ordinary line.

Today, extensions to T-1 services include bandwidths very much higher than originally proposed. American Airlines, Merrill Lynch, and Internet service providers are among the many firms capitalizing on T technology. Many companies today use T-3 services, and its use is expected to expand rapidly in the future. Table 6.1 displays the current structure of extended service in North America. In addition, fractional T-1 service, which permits customers to use a portion of a T-1 link, is attracting widespread interest.

TABLE 6.1 T Services

Digital Facilities	Transmission Rate	Number of T-1 Equivalents
T-1	1.544m bps	1
T-1C	3.125m bps	2
T-2	6.312m bps	4
T-3	44.746m bps	28
T-4	274.176m bps	168

Electronically sophisticated multiplexers handle T lines. These devices manage traffic, detect and isolate defective lines, and provide redundancy, thereby improving reliability. Multiplexers become more complex as the functional capability of their logic and memory circuitry continues to grow as it has in computer technology. Improved technology allows compression of voice signals, improving throughput while increasing reliability through error detection, correction, and redundancy techniques. These improvements in telecommunication services reduce cost and increase demand.

T technology is not confined to the private sector. The Federal Reserve System is consolidating 12 separate networks into Fednet, a new network based on fractional and full T-1 capability. Fednet consolidates the Fed's 12 data processing centers into three and permits depository institutions with similar equipment to connect to the Fed system. The network carries more than $1.25 trillion in transactions daily.

Voice Signal Compression

Is it possible to increase the number of phone conversations beyond 24 over a T-1 line while maintaining satisfactory quality? Is there some way to increase the effective capacity of a line for digitized voice information?

The first step in improving throughput was recognizing that the traditional sampling rate of 64,000 bits per second is overly conservative for voice conversations and that 32,000 bps is quite satisfactory. The technique that uses the 32,000 bps sampling rate is known as Adaptive Differential Pulse Code Modulation (ADPCM). Decreasing the bit rate by 50 percent to 32,000 bps through ADPCM technology doubles the number of voice messages a T-1 line carries from 24 to 48. Other more sophisticated speech compression techniques that encode voice information based on speech characteristics have also been developed. GTE developed one such technique, which increases the effective T-1 capacity by another factor of two, raising the number of simultaneous voice conversations to 96.

Techniques that detect short, silent pauses between words and sentences offer additional room for significant improvement in throughput. One such algorithm, called Time Assignment Speech Interpolation (TASI), can add another factor of two or three to the throughput of a T-1 line. Thus, under optimum conditions using current technology, up to 288 voice messages can be transmitted concurrently on a single T-1 line. Implemented in computer-like digital signal processors, digital compression techniques greatly increase the phone network's capacity and give scientists great incentive to develop compression technology for video signals.

Fiber-based Advanced Transport Technologies

High-speed networks with data rates of 45 mbps or higher frequently employ microwave or fiber-optic transmission technologies. Because of cost, size, weight, and flexibility, and because many nets are being installed in large buildings in large cities, most high-bandwidth networks are being constructed with fiber-optic media.

Fiber-optic systems created an explosion of bandwidth in wireline networks. Driven by economics and technical considerations not by supercharged promoters or government policy, its continuing explosive growth is inevitable. Fiber costs less to manufacture, install, and maintain than copper; its costs decline while copper's costs are likely to remain steady or increase. As a result the installed base of fiber cables will grow, gradually replacing copper wire and coaxial cables as the medium of choice. Today, for example, telephone companies and others install fiber at the rate of 4000 miles daily.[7]

The decision to install fiber where new communication links are required and to replace copper to reduce maintenance costs in presently installed links is independent of current or anticipated applications for the increased bandwidth. Today, optical fiber is so cheap that it is being installed even without immediate need. For example, strands of unused fiber, called dark fiber, are placed in the center of power cables in newly constructed power lines—electricity in the metal cable does not interfere with light passing down the fiber. Fiber cables are installed in trenches with new interstate natural gas or petroleum pipelines, and they follow railroads or other such projects under construction. Current fiber costs and its potential justify investments by firms owning rights-of-way.

Scientists have yet to reach the transmission limits of fiber lines, the inherent bandwidth of which is about 25,000 gigahertz. With improved materials, optical amplifiers, and lightwave multiplexing, a reasonable expectation is that fiber lines can transmit 2000 gigabits per second.[8] This is equivalent to about 200 million simultaneous phone conversations. At these rates, all phones in Europe can be connected to phones in North America simultaneously—over one fiber line.

Growth in the fiber network and in our ability to transmit information through it yields bandwidth growth rates of about 100 percent per year. Doubling in capacity every year means that network bandwidth will increase by a factor of more than a thousand in a decade.[9]

One important standard for high-speed networks using fiber-optic technology is Fiber Distributed Data Interface (FDDI). The American National Standards Institute developed the FDDI standard and designed it for organizations that need flexible, high-performance networks. Basic FDDI performance is defined as 100 megabits per second. FDDI supports 500 stations extended over a maximum fiber length of 200 kilometers. Station spacing must not exceed two kilometers.

FDDI accommodates future growth in the number of network stations and in traffic loads resulting from client/server operations and complex applications such as computer-aided design and graphics manipulation. By employing fiber-optic technology, FDDI transport rates provide substantial improvements over typical 4 to 16 mbps LAN implementations; thus, FDDI is useful for linking lower-speed LANs to configure internets.

FDDI architecture employs two fiber rings. In each, data flows in opposite directions. Figure 6.11 shows this configuration. The secondary ring is for backup and reliability. The dual ring concept aids in network initiation and reconfiguration, and provides recovery capability in case of ring failure. If a network device fails or a fiber fault occurs, the network maintains operations because the secondary ring can restore the communication path. This inexpensive redundancy greatly increases the fault tolerance of FDDI networks.

FDDI networks can be configured in several ways to support high-speed workgroups, interconnect separate LANs, or provide a backbone net for multiple buildings or campus networks. In Europe, for example, the Italian government installed an FDDI backbone network connecting over 300 toll stations on its highway system. The system gathers traffic statistics, updates weather and road conditions, and manages toll collection. In the U.S., phone companies have installed FDDI rings that provide competitive service to established local providers.

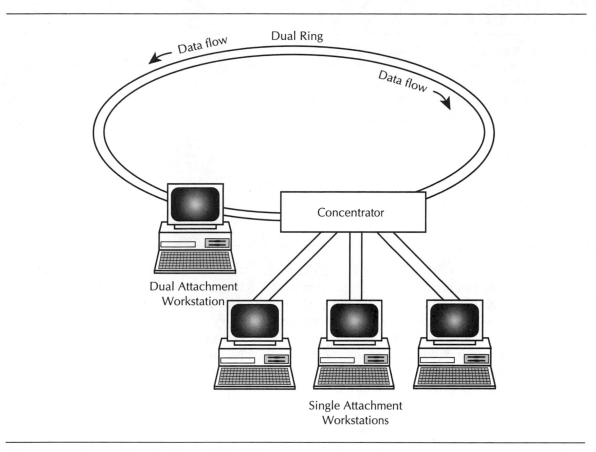

FIGURE 6.11   FDDI Network Configuration

Metropolitan Area Nets (MANs) will benefit considerably from FDDI and its extensions. FDDI and its enhancements, FDDI-II, FDDI Follow-On, and its kin, Distributed Queue Dual Bus (DQCB), will increase network throughput, improve reliability, and form the base for growth in fiber-optic networks.

At very high data rates over optical fibers, Synchronized Optical Network or SONET is the future standard. Its performance base starts at approximately 51.84 megabits per second. This rate is called Synchronous Transport Level 1 (STS-1). The STS-1 data rate is multiplexed to higher transmissions and is defined to a rate of 2.4 gigabits per second. Actually, STS-n defines multiplexing, where n takes values of 1, 3, 9, 12, 18, 24, 36, and 48 only. Table 6.2 shows the STS hierarchy.

TABLE 6.2   STS Line Rates

STS Level	Line Rates (mbps)
STS-1	51.84
STS-3	155.42
STS-9	466.56
STS-12	622.08
STS-18	933.12
STS-24	1244.16
STS-36	1866.24
STS-48	2488.32

The STS-1 SONET signal is a 9×90 byte data block that repeats every 125 microseconds, or 8000 times per second. The block contains 6480 bits (9×90×8); thus, the data rate is 51.84 mbps (8000×6480). Extensions to the data rates above STS-48 are expected to yield transmission speeds greater than 10 gigabits per second.

Rapid advances in SONET technology are expected during the next decade in conjunction with advances in digital switch interfaces, optoelectronic switches, and photonic switches. These and other developments are likely to make SONET technology the preferred choice for high-speed data transmission in the next 20 years.

Other Advanced Transport Technologies

Most communication technology is either circuit switched, as in the narrow-band phone system, or packet switched for data traffic. The main inefficiency of circuit switched voice or fax is wasted dedicated bandwidth during breaks in the voice or fax communication. Most voice communication is at the rate of one word per second, slow in comparison to the dedicated line's capability. Packet switching buffers data into packets before transmission and several data sources share one circuit. Thus, packet switching uses available bandwidth more fully than circuit switching. Circuit sharing causes slight immaterial delays in data transmission, which are unacceptable in voice communication. Advanced transport technologies using broadband switched nets can use packets for virtually all types of communication.

Two advanced transport technologies are frame switching (frame relay) and cell switching (asynchronous transfer mode or ATM). Both technologies use a star-wired network with dedicated lines connected to a center switch (like the phone net). Thus, ports can communicate with each other at full line speed. The ATM and Frame Relay information packets' construction differs. Figure 6.12 illustrates their differences.

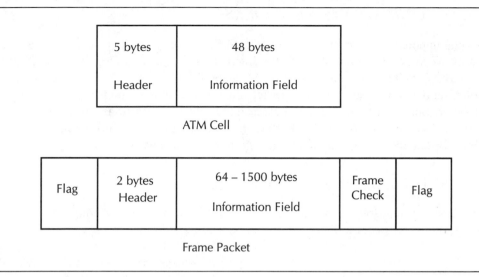

5 bytes	48 bytes
Header	Information Field

ATM Cell

Flag	2 bytes Header	64 – 1500 bytes Information Field	Frame Check	Flag

Frame Packet

FIGURE 6.12   ATM Cells and Frame Relay Pockets

The frame contains an information field of variable length and packet data that ensures integrity of the information. The information field can vary from 64 bytes to 1500 bytes in length. The ATM cell structure is a fixed 53-byte length and assumes network integrity to ensure information transmission accuracy (a good assumption for advanced networks). With frame switching, switch management also differs. Here, the circuit is established for each packet. But with ATM, the initiating station requests the destination and the bandwidth. The switch establishes dedicated bandwidth to the message for its duration.

ATM equipment merges all forms of information (voice, data, video, etc.) for transmission to the switch where it is dispatched to its destination, which also contains ATM equipment. Because packets are small and fixed in length, traffic is very predictable. Voice data can be given a fixed number of cells to ensure high quality and the remaining time allotted to data traffic. As this method suggests, ATM is a sophisticated form of time-division multiplexing occurring over broadband lines. Switched multi-megabyte services such as ATM can deliver data rates exceeding 100 megabytes per second, limited by line, switch, and equipment bandwidth. ATM seems to be the technology of the future as facilities and equipment costs decline.[10]

Desktop communication capability is growing very rapidly, too. Transmission speed at the desktop was about 1.0 mbps in 1980, 50.0 mbps in 1990, and projected to be 10.0 gbps by the year 2000. This phenomenon poses great challenges to microprocessor developers. PC processor speeds have increased very dramatically with advantages for dedicated computing, but managing bandwidth becomes increasingly important as PCs evolve into communication centers. By the year 2000, PCs will process several hundred MIPS *and* handle data traffic at the rate of gigabytes per second. Enormous computation and communication capability will be available to office workers in the near future.

Several important organizations establish standards that coordinate the details of data communication networks. Founded in 1946, the International Standards Organization (ISO) has very effectively instituted standards on a broad range of topics. Mostly, its members are standards organizations representing nations participating on a voluntary, non-treaty basis. The ISO has authored more than 5000 standards designed to facilitate international commerce and economic activity. One of its most important efforts has been in data communications. It established the Open System Interconnection model, a very influential standard.

Open System Interconnection Model

The Open System Interconnection model (OSI) creates an open or nonproprietary data communications architecture intended to promote end-to-end compatibility between communications sub-systems. The extremely complex and detailed model requires many thousands of pages to describe. It has been under development for nearly two decades by individuals representing nearly every facet of the communications industry. The OSI model represents an important, critical development in data communications technology.

The model describes how the network operates and the protocols it uses. It consists of a seven-layer architecture in which each layer views the one below it as if it were the actual path to the other communication system. The model achieves compatibility between communication systems independent of specific application coding. The layering's effect is to achieve modular communications systems through carefully defined rules and terminology. In addition to accomplishing this important task, the model has had a major influence on our perceptions of network communications. Figure 6.13 shows the layered OSI model.

The OSI model describes protocols governing data flowing from one user's application program, to the network media, along the media, then back to the receiving application. The path through the model is bi-directional. Each layer specifies or provides services and has at least one protocol or definition of how the applications establish communication, exchange messages, or transmit data. The layered architecture defines divisions of responsibility and permits changes within one layer without disrupting the entire model. Thus, OSI modularity simplifies complexity. OSI is important because it provides an organized, logical solution to a very large communications problem.

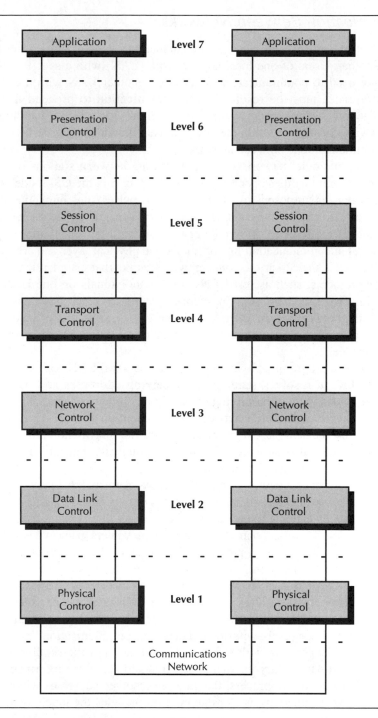

FIGURE 6.13 Open System Interconnect Model

Communication Between Networks

Communication devices, and the data formats that they support, may differ from one user network to another. Users of one local area network (LAN) with a specific communication data format may need to communicate with users on another LAN with a different data format. These differences must be resolved for communication to proceed from one network to another. The lowest three layers in the OSI model provide rules for handling these incompatibilities. Devices operating within the physical, data link, and network layers of the OSI model manage communications between networks.

Extending a single network, or communicating between stations on different networks, requires specialized facilities operating at layers 1, 2, or 3 in the OSI model. One such facility is a signal repeater. Its purpose is to retime or reshape pulses, thus mitigating the effects of signal attenuation or distortion and extending the network's length. In local telephone and cable networks, repeaters are inserted every mile or so to forward the signal after improving its quality. Repeaters on the communication line operate at the physical layer and are called layer 1 relays.

Many times it's necessary to connect two networks that use identical data communication formats or protocols, such as two LANs, so that individuals on one LAN can communicate with individuals on another LAN. For communication to occur, the LANs must be electrically connected, the networks must share knowledge of device addresses, and disciplines must be followed to prevent disruption to network operations. The facility to accomplish this is called a network bridge.

A bridge is a programmable facility that connects two LANs operating with identical protocols. The bridge resolves sending and receiving addresses and passes packets of data between the LANs without modifying or adding anything to the packet. The bridge intercepts and analyzes information on a network communication line. If the sending and receiving addresses are on the same LAN, the bridge passes the information along. If the addresses are for different LANs, it routes the information through its data link layer to the connecting LAN. Thus bridges are layer 2 facilities.

A router/gateway is a device used to connect two networks that may or may not have similar protocols. If the networks are similar, the device is usually called a router. If they are not similar, the device is usually called a gateway. (Definitions in the literature are inconsistent.) A collection of communication networks connected together with routers/gateways is called an internet. An internet with a complex topology presents a significant challenge to routers and gateways, so their capabilities must be quite sophisticated. The gateway employs an internet protocol or, in other words, has knowledge of each network's operating protocol and can translate messages from one protocol to another. It also resolves addresses and establishes routes or paths.

During operation, routers/gateways determine if the message is intended for a user on the same LAN. If so, it passes the information down the communication line. If the message is for a user on a LAN operating with a protocol different from the sending LAN's protocol, the gateway makes the necessary internet conversion and sends the information down the line of the receiver's LAN. In the process, the gateway resolves addresses within its data link layer. The router/gateway functions as a bridge but also includes the much more complex functions of protocol conversion.

Because multiple paths can exist in the internet, gateways must resolve path routing. For example, a user on one net may want to communicate with a user on an adjacent net, but traffic congestion or equipment failure may dictate that the communication be routed indirectly through other interconnecting nets. Because nets comprising the internet usually contain different protocols, gateways must be able to resolve routing problems and protocol differences. These complexities, along with the usual network problems and human errors and the need to satisfy accounting, security, and a host of other demands, make internet operations very complex. The OSI architecture organizes this complexity very effectively.

Vendor Network Architectures

Several organizations developed network architectures that have gained popularity and been widely adopted. The U.S. Defense Department sponsored Transmission Control Protocol/Internet Protocol (TCP/IP), while Digital Equipment Corporation and IBM initiated Digital Network Architecture (DNA) and Systems Network Architecture (SNA), respectively.

Introduced by IBM in 1973, Systems Network Architecture (SNA) has secured a large user base. SNA, an IBM proprietary product, is considered a *de facto* standard for IBM customers worldwide. SNA is a hierarchical, layered model similar to OSI that supports distributed mainframes, token-ring local area networks, and peer-to-peer communication. DNA is DEC's proprietary network standard. It, too, has a layered architecture roughly comparable to ISO/OSI. Today, with the Internet's explosive growth, TCP/IP is the most widely used communication standard.

The U.S. Defense Department developed TCP/IP as a set of protocols for its ARPAnet network—therefore, it is nonproprietary but supported by many vendors. UNIX and ethernet environments use TCP/IP, and IBM, DEC, and Apple Computer support TCP/IP products. TCP/IP supports network interconnectivity and is widely used on the Internet. Much data transported on the Internet using TCP/IP consists of Web pages encoded in Hypertext Markup Language (HTML). The protocol or communication method servers and clients use to exchange HTML documents over the World Wide Web is the Hypertext Transfer Protocol (HTTP). Thus, using Purdue University as an example, Web site addresses are written as http://www.Purdue.edu.

Demand for voice and data communications in government operations and in commercial businesses continues to grow rapidly as communication costs drop and as organizations learn to apply new technology advantageously. Options available to network implementors have also increased with the evolution of network standards such as OSI, ISDN, SONET, and others. Many other options for network planners are being developed, thus spurring growth in local area and wide-area nets for voice, data, and video communication. New architectures, powerful network operating systems, and sophisticated network management systems are required to attain full advantage of the network and computing technology.

The cable network in the U.S. was initiated on a limited scale in 1949. Now, approximately 90 cable networks service more than 60 percent of U.S. homes and many businesses. Cable passes near another 30 percent of non-subscriber homes. In its penetration of households, cable ranks third among wirelines behind power and phone. Its growth has paralleled the growth of broadcast television in the U.S.

Cable is characterized as direct wireline (non-switched), generally not interactive, broadband, and private. Cable providers expect that subscribers will select one of several tens of signals available on the line and that all subscribers may want to use the line simultaneously. In contrast with the phone system's star architecture, cable network architecture is tree and branch. All signals originate at the headend and are cablecast unidirectionally to subscriber equipment. Although the cable networks' capacity is high compared to phone networks', cable providers did not initially plan to use the network for bidirectional communication. Considerable activity is now underway, however, to make the cable TV network interactive.

Cable Network Architecture and Headend

The point at which signals are placed on the local cable network is called the headend. The headend connects trunk lines that form the primary distribution network. Subscriber loops tap the distribution trunks to bring signals into individual homes or businesses. Thus, the network is called tree and branch. Figure 6.14 shows this architecture.

The cable network headend contains extensive sophisticated equipment needed to obtain signals from various sources and to place them on distribution cables. The headend may obtain radio or television source signals from satellite or microwave transmissions or from other distribution sources over landlines. Headend sites include towers and antennas needed to obtain signals or to retransmit locally-generated material.

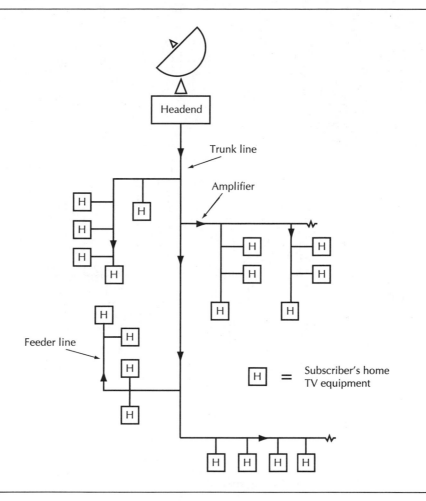

FIGURE 6.14   Cable Network Architecture

The headend also contains production equipment such as VCRs, video cameras, keyboards, control panels, and computers used to produce local programming and to control video signal distribution. Signal processors clean snow and hiss from incoming signals, filter interference created by unwanted signals, maintain signal strength, and, in some cases, convert the signal's frequency before passing it to the modulator. The modulator places the signal on the cable for transmission. In addition, the headend may transmit some signals to neighboring headends for retransmission to subscribers on adjacent cable networks. Figure 6.15 shows the configuration of headend equipment.

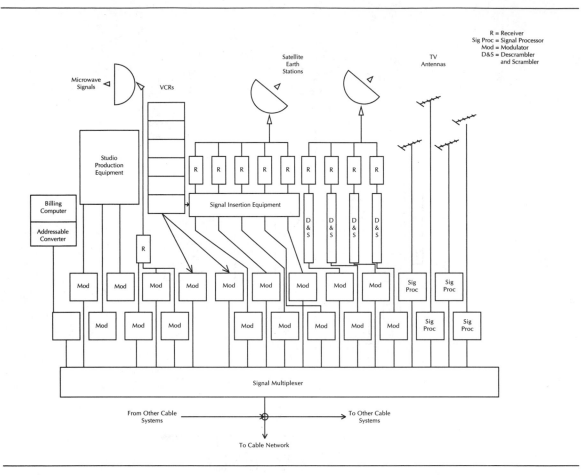

FIGURE 6.15 Cable Headend Configuration

The headend also contains computer equipment for coding the signals so that only sites subscribing to the signal (paying for it) can receive it. Converter equipment at the subscriber end decodes the signal, thus enabling reception and viewing. Headend computers also handle billing.

The cable headend's purpose is to acquire signals, maintain or enhance their quality, place them on distribution trunk lines, and provide administrative network control. Highly automated and intensely electronic, the headend comprises only six percent of the network's costs. The distribution system accounts for the remaining costs.

Cable Distribution Systems

The cable network distribution system consists of trunk lines beginning at the headend, travelling through the service area, branching at various points, and terminating at the service area boundary. At points along the trunk lines, feeder lines branch to individual subscriber locations.

Figure 6.16 shows the trunk and feeder line system. In a typical cable network, the trunk system accounts for 19 percent of the cost and the feeder system (the last mile) linking sites to trunk lines accounts for 75 percent.

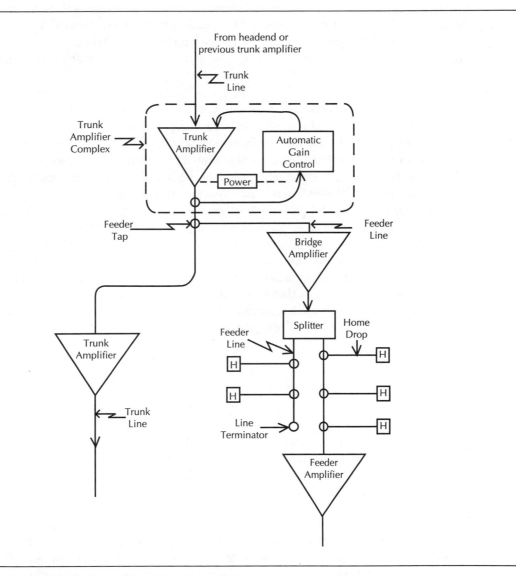

FIGURE 6.16   Cable Trunk and Feeder Network

The trunk line is a coaxial cable about three-quarters of an inch in diameter designed to carry up to 75 analog video signals with minimum signal loss or attenuation. Even with large cables, amplifiers and gain control loops must be placed along the trunk every one-quarter to one-half mile to maintain uniform signal strength and quality. As Figure 6.16 shows, taps into the trunk permit connections to the feeder system at various points.

The feeder system is constructed of smaller cable (three-eights to one-half inch in diameter) that transmits quality signals over the shorter distances from trunks to individual homes or businesses. The feeder system contains line splitters (multiline connectors), taps for home drop lines, and feeder amplifiers, if feeders extend beyond 300 to 400 feet or so. In addition to video signals, trunk and feeder lines must carry electrical power for the trunk and feeder amplifiers and maintain interfaces for use by line technicians.

The cable network becomes more complex as operators attempt to improve system reliability, security, function, and performance. In some instances, the network consists of dual cables to increase system capacity; and in other instances, operators have added cable to provide special functions for individual companies or institutions. For example, to assist telecommuters, Digital Equipment Corporation links some employee homes to their office locations via cable. Other firms use cable for teleconferencing—maintaining interactive video between conference rooms in individual office buildings. Some cable systems include devices within the feeder system to maintain signal security and prevent unauthorized use. Driven by many forces and responding to many opportunities, the cable network is rapidly becoming complex.

Cable services provide one-way transmission of video content (cablecasting) that requires end-user selection or intervention to obtain the desired program. Defined and delimited this way, cable services are not regulated as common carriers or public utilities but are franchised by local governments. Thus, cable networks are deemed private, not public, utilities. As they depart from delivering basic cable services, however, cable systems come under governmental surveillance at several levels.[11] Today, they attract increased attention.

Cable System Capacity

Cable networks are broadband. In contrast with the 4000 hertz telephone local loop, cable feeders carry many frequency-division multiplexed channels, each with a bandwidth of approximately six megahertz. The information carrying capacity of a single feeder is 50,000 times that of the currently configured phone access line. However, new developments in digital signal processing will allow feeder cables to carry even more video signals. These developments in cable engineering, along with improvements to the telephone network, will also increase the capacity of local phone access lines, perhaps enough to enable the phone network to carry digital TV signals.

Signal amplifiers' frequency response currently limits cable TV channel capacity. The frequency spectrum of modern trunk lines extends from 54 megahertz to about 550 megahertz with several small gaps to allow for system control, digital information, and FM broadcast signals. Dividing this spectrum into the six megahertz (including video guardbands) subdivisions required for each channel permits 80 signal channels. In practice, most systems contain 50 to 54 channels. New technology in the form of fiber-optic links and digital signal compression may add 450 additional channels to the cable network. With additional capacity, cable operators anticipate interactive, multimedia applications.

Today's standard analog television broadcast contains about 100 million bits of information per second. By digitizing the analog signal, digital compression techniques can reduce the bandwidth needed to transmit the signal by a factor of ten without noticeable signal degeneration. Some techniques mathematically smooth color details not visible to the human eye, while others eliminate information that does not change from frame to frame. Video signal compression requires specially designed, high-speed processors. The notion that video signals can be compressed by a factor of 10 quickly led to the belief that 500-channel TV is in the near future.

WIRELESS SYSTEMS

In 1991, George Gilder predicted:

> "What currently goes through wires, chiefly voice, will move to the air; what currently goes through the air, chiefly video, will move to wires. The phone will become wireless, as mobile as a watch and as personal as a wallet; computer video will run over fiber-optic cables in a switched digital system as convenient as the telephone is today."[12]

Today, radio and wireless communication play an important role in AM, FM, and TV broadcast systems and in a wide variety of communications for governments, military organizations, private corporations, and others. In the next 10 years, however, wireless communications networks will link workers in offices, distribution centers, and field locations with voice, fax, data, and video, providing abundant information for the most demanding 21st century job. Through wireless transmission, everyone who wants to can be online, everywhere, all the time. Transitions occurring between forms of communication will profoundly change world communication.

Many observers have noted that these communication transitions are occurring with revolutionary speed. Cellular voice and data services and mobile data networks are growing 20 to 30 percent annually, and the market is just opening in many places. In the U.S., mobile data networks and the convergence of wireless communications and portable computing are growing rapidly. Today, most IS and telecom managers use wireless products—those who don't are actively considering acquiring wireless equipment. Advancing technology, declining costs, and immediate benefits drive increased customer demand.

Land-Based Systems

Early cellular technology was based on a standard transmission protocol for analog phones, developed by Bell Labs, called Advanced Mobile Phone Services (AMPS). Digital technologies are replacing AMPS, promising to bring many efficiencies to the cellular network just as digitization brought many advances in wired systems. Tiny microprocessors within cellular phones will perform digitization and multiplexing and control radio operation, too, enabling users to send voice, fax, and data from a single cellular device. For the next several years, however, analog and digital systems will coexist. Eventually, digital systems will dominate.

Cellular systems are designed as a dense network of low-power, broadband radio transmitters and receivers. Figure 6.17 illustrates the architecture of the cellular network.

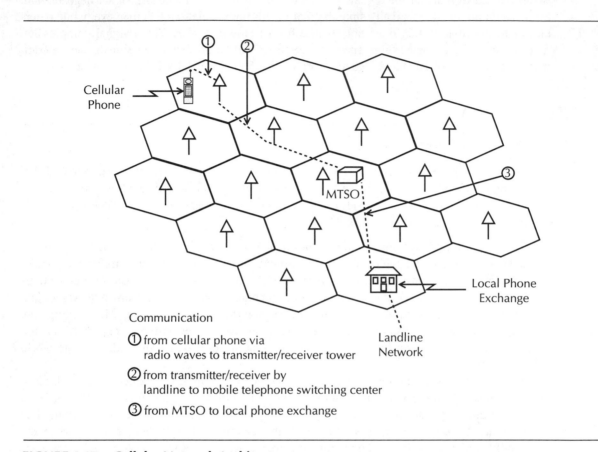

Communication

① from cellular phone via
radio waves to transmitter/receiver tower

② from transmitter/receiver by
landline to mobile telephone switching center

③ from MTSO to local phone exchange

FIGURE 6.17   Cellular Network Architecture

Operating a network depends on a combination of radio and landline communication. Figure 6.17 shows how a transmitter/receiver located in the cell site receives the call originating from a cellular phone. Each cell or geographic region contains one transmitter/receiver that controls traffic within the cell and incoming and outgoing traffic. The message moves from the cell site via wires or radio to the Mobile Telephone Switching Office (MTSO) to the local exchange, where it proceeds as a conventional call. If the call is to another cellular phone in the area, the MTSO broadcasts the number to all its cell sites. Upon confirming the location of the called phone, the MTSO completes communication between the two local cells.

If either the originating caller or the called party is traveling and the transmitter/receiver detects a weak signal, it transfers the call to equipment in an adjacent cell. If a caller leaves the local service area, another MTSO completes the call. This is called roaming. Computers and computerized digital switches control and monitor the operation of this complex, high-tech network made possible by microchip technology.

In contrast with standard AM and FM broadcast stations, which increase their service area by increasing their signal strength, cellular systems depend on low-power, lightweight, and easily-transported devices. Thus, the service area, or cell, must be relatively small and the transmitter/receiver network relatively dense.

According to Shannon's theory (see note 4), if the signal bandwidth is high enough, it can maintain signal rate even when signal power declines below the noise level. Thus, small, lightweight, very low-powered cellular phones can communicate effectively in the presence of noise if sufficient bandwidth is allocated to the transmission. Cellular systems rely on a rather wide portion of the radio spectrum and a dense network of transmitter/receivers.

Several new digital technologies are replacing the analog AMPS system. Cellular Digital Packet Data (CDPD) places digital message packets in unused or underused portions of the radio spectrum or during conversational pauses. This speeds data transmission by a factor of about eight over the speed of conventional cellular transmission.

Another method, Code Division Multiple Access (CDMA), tags each message with a code and spreads it over a wide spectrum. With CDMA, more low-power messages (calls) can occupy the wide channel. Some claim that CDMA offers nearly a 20-fold improvement over conventional AMPS. Another system based on time-division multiplexing called Time Division Multiple Access (TDMA) divides the channel into time slots and multiplexes three to seven conversations in a manner similar to time-division multiplexing in wired systems. Not as effective as CDMA, TDMA deployment is more advanced.

Cellular systems have great potential but are limited to land areas and are less economic in sparsely settled regions. To overcome these limitations, several firms are building or planning to build cellular systems that replace stationary cells with mobile cells located on low-earth-orbit satellites. The system is a truly global cellular network, but its technology is exotic, expensive, and risky.

Satellite Cellular Networks

The concept of a global cellular network is based on precisely locating numerous low-orbit satellites so that any point on earth can communicate with at least one satellite at all times. The satellites can communicate with each other and with ground stations so that any two people with suitable cellular phones can communicate, regardless of their locations. The concept includes an electronic, international directory containing a unique phone number for each cellular phone in the system. In principle, an individual on a trans-Pacific cruise can maintain phone contact with the home office 24 hours a day using voice, fax, or even video communication. At least eight separate systems aimed at capturing this business opportunity are in the conceptual or developmental stages.

One system, conceived by Motorola in 1987 and estimated to cost $3.4 billion, uses 77 satellites in polar orbit 500 miles above the earth and 20 ground stations.[13] Satellites at this altitude move about 4.5 miles per second relative to a fixed point on the earth's surface and have a viewing area (cell size) of about 400 miles in diameter. Thus, an individual using the system for more than a brief period needs several satellites' services. The system automatically transfers an individual's message to the appropriate satellite to maintain the communication path. For long-distance communication, several satellites in turn may serve the call originator, while several others may serve the recipient. The system switches and processes calls in space, but may also involve one or more ground stations.

Transferring traffic from satellite to satellite is akin to transferring traffic from one land-based cellular station to another, as mobile callers move from one station's jurisdiction to another's. For obvious reasons, the satellite process is much more complex.

The Motorola system, termed Iridium (the element iridium has atomic number 77), differs considerably from current communications satellite systems. Traditional communications satellites operate in geosynchronous orbit about 22,300 miles above the earth. At this altitude, a satellite remains nearly stationary over a fixed point on land where it can view about one-fourth of the earth's surface. (Home satellite TV systems are of this type.) The distance from satellite to earth delays electromagnetic transmission to the satellite and back by about one-fourth second, insignificant to TV viewing, but bothersome for phone conversations. To cover the distance, cellular phones using geosynchronous satellites require bulky battery packs. In contrast, the Iridium system experiences imperceptible communication delays once the call is established and can maintain transmission with lower-power cellular phones.

The real issues facing the Iridium project and others like it center around demand for the system and the developer's ability to market it at its anticipated costs. In Motorola's case, the phone device is expected to cost about $2500. Monthly service fees are projected at $50, and a five-minute call will cost about $15. As usage develops, however, these costs can decline considerably. In comparison, land-based cellular service in the U.S. averages about 50 cents per minute, monthly charges are $20 to $30, and the phone device costs $100 or so. Advocates claim that business and personal traffic volume in less developed countries and in sparsely settled regions will make the Iridium and similar systems profitable. By the year 2000, developers hope to profit by capturing one to two percent of the estimated market of 100 million cellular users.

To some people, these projects indicate communication overkill. They believe brilliant technology deceives developers who overlook lower-tech, lower-cost alternatives. Regardless of the merits of the arguments on either side, satellite cellular systems, land-based cellular networks, land and undersea cable systems, and conventional satellites greatly increase global communication capability.

SUMMARY

Global telecommunications is one important result of advanced information technology. Just as in computer systems, semiconductor technology and programming are fundamental to the development and implementation of telecommunications systems. Today, telecommunications hardware and software are being integrated with data processing equipment; and voice, image, and data communications are being integrated on a large scale.

Advanced hardware and software to serve customer needs are available for communication between individuals and businesses with LANs or internets using wireline or wireless technology. Most business organizations in the U.S. find these capabilities attractive and have adopted them. On a larger scale, telecommunications systems serve national or international needs. Multinational firms and others engaged in international commerce are rapidly implementing these technologies. Thus, the application of telecommunications technology to business and industry is critically important to information technology management.

Review Questions

1. What is telecommunications and what defines a telecommunications network?

2. What determines the architecture of a network system?

3. What are the most important characteristics of the U.S. telephone system?

4. What are the major elements of the U.S. telephone network system?

5. Describe the function of the local exchange. Why is the local access line called the bottleneck in the phone system?

6. What forms of media are found in telephone networks?

7. Describe the various forms of multiplexing discussed in this chapter. Which types do not require digital signals?

8. Describe T-1 service.

9. Describe some conditions that make voice signal compression feasible. What additional factors apply to video signals?

10. What are the advantages and disadvantages of ISDN?

11. Why is the International Standards Organization OSI model so important?

12. What are the main features of the OSI model?

13. Distinguish between repeaters, bridges, and gateways. What are some uses for these devices?

14. What is the theoretical ATM packet rate on an STS-1 SONET line? What are some implications of this rate?

15. What are the characteristics of the U.S. cable TV network?

16. Differentiate between AMPS and CDPD. How does TDMA differ from CDMA?

17. Describe the operation of the Iridium satellite system.

Discussion Questions

1. Describe the operation of the phone network in a call going from New York to Los Angeles. What possible forms of media may be involved in this call?

2. Discuss the steps required to make the phone system a digital network end-to-end.

3. Discuss the process for digitizing voice signals. What are the prerequisites for signal digitization, and how does it enable multiplexing, switching, and advanced transmission techniques?

4. Discuss advantages and disadvantages of coaxial cable and fiber-optic cable. Based on what you have read or heard, which media will become dominant, and why?

5. According to some, ISDN has been implemented rather slowly in the U.S. Discuss reasons why this might be so. What are some consequences for IT management?

6. Discuss the difficulties CIOs and their firms face in making network choices from today's menu. What management principles are useful in resolving these difficulties?

7. Discuss the similarities and differences between the phone network and the cable network in the U.S. today. What modifications to each can make them more alike?

8. Assume you are the CIO of a U.S. firm. Discuss the possible uses of the TV cable network in your business.

9. Compare and contrast the existing land-based cellular system with the developing Iridium system.

10. George Gilder (note 7) said "The new rule of radio is the shorter the transmission path, the better the system. Like transistors on semiconductor chips, transmitters are more efficient the more closely they are packed together." Discuss the significance of his statement.

Assignments

1. Visit a local telephone switching center in your area and investigate the types of switches in use. Write a two-page report describing the center.

2. Obtain a detailed description of cable system headends from library references to assess the difficulties involved in providing video on demand.

3. Obtain a description of the standards pertaining to one type of LAN network and summarize its main features on one or two pages.

4. Write a brief description of one of the systems competing with Motorola's Iridium network. What are the relative advantages and disadvantages of each?

[1] Vinton G. Cerf, "Networks," *Scientific American*, Special Issue Vol. 6, No. 1, 1995, 44. Cerf led the Defense Advanced Research Projects Agency (DARPA) team that developed the TCP/IP protocol, widely used today for connecting packet-switching computer networks that comprise the Internet.

[2] Heavy Internet activity is invalidating this assumption and causing difficulty for network providers. Today, about 85 percent of the long-distance network is in use at any given time, up from an optimum 65 percent according to Salomon Smith Barney. Steve Rosenbush, "Network Busy Signals?, *USA TODAY*, December 2, 1997, 3B.

[3] Analog signals are also characterized by phase, a technical notion that's important in some cases. A simplified description of analog signal carrying capacity, the baud rate, can be found in Scott Woolley, "This Pasture Looks Greener," *Forbes*, October 20, 1997, 294.

[4] Shannon showed that transmission speed in bits per second (bps) is related to bandwidth, signal power, and the amount of power in the noise according to the following equation:

$$bps = B \log(base\ 2)(1 + (S/N))$$

where bps is the transmission speed in bits per second, B is the transmission line bandwidth, S is the signal power in watts, and N is the signal power of the noise in watts.

[5] Hitachi has achieved a data transmission rate of 40 gigabits per second (40,000,000,000 bps) using an optical fiber—the practical equivalent of about 600,000 simultaneous phone conversations on one fiber.

[6] Kim Girad, "Pending ADSL Services Will Speed 'Net Access," *Computerworld*, September 23, 1996, 28.

[7] George Gilder, "Telecosm: Feasting on the Giant Peach" *Forbes ASAP*, August 26, 1996, 85.

[8] Emmanuel Desurvire, "Lightwave Communications: The Fifth Generation," *Scientific American*, Special Issue, 1995, 54. Scientific results published in early 1996 have confirmed 1,000 gigabit transmission speeds on fiber using 50 wave-division multiplexed signals.

[9] Gilder, 90.

[10] Based on a survey of 350 sites in October 1994 by the Business Research Group, 10 percent planned LAN implementations and 21 percent planned WAN implementations of ATM. Stephen Klett Jr., "Users Anticipate Benefits From ATM", *Computerworld*, November 14, 1994, 2.

[11] The 1996 Telecommunications Act and various other regulations apply to cable systems.

[12] George Gilder,"Into the Telecosm," *Harvard Business Review*, March-April 1991, 154.

[13] "Phone Calls From Anywhere," *The Gazette Telegraph*, August 30, 1992, E1. The status of Motorola's satellite projects can be monitored by browsing www.Mot.com.

7

Legislative and Industry Trends

The Smallest Baby Bell Grows Much Bigger

After the breakup of AT&T in 1984, SBC Communications (formerly Southwestern Bell) became the new Regional Bell Operating Company (RBOC) serving Kansas, Missouri, Arkansas, Oklahoma, and Texas. SBC, the smallest of the new Baby Bells, relentlessly pursued an orderly, profitable growth path. In the 10 years ending in 1995, free cash flow grew more than 23 percent annually and common shares dividends increased, growing 5.1 percent annually.

SBC executed its growth strategy skillfully. In December 1990 in a major move south of the border adjacent to its Texas market, a consortium led by an SBC subsidiary purchased voting control of Telefonos de Mexico (Telmex) from the Mexican government.[1] The consortium included a subsidiary of France Telecom and a diversified Mexican business alliance led by Grupo Carso. Including an option to buy additional shares, the purchase price to SBC was $486 million. In September 1992 SBC exercised its option and invested $467 million for more than 530 million Class L shares. Today, SBC's Telmex investment value is twice the purchase price.

In a package called The Works, SBC bundles Call Waiting, Call Forwarding, Three-Way Calling, Caller ID, Priority Call, Call Return, and Call Blocker for a single monthly charge. Since its introduction in 1994, nine percent of SBC's customers have adopted this package. In an initiative other RBOCs now mimic, customers receive discounts for using a recently introduced SBC credit card to pay their phone bills. And, on the technology side, SBC accelerated the buildout of its ISDN capability, making it available to nearly all customers by the end of 1996.

Through expansion and alliances, SBC increased its U.S. cellular business to more than four million customers in 65 markets. To enhance growth, it developed a marketing alliance with GTE, gaining access to 14 million potential customers in 30 Texas markets. Today, 11 percent of its landline customers also purchase cellular service from SBC, a percentage significantly higher than many of its peers. The company also holds interests in wireless systems in upstate New York, Boston, Chicago, and parts of central Illinois.

During the past several years, SBC expanded its international business by obtaining a 15.5 percent share of a South African cellular company and by adding ownership interests in Chilean and South Korean cellular firms. SBC also obtained a 10 percent ownership position in one of France's nationwide cellular systems, and it's expanding cellular business in Mexico through Telmex.

Meanwhile, Pacific Telesis, the RBOC serving California and Nevada experienced financial difficulties. Having divested itself of its rapidly growing cellular business, AirTouch, Pacific Telesis faced tough competition in its local markets and demands on its resources for expansion and modernization. Challenged by deteriorating financial conditions and the prospect of reducing its dividend or worse, the company began searching for a merger partner, one with pockets deeper than its own. After rebuffs by several firms, Pacific Telesis agreed to be acquired by SBC in April 1996.

Given Pacific Telesis' situation, a merger made sense and perhaps could even be called essential. With few anti-trust implications, the Justice Department approved, and the merger was consummated in April 1997. The acquisition more than doubled SBC's phone access lines. Aside from increased size, what's in this for SBC? Do its shareholders and customers stand to gain from the acquisition?

Given the geography of the new company, potential exists for improved customer focus on Mexico. More than half of the calls from the U.S. to Mexico originate from the new SBC territory, as do one-fifth of the calls to Asia.[2] Thus, after the merger countries in Latin America and Asia may be better or more efficiently served.

Additional potential arises when the firm receives permission to enter the long-distance market because the new company will operate in seven of the 10 largest metropolitan areas in the U.S. With competition from the traditional long-distance suppliers and GTE, who also serve these markets, acquiring inter-regional traffic will be difficult. However, this gives the new company an opportunity to develop its PCS business and capitalize on its licenses in San Francisco and Los Angeles. It also has an opportunity to expand services overseas and develop a revenue stream from new long-distance business. Given SBC's track record for managing operations, this venture will probably succeed.

In announcing second quarter 1997 results, CEO Whitacre stated, "We are more convinced than ever before that the merger will be extremely positive for our customers and shareowners, and we're working hard to unlock the full value of the merger as quickly as possible." Preparing for what many believe will be a bright future, during 1997 SBC spent $2.3 billion to restructure its business to absorb Pacific Telesis and to close some redundant and unprofitable operations.

INTRODUCTION

The information technology industry is large, competitive, highly complex, and rapidly growing. Now one of the world's largest industries, its 1997 revenues exceeded $600 billion.[3] In the span of barely 50 years, the revenues of the U.S. electronic computing industry grew to several 100 billion dollars; and thousands of firms now employ several million individuals worldwide.[4] Some firms develop and manufacture components such as semiconductor chips or disk storage devices. Others assemble and market systems constructed from various components they manufacture themselves or purchase from others. Many firms develop the tens of thousands of application programs that enable hardware to do useful work. All attempt to capitalize on technology; some invent it and others develop its usefulness.

The U.S. telecommunications industry is much older than and about as large as its computer counterpart. It, too, develops and capitalizes on new technology. Many breakthroughs depend on inserting semiconductor and computing technology in communications devices such as switches and multiplexors; others stem from developments in new media or devices such as optical fiber or satellites. In addition to developing and implementing new technology and providing a rapidly growing array of new services, the industry responds to dramatic, highly significant regulatory and environmental changes in the U.S. and most other nations.

THE SEMICONDUCTOR INDUSTRY

Semiconductor devices are fundamental building blocks of information-age products. More than 100 companies compete for a share of the $115 billion worldwide semiconductor market. Most of the semiconductor industry's outputs are commodity products such as memory chips or microprocessors, chips for consumer electronics or industrial products, or devices for automotive, communications, or military applications. Product development and manufacture is very expensive in this highly capital-intensive industry.

Major companies around the world participate in the semiconductor business. Japanese firms capture about 40 percent of the market; North American companies hold 41 percent; European and other firms secure the remainder. Table 7.1 lists the world's largest semiconductor suppliers.

TABLE 7.1   The World's Largest Semiconductor Suppliers

1.	Intel	U.S.	6.	Texas Instruments	U.S.
2.	NEC	Japan	7.	Samsung	Korea
3.	Toshiba	Japan	8.	Mitsubishi	Japan
4.	Motorola	U.S.	9.	Fujitsu	Japan
5.	Hitachi	Japan	10.	Matsushita	Japan

Because the industry is so capital intensive, large companies dominate its market. Intel, for example, is expected to generate about $36 billion in 1998 revenue.[5] Nevertheless, many small firms have found niches making specialty products. In addition to the firms listed above, many manufacturers also build semiconductors for their own use. IBM, Hewlett-Packard, and Digital produce products for internal use worth billions of dollars. AT&T's former Micro-electronics division, now the independent NCR, produces chips for its computer and telecommunications products and also sells to others. In addition, many firms share design and manufacturing capability with larger companies.

Worldwide consumption of semiconductor products is primarily for computer/data-communications systems (47 percent) and consumer products (21 percent). Communications systems consume 16 percent of the output, industrial products 10 percent, and automobile and other products the remaining six percent. Tiny, powerful semiconductor devices are rapidly becoming ubiquitous in modern society.

The electronic computer industry, barely 50 years old, is one of the world's largest. Hardware, software, and service products are rapidly establishing markets throughout industrialized nations, permeating offices, businesses, and individual residences. About 70 percent of the market is in the U.S. and Western Europe but four Japanese firms are among the top eight information technology suppliers.

Although government agencies used the computers first, businesses quickly invested in them, too. Today, desktop and mobile PCs used for a broad range of individual and business purposes drive industry growth. Nearly all U.S. businesses use PCs extensively. Forty-one percent of U.S. households now have one or more PCs; however, PC sales growth rates are higher outside the U.S. Modem sales everywhere are growing most rapidly of all, indicating spectacular growth in connectivity.

As the industry grows and matures, companies within it prosper unevenly; some early entrants departed rather quickly—General Electric and RCA, for example. Others with more staying power grew rapidly. Some struggled, then merged, and ultimately survived as part of another company. For example, the 1986 merger of Burroughs and Sperry Corporation formed Unisys. Agile competitors, weak strategies, and the rapidly changing environment eventually overcame still others whose formulas seemed winning. Humbled in the early 1990s, Digital Equipment Corporation and IBM are presently on the mend; however, Apple Computer still experiences extreme distress. Thousands of smaller firms prosper by developing and marketing an almost endless stream of hardware or software products. Today, numerous, successful firms provide specialized information systems services, from consulting to disaster recovery to systems integration.

Hardware Suppliers

Mobile personal computing, with growth rates estimated to be 25 to 40 percent for the next several years, is the most rapidly growing portion of the computer industry. Market leaders in portable PCs, notebooks, and subnotebooks include Toshiba, Compaq, Apple, and IBM, but many others compete in this market.[6] The largest supplier to the U.S. is Toshiba with about 18 percent market share, followed by Compaq with a 15 percent share. Nevertheless, firms other than the top four suppliers sell more than 46 percent of all notebooks. Rapid growth in mobile computing foretells parallel growth in wireless data communication and networking.

Desktop computers and workstations connected to servers are the norm in modern firms. Most businesses invest heavily in personal computers for their employees, and many network them to servers and to enterprise systems in client/server and intranet configurations. Of the many firms actively supplying products to this market, Apple, IBM, Compaq, NEC, and Dell lead; the top 10 firms capture about 50 percent of the market. The workstation/server market is more concentrated: Sun MicroSystems secures about 40 percent, Hewlett-Packard has 20 percent, and DEC nearly 11 percent.[7] IBM, Hewlett-Packard, and Digital garner nearly three-fourths of the market for mid-range systems, usually networked and often serving client systems.

IBM dominates the worldwide market for large-scale, enterprise systems with an 85 percent share—Amdahl's share is 13 percent and Hitachi's about two percent. Predicted to decline as businesses decentralized their computing and rebalanced with personal and networked computing, mainframe MIP shipments doubled between 1993 and 1995. IBM's fourth generation of S/390 processors, shipped in June 1997, offer five times more processing power than mainframe systems shipped three years earlier and are in great demand by large customers. In addition to processors, many hardware suppliers also market printers, storage systems, video display devices, tape storage, and software products.

The Software Business

The software business grew rapidly as many firms competed for software sales. The top 100 *pure* software vendors, led by Microsoft, Oracle, and Computer Associates International, generate more than $30 billion annually. IBM, whose total annual revenue exceeds $80 billion, is the second largest software vendor with annual software sales exceeding $11 billion—other hardware vendors such as DEC, Unisys, and Hewlett-Packard also support their products with software.

Microsoft leads the software industry and has become the third largest U.S. firm in market capitalization by successfully capturing a dominant share of PC operating system sales. Windows 95 (100 million copies sold worldwide) and Windows NT provide primary software support for Intel microprocessors, which lead the market for workstations, PCs, and notebooks. The combination is so popular it's named Wintel, for Windows and Intel. Windows NT, which supports Intel, Power PC, and Alpha microprocessors, and Windows 98 products indicate that Microsoft will dominate operating systems for some time. Interestingly, while noted for its operating systems, Microsoft earns two-thirds of its revenue from applications software such as Microsoft Office, MS-Word, Microsoft Works, and many others.

Many other firms supply important software. Among the largest, Computer Associates, Oracle, Novell, and Lotus Development Corporation (now part of IBM), are noted for their many general business applications, database systems, networking systems, and spreadsheets.

Industry Dynamics

The computer industry is in a state of flux as many its major players adjust their strategies to cope with competition, changing market conditions, and technology advances. Mergers, acquisitions, and joint ventures are common among smaller hardware, software, and peripheral manufacturers attempting to marshal resources and marketing clout to ward off stronger competitors. But the phenomenon occurs among larger players, too. After decades of priding itself on self-sufficiency, IBM built thousands of alliances to strengthen its position. After years of going it alone, severely weakened Apple struggles for survival—it recently received an infusion of cash ($150 million) from its chief rival Microsoft. The trend is unmistakable. The business is so dynamic and so volatile that nearly all firms seek safety by joining with others.

The trend is not confined to the U.S. but includes major firms in countries around the world. Walter Wriston convincingly describes the situation:

> During the 1980s, high-technology firms, such as Siemens (Germany), Philips (Benelux), GGE, Bull, and Thomson (France), Olivetti (Italy), AT&T, IBM, Control Data, Fujitsu, Toshiba, and NEC (Japan), each forged numerous foreign alliances; some of them formed dozens. A chart of Siemens international cooperative agreements is a genealogist's delight, including, among others, Ericsson, Toshiba, Fujitsu, Fuji, GTE, Corning Glass, Intel, Xerox, KTM, Philips, B. E., GEC, Thomson, Microsoft, and World Logic Systems. IBM has so many alliances in Japan that there is a Japanese book on the subject called *IBM's Alliance Strategy in Japan*.[8]

But dynamism is not confined to the semiconductor and computer industries; the telecommunications industry here and abroad experiences considerable turbulence, too. As developers combine semiconductor logic and memory chips with mass storage and software to build computers and consumer products, telecommunications engineers also use these building blocks to fabricate switches, multiplexors, and other devices to control transmission media and digital traffic. Combining advanced transmission technology with hi-tech electronics, they're rapidly building the global information infrastructure.

THE INFORMATION INFRASTRUCTURE

In the U.S., the electronic information infrastructure consists of many elements, including broadcast radio and TV, cable and telephone wireline and wireless networks, the Internet, and many other networks owned or operated by businesses and government agencies. Collectively these electronic information pathways form the National Information Infrastructure or the Information Superhighway. Telephone systems, cable networks, and the Internet provide rapidly expanding electronic access to vast communications systems linking individuals here and abroad. This bounty of communication capability defines the information age.

Companies that operate phone and cable systems invest large sums to fabricate this infrastructure. They also devise strategies to use the infrastructure in new, profitable ways for the consuming public. Figure 7.1 illustrates the factors driving these investments. It shows the critical contrasts between these networks' transmission bandwidth and switching capability.

The switched phone network in the U.S. permits individuals to reach businesses, 93 percent of households, and most other countries through direct-distance dialing. Rapid, trouble-free switching and broad deployment are the phone networks' most important strengths. The extremely narrow bandwidth of the local loop, however, makes it inadequate for delivering hi-fi sound or video.

In contrast, cable delivers many video channels over wideband links broadly deployed in urban areas. But lacking switching capability and two-way communication, cable cannot deliver narrowband voice messages between customers over its broadband system. To develop a single broadband national network, cable networks must install broadband switching, and phone systems must greatly increase the effective bandwidth of the local loop in their switched networks. Figure 7.1 depicts industry positions and trends as we move toward the network of the future.

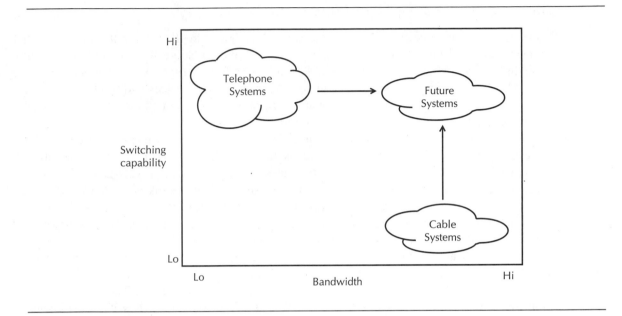

FIGURE 7.1   Switching vs. Bandwidth

In practice, phone and cable firms are developing their networks and investing in building the infrastructure. Using telephone, video, and computer technologies, they strive to provide businesses and individual consumers with many new communications options. Technology advances and new business ventures will help develop tomorrow's network, but legislative and regulatory actions must also address issues like consumer protection and access equity.

REGULATING TELECOMMUNICATION SERVICES

The court-mandated breakup of AT&T in 1982 (implemented in 1984) and the 1996 Telecommunications Law played a major part in shaping today's telecommunications service industry. Among other stipulations, the 1982 judgment mandated competition in long-distance and created seven Regional Bell Operating Companies (RBOCs) from AT&T's 22 operating companies. Since 1984, the number of long-distance suppliers has grown from 108 to more than 800. RBOCs maintain a near monopoly on local service within their regions, however, as the number of smaller local phone companies declines.

The 1996 Act's intent is to increase competition once again by permitting RBOCs to offer long-distance service in their regions, provided they open their service areas to effective local competition. In the last two decades, Court decisions, FCC actions, and Federal legislation profoundly altered the U.S. telecommunications regulatory landscape.

The Court Orders the Divestiture of AT&T

Many legal actions targeted AT&T since its operations began in 1885, but not until 1949 did the Department of Justice file an anti-trust suit to curb the phone giant's growing power and expanding scope. After much wrangling, the suit was settled in 1956. A consent decree confined AT&T's activities to furnishing regulated communication services and producing phone equipment. The decree also required company to license its patents to others; still, it remained integrated and regulated.

Nevertheless, the company continued to experience stress from competitors and from advancing technology. In 1968 over strenuous objections, the court directed AT&T to permit attachment of non-Bell equipment, such as answering machines, to its network.[9] One year later, again over AT&T's vigorous objections, the FCC let MCI establish a microwave link between Chicago and St. Louis for telephone communications. Growing competition, advancing technology, and concern for service to users prompted many other actions. The most important occurred in 1974, however, when the Department of Justice filed its second anti-trust suit against AT&T.

In 1982 litigants settled the anti-trust case by agreeing to a Modified Final Judgment (MFJ). The settlement stipulated that AT&T must divest its local telephone companies and compete in the long-distance and data communications markets. The 22 AT&T local service companies reorganized into seven Regional Bell Operating Companies (RBOCs) and intense competition from growing numbers of new long-distance companies commenced.

The 1996 Telecommunications Law

Just as several decades of advancing technology, growing competition, and rising consumer demands led to AT&T's divestiture, similar pressures once again provided motivation for the 1996 Telecommunications Law.

During the early 1990s, cable and phone company executives operated in an uneasy atmosphere. Advancing technology and consumer preferences quickly turned these executives into competitors. Cable firms that install phone-company-like switches in their networks can threaten phone companies by bringing consumers conventional and advanced services like video-phones and Internet services. Conversely, phone companies that install broadband fiber-optic cables to home can threaten cable firms by delivering video-on-demand and other services. Facing these realities and unwilling to wait for a green light from legislators, regulators, or the courts, industry executives asserted themselves.

In February 1993, Southwestern Bell's agreement to buy two cable systems in the Washington, D.C., area, was the first telephone and cable TV deal in the U.S. Just three months later, US West announced its plan to buy 25.5 percent of Time Warner (TW) for $2.5 billion. This purchase gave US West a 50 percent partnership in TW's cable system, the second largest in the U.S. In October 1993, however, came the announcement of the largest and most sensational deal. Bell Atlantic, the second largest RBOC, announced plans to purchase Denver-based Tele-Communications, Inc., the world's largest cable system, in a deal worth about $33 billion. Industry giants were dramatically maneuvering to position themselves for the technology-driven future.

Stunning business events like these and activities within the FCC and Congress, including hearings by Congressman Ed Markey's House Telecommunications Subcommittee, increased pressure for Congressional action. Facing overwhelming, relentless external forces, Congress responded with the 1996 Telecommunications Act, which President Clinton signed into law on February 8, 1996.[10]

After years of Congressional study and with strong industry support, the Telecommunications Act passed the House 414 to 16 and the Senate 91 to 5. "We've been moving toward the sunny seas of competition for some time," said the Federal Communication Commission's Chairman Reed Hundt. "This bill puts an engine on the boat."[11]

The 1996 act largely frees industry participants from more than 60 years of outdated federal and state oversight and regulation. It supersedes much of the depression-era Communications Act of 1934 and the 1982 Modified Final Judgment that mandated AT&T's break-up. The 1996 Act's intent is to unleash powerful competitive forces among the seven regional Bell companies, AT&T and other long-distance suppliers, and cable TV companies. The Act fundamentally shifts the way communication services are delivered—it initiates a new telecommunications era in the U.S.

The Act permits and encourages competition in the local exchange market where RBOCs reign virtually unchallenged, and it also fosters competition between RBOCs and long-distance suppliers. But it does much more. The Act affects TV-station ownership, alters cable rate regulation, lets phone companies sell video over phone lines, permits cross-ownership between cable and phone companies in small communities, and guarantees phone service everywhere, including remote areas. Designed to leverage free market forces for consumer advantage, the 1996 Act is accurately called the Telecommunications Competition Act of 1996.[12]

THE TELECOMMUNICATIONS SERVICES INDUSTRY

Local Service Providers

Beginning January 1, 1984, AT&T's divestiture created seven RBOCs from the its 22 wholly-owned subsidiary telephone companies. AT&T also transferred its cellular facilities to RBOCs, along with a one-seventh interest in Bell Communications Research Corporation. Before divestiture, Bell companies served approximately 87 million access lines; independent companies owned about 22 million.

The provisions of the Modified Final Judgment (MFJ) carefully controlled activities. The judgment required them to provide local access to all long-distance carriers equal in all respects to that provided to AT&T or its affiliates. Additionally, the MFJ prohibited RBOCs from providing long-distance services or manufacturing telecommunications products or customer premises equipment.

Today, several hundred million telephones and other communications devices obtain local service from RBOCs, from GTE, or from more than 1300 smaller companies. Nearly 100 tiny companies own fewer than 1000 access lines each, mostly in less densely populated rural areas. Altogether, these local telephone companies serve nearly 200 million access lines. About 25 companies account for 90 percent of the local and long-distance traffic. In addition, the largest local companies also provide wireless services in their operating areas.

Business Developments

Local operating companies invested nearly $300 billion in plants and equipment and employ about 600,000 workers to serve U.S. businesses and households. Seven RBOCs and GTE (larger than any RBOC but always independent of AT&T), are projected to earn nearly $140 billion in 1998. They plan additional investments of nearly $100 billion over the next five years, modernizing and expanding their capabilities as part of the information superhighway expansion.

The local telecommunications service industry changes rapidly as formerly closed markets experience competition from cable operators, competitive access providers (CAPs), and long-distance companies. Cable operators expect to capture a share of the residential phone market as CAPs woo the local companies' high-volume business customers. In addition, local companies entering the cable business try to satisfy conditions of the 1996 Act so they can compete for long-distance business within their regions.

To protect their markets, local companies are reengineering their operations to reduce costs and offer new services to retain customers. Major players are rapidly expanding their wireless operations and have introduced interactive multimedia networks (with little success so far), frequently in joint ventures or alliances with others. They are beginning to enter the promising, potentially huge, personal communications services (PCS) market, too. In addition, most are expanding their overseas operations, hoping to exploit their strengths in less competitive, high growth potential areas.

By specifically establishing an environment fostering competition, the 1996 Act caused turmoil among the largest local service providers. Their goal, now possible under the new rules, is to become one-stop suppliers of telecom services—local wireline, long distance, wireless cellular and personal communication services, and cable TV, too. For most, the highest priority target is the lucrative long-distance market within their regions. Now collecting fees from long-distance companies who access their networks, local service providers would gladly trade these for a chance to provide more profitable in-region, long-distance service, the first step toward their goal of becoming one-stop suppliers.

The specific intent of the 1996 Act, however, is to bring competition to the local market, since the MFJ already greatly increased long-distance competition. To this end, the Act requires RBOCs to satisfy a rigorous 14-point checklist that guarantees effective local competition before permitting them to offer in-region long-distance. (The appendix to this Chapter summarizes the checklist.) Interpreting the checklist strictly, the first three RBOCs that sought FCC permission failed. With widely dispersed operations and always independent of AT&T, GTE is exempt from meeting the requirements of the checklist.

The 1996 Act's passage and the MFJ's negation led to several other important developments. One, noted in the Business Vignette, is the merger between SBC Communications and Pacific Telesis concluded in April 1997. And, within months of the Act's passage, Bell Atlantic and NYNEX announced merger plans. The consummated mergers reduced the seven RBOCs to five, rearranged the landscape leaving Ameritech, US West, and BellSouth unchanged, and caused some to wonder whether the 1996 Law really is pro-competition.

By any measure of size, the newly created firms head the pack. Table 7.2 shows how RBOCs and GTE ranked by revenue following these mergers. Other measures, such as number of employees or access lines, correlate with revenue.

TABLE 7.2   Phone Company Ranking

Company	1998 Revenue (estimated)
Bell Atlantic-NYNEX	$31.5 billion
Pacific Telesis-SBC Communications	26.8 "
GTE	25.0 "
BellSouth	21.4 "
Ameritech	17.5 "
US West (Communications & Media Group)	17.1 "
AT&T (for comparison)	55.1 "

Unlike the SBC-Pacific Telesis deal, the marriage of NYNEX and Bell Atlantic was one of convenience. The newly merged company controls most local services from Virginia to Maine, including its already well-developed alliance in wireless, Bell Atlantic NYNEX Mobile. The companies also formed a four-way partnership with AirTouch and US West called PCS PrimeCo, which operates PCS systems and provides PCS services.

The merger of Bell Atlantic and NYNEX created an entity with a dominating presence on the East coast. Owning more than 30 percent of U.S. local lines, it will be the termination point for more than a third of America's international traffic.[13] The merger generated considerable resistance from state regulators, though little from the FCC or the Justice Department. After all, Congressional intent in passing the 1996 Act was to promote competition and to secure lower prices and higher quality services for consumers. The partners explained to many inquiring groups how their merger benefits consumers in their region.

Lured by $20 billion worth of long-distance calls originating in the 12-state area, participants remained fixed on their objectives. The decision finally rested with the Department of Justice, which determined that the deal violated no regulations and satisfied other legal conditions such as the anti-trust laws. The Department of Justice gave the merger final approval in 1997.

Ameritech (AIT) serves the upper Midwest in Wisconsin, Michigan, Illinois, Indiana, and Ohio, having originated from former Bell companies operating in these states. In the 14 years since divestiture, AIT business has grown by offering increased and improved services in its core area, enlarging its presence outside its region and expanding aggressively overseas. It devotes an increasing share of its $2 billion in annual capital expenditures to developing cable TV, security monitoring, and other networks.

In addition to constructing cable TV networks, AIT joined a project with The Walt Disney Company, GTE, BellSouth, and SBC Communications to produce *Americast* programming. Ameritech expects this venture to grow 11 percent per year.

Ameritech's cellular and paging volumes grow at double digit rates, too. To increase its wireless presence, Ameritech obtained PCS licenses in Indianapolis and Cleveland and makes investments in those markets. Providing more functions to current customers and developing new markets are two of its strategic thrusts. Investing internationally is another.

In a joint venture with Deutsche Telekom, AIT owns 67 percent of the privatized Hungarian telecommunications system, MATAV. The system increases the number of access lines by more than 15 percent per year and also serves a rapidly growing base of cellular customers. Ameritech holds a 24.8 percent stake in New Zealand Telecom, which it intends to sell. This investment has grown three-fold since 1990. In Poland, AIT owns 24.5 percent of Centertel, the first company to provide cellular service, and it owns 25 percent of NetCom, the first competitive cellular company licensed in Norway. In 1995 Ameritech led a consortium selected as a partner for Belgacom, which supplies local, long distance, and cellular for 10 million Europeans. AIT's share of Belgacom is 17.5 percent.

With its strong position in the Midwestern market, AIT intends to expand its multimedia activities and to capture a large share of its $8.5 billion intraregional long-distance market. Sixty percent of all long-distance traffic in AIT's region is intraregional. Anxious to satisfy the 14-point checklist, Ameritech also intends to protect its local exchange business from erosion. Ameritech's U.S. strategy is to flourish by coddling customers and cultivating commerce. Seeking prosperity while managing risk characterizes this strategy.

BellSouth serves the rapidly growing Southeast, including the states of Alabama, Florida, Georgia, Kentucky, Louisiana, Mississippi, North Carolina, South Carolina, and Tennessee. With more than 20 million lines, BellSouth is the largest U.S. local phone company by this measure. Wireless and international activities account for 12 percent of its revenue, and network and related services for 72 percent. One of BellSouth's most rapidly growing markets is cellular service, now provided to more than two million customers.

In the U.S., BellSouth participates in a joint venture with The Walt Disney Company, Ameritech, and SBC Communications to develop, market, and deliver innovative video programming to consumers. The venture expects to invest $500 million in video markets over the next five years. BellSouth has many other domestic initiatives, too.

Overseas, BellSouth operates or shares operations in a total of 16 countries throughout Latin America, Asia, Western Europe, the Middle East, and the Pacific Rim. These operations include cellular service, long-distance service, private data networks, and mobile data nets. BellSouth also has alliances with two Chinese companies to develop advanced telecommunications technologies for rapidly expanding Chinese markets. BellSouth is positioning itself to grow its markets in the U.S. by offering advanced services and to expand rapidly in selected areas around the globe.

GTE is the eighth largest telecommunications company in the world, the third largest U.S.-based local phone company, and the fourth largest U.S. cellular provider. Operating nationally, GTE provides wireline or wireless service in all 50 states and has a wireline presence in 164 metropolitan areas. GTE's wireless business doubled in the past two and one-half years and is expected to grow at a similar rate during the next several years.

Overseas, GTE has operated the Dominican Republic's phone system for 44 years and also manages Venezuela's system. Both systems include wireline and wireless facilities. Cooperating with others, GTE provides cellular service to Argentina and plans to expand operations in Latin America where its presence is established. In Canada, GTE owns a majority share of BC TEL and operates the system serving British Columbia and Vancouver and the network serving the province of Quebec.

Among many other initiatives, GTE is expanding rapidly into the data services business. In 1994, the company introduced its World Class Network built on SONET and ATM switches. By the end of 1995, this service was available in 14 major business markets. GTE plans to expand this service to 275 SONET rings in 1998.

Serving 14 western states from Minnesota to Washington and from Montana to Arizona, US West is the largest RBOC geographically, even after the two recent RBOC mergers. Its approach to market expansion also differs considerably from any others'.

In 1994, USW purchased Wometco Cable Corporation and Georgia Cable Holdings for $1.2 billion and invested $2.5 billion for a 25.5 percent share of Time Warner. US West's investments in cable firms indicate its belief that cable may provide it an opportunity to offer competitive phone service in RBOC regions and that content (programming) will be profitable as cable expands toward 500 channels.

In late 1995, US West created two classes of stock separating its services into basic wireline phone, US West Communications, and all others, US West Media Group. In October 1997, USW announced that it would split into two firms, confirming the failure of its strategy to find synergy by combining phone and cable systems. USW's actions indicate to many that merging cable and phone systems present greater obstacles than originally believed, thus postponing what some thought was a natural development. Analysts expect that USW will find a buyer for all or part of its cable assets (US West Media Group) or US West Communications may merge with or be acquired by another phone company. The company retains many other options and initiatives.

Overseas, USW holds 50 percent of TeleWest Communications plc, a U.K. cable television/telephone business, and 50 percent of Mercury One-2-One, a personal communications joint services venture also in the U.K. As its telephone business grows rapidly, its cable services grow more slowly than expected. USW also has wireless joint venture operations in the U.K., Hungary, the Czech Republic, Slovakia, and Russia, and holds interests in other European cable television ventures, too.

Some RBOCs are considering separating their traditional wireline operations from their wireless and cable operations to gain more flexibility in these new endeavors. In April 1994,

Pacific Telesis spun off its domestic and international cellular, paging, and other wireless operations in a stock distribution that created a new company, AirTouch. Later, AirTouch partnered with US West Cellular, gaining a major presence in the U.S. cellular business.

Cable systems and programming captured the attention of RBOCs and others struggling to maintain a competitive posture in the exploding telecommunications marketplace. In addition to the US West purchases, SBC Communications, BellSouth, and Ameritech have a joint venture with The Walt Disney Company. Bell Atlantic, NYNEX, and Pacific Telesis initiated a joint production venture using Creative Artists Agency as a consultant. And Sprint is part of a joint venture with Tele-Communications Inc., Comcast Corporation (now partly owned by Microsoft), and Cox Communications. As the information infrastructure develops, these and many other business ventures play increasingly important roles.

LOCAL AND LONG-DISTANCE CONSIDERATIONS

The MFJ prohibited RBOCs from offering long-distance service and required equal access to RBOC facilities for all long-distance companies. This means that long-distance messages, using higher-level exchanges shown in Figure 6.2, pass from the local company to the long-distance company, then to a local company again to complete the call.

Geographically, RBOC service areas are divided into Local Access Transport Areas (LATAs) within which the local company provides service. The approximately 200 U. S. LATAs are mostly centered around metropolitan areas. Local companies serving the LATA may charge tolls for some calls, but the MFJ required that a long-distance carrier handle all interLATA traffic.

For example, consider the cities of Houston and Dallas, each in separate LATAs within SBC's region. Houston customers can make local calls within a defined area with no extra charges on their monthly bill, but they pay toll charges for calls within the LATA and beyond this defined area. However, a long-distance carrier such as AT&T or MCI must handle interLATA calls from Houston to Dallas. The long-distance carrier pays a fee to access SBC's network, and SBC bills its customers for long-distance calls and remits the charges, minus access fees, to the long-distance company.

This arrangement will change, however, because the 1996 Act permits RBOCs to provide interLATA service after ensuring that effective competition exists in the local market. After satisfying the requirements of the 14-point checklist, SBC and long-distance providers such as AT&T and others, for example, may all compete for both local and long-distance service in Houston, Dallas, and other markets.

The environment is more complicated in some regions of the country, however, where independent local companies operate within RBOC regions, and are interconnected with RBOC and long-distance equipment.[14] These non-Bell companies are an important part of the U.S. phone system.

LONG-DISTANCE SERVICE PROVIDERS

AT&T (with about 56 percent market share), MCI (20 percent), Sprint (10 percent), and about 800 smaller companies (14 percent) provide long-distance service in the U.S.. Table 7.3 shows the largest U.S. firms ranked by 1996 revenue.

TABLE 7.3   Largest U.S. Long-Distance Companies[15]

Company	Revenue (billions, 1996)	Lines served
1. AT&T	$45.60	100.4 million
2 MCI	16.80	22.9 "
3. Sprint	8.30	11.8 "
4. WorldCom	4.50	4.5 "
5. Frontier	1.89	1.0 "
6. Excel/Telco	1.78	
7. LCI	1.10	
8. Tel-Save	.32	

In the last decade, long-distance service has grown annually an average of 10.5 percent as compared with local access lines' growth of about three percent. The number of local companies steadily declined during this time as the industry consolidated, while the number of long-distance providers grew to more than 800. Although some new long-distance suppliers such as RBOCs may emerge, local industry consolidation continues.

Developments in the Long-Distance Market

During the past several years, mergers and acquisitions reshaped the global competitive environment as firms positioned themselves for the rapidly-growing, worldwide long-distance market. Nearly every major company in the U.S. and abroad has engaged in this activity.

To prepare for the global, competitive future, AT&T divided itself into three independent, publicly-held companies—the new AT&T, Lucent Technologies, and NCR. The new AT&T earns revenues of $55 billion, Lucent Technologies about $22 billion, and NCR more than $8 billion. These firms' total employment is 296,000 people.

Lucent Technologies manufactures and distributes telecommunications hardware and software systems, high-performance integrated circuits and optoelectronic components, and designs, manufactures, installs, and maintains business communications systems in more than 90 countries. AT&T Global Information Solutions, the semiconductor and computer arm of AT&T, assumed its old name, NCR, before once again becoming an independent company. It hopes to lead in high-end computer systems and in transaction processing hardware and software for retailers and financial institutions with data warehousing, scanners, and ATM machines.

The new AT&T remains one of the world's largest telecommunications companies. AT&T's huge network includes more than 100,000 miles of underseas fiber-optic cable linking the continents of the globe. AT&T Wireless Services operates in more than 100 major cities and expands rapidly in the U.S. and abroad. AT&T's goal is to become a worldwide provider of wireless under its well-known brand name. As it expands its wireless operations, the company hopes to generate more traffic for its long-distance network, thus building revenue and saving access charges, too. The numbers are compelling: Eliminating access charges at both ends of a long-distance call increases revenue about 80 percent with little or no cost increase.

Commenting on reasons for restructuring, CEO Robert Allen said, "Our decision to go this route reflects our determination to shape and lead the dramatic changes that have already begun in the worldwide market for communications and information services—a market that promises to double in size before we ring in the new century."[16] Recognizing that its financial strength would not likely increase as it faces growing competition, Allen wanted to restructure to be more competitive while conditions were favorable.

The 1996 Telecommunications Act also lent urgency to AT&T's actions. Ownership of 125 million local access lines gives RBOCs and GTE a major advantage in their quest to invade the long-distance market that AT&T, MCI, and Sprint now dominate. More than half of all long-distance calls begin and end within the local exchange companies' territory—making them juicy targets when the Bells become free to attack. Although the FCC and others will determine many ground rules for competition in this area, the Act appears to favor the Baby Bells.

Things heat up on the international front, too, with British Telecom's (BT) purchase of 20 percent of MCI creating Concert Global Communications PLC, and France Telecom's and Deutsche Telekom's acquisition of 20 percent of Sprint.

With revenue exceeding $42 billion from more than 40 million customers in 70 countries, Concert ranks with AT&T in size and global reach. MCI Communications gives Concert a strong entry into U.S. markets. It operates more than 3600 million circuit miles of microwave and fiber-optic network, providing toll-free (800) service, worldwide direct dialing, and other functions. It plans to invest $2 billion in fiber rings and switching systems in 20 major U.S. cities and eventually reach 200 urban markets. An aggressive, rapidly growing U.S. company combined with a slow-growth, monopolistic British firm spells trouble for European and American firms alike.

However, events took a dramatic turn. In bidding to acquire all of MCI, BT offered $24 billion for the 80 percent of MCI it did not own. The bid appeared to be successful, but, when large, planned investments in infrastructure reduced MCI's earnings, BT lost its nerve and reduced its bid by $5 billion to $19 billion. Sensing an opportunity and recognizing MCI's long-term value, the ever-aggressive WorldCom made a $30 billion all-stock offer for the 80 percent of outstanding MCI stock and $7 billion for BT's 20 percent. A $28 billion all-cash bid from GTE for 80 percent of MCI soon followed WorldCom's offer. With several alternatives to consider, including the possibility of yet another bid, MCI paused briefly to evaluate its position, then accepted WorldCom's proposition.

Although this merger appears headed for consummation, it leaves many options for the remainder of the industry including BT (estimated 1998 revenue of $25.8 Billion), GTE, and most other major international players. Two firms most affected by the impending merger are BT and AT&T. BT's market share is eroding at home and internationally, and its poor handling of its MCI bid inspired no confidence from other potential suitors or partners. For AT&T, another strong competitor in local, long-distance, and Internet services markets means pressure on market share and on profits. It, too, may need to seek safety in a merger partner.

Sprint, third largest U.S. long-distance supplier, serves about eight million customers and provides voice service to all the world's direct-dial countries, more than 290 countries and locations. Its U.S. network was the first to be 100 percent digital fiber-optic.

In 1995 France Telecom and Deutsche Telekom took a 20 percent interest in Sprint, forming a tight alliance called Global One. With a dense fiber network in Germany, years of expertise in global data service from France Telecom, and satellite capability from many sources, Global One is a formidable competitor to AT&T's Worldpartners and to other international suppliers. Sprint now provides local service to 6.4 million local lines in 19 states, giving the alliance a strong entrée into the U.S. local market, too.

As privatization continues and economic ties among nations strengthen, global networks become increasingly vital to international trade, and competition continues to grow. The world's major telecommunications providers are positioning themselves for global competition in various ways exemplified by the mergers discussed earlier. Table 7.4 lists the top 10 suppliers after consolidations.

TABLE 7.4   The World's Largest Phone Companies

Company	1998 Revenue (estimated)
1. NTT * (Nippon Tel and Tel)	$75.6 Billion
2. AT&T	55.1 "
3. Deutsche Telekom	39.6 "
4. Bell Atlantic/NYNEX	31.5 "
5. WorldCom (including MCI)	30.9 "
6. France Telecom	27.0 "
7. SBC/Pacific Telesis	26.8 "
8. British Telecom	25.8 "
9. GTE	25.0 "
10. BellSouth	21.4 "

* NTT is considering dividing in two.

For many reasons, the landscape of global telecommunications is certain to change in the years ahead. What we have seen so far is only a prelude. From today's evolving base, two billion new customers will obtain wireline or wireless phone service in the next decade.

CELLULAR AND WIRELESS OPERATORS

Current U.S. regulations require at least two cellular providers in each metropolitan and rural area; one is usually the local wireline operator. Except for Pacific Telesis, each RBOC provides cellular service as do a large number of other firms.

Capital investments made by cellular providers exceed $14 billion or about $875 per subscriber. To improve cellular operations for subscribers as they travel (roam), 15 providers are developing MobilLink, a seamless service that covers 90 percent of the U.S. and 80 percent of Canada. Cellular companies know the importance of ease-of-use in the cellular business and in computing.

On a global basis, cellular service is a growth business. Dataquest Inc. predicts that the global cellular market will grow from 195 million subscribers and $125 billion in revenues in 1997 to 550 million subscribers and $275 billion in revenues in 2001. In the U.S., according to The Yankee Group, the number of subscribers will grow from 50.4 million in 1997 to 90.4 million in 2000, while revenues will nearly double to $46.5 billion.[17] Consumer demand fuels market growth (especially in less-developed regions) and declining costs.

Wireless rates are expected to drop dramatically as the industry scales up and technology costs rapidly decline. According to predictions, wireless cost trends will exert enormous pressure on traditional wireline suppliers as consumers migrate from wireline to wireless service. Technology Futures Inc. predicts that wireless cost per minute in 2007 will be three cents as opposed to 15 cents in 1997; by then, 57 percent of cellular subscribers will have stopped using wireline service.[18] Obviously, digital cellular providers pose great threats to analog cellular and traditional wireline companies.

Communications suppliers face other threats, too. Wireless service will move from land-based systems to space platforms, offering consumers more options.

Satellite Cellular Networks

Tempted by the potential of a global cellular network and by Motorola's advances in this area, at least eight separate systems aimed at capturing this business opportunity are in the conceptual or developmental stages. If completed as planned, these systems will add 1200 low-earth-orbit (LEO) satellites to the approximately 200 already in geosynchronous orbit.[19]

Many communications satellites are parked in geosynchronous orbit about 22,300 miles above earth. The distance delays communication because electromagnetic transmission to the satellite and back takes about one-fourth second. The distance also creates relatively high power demands. LEO systems overcome these difficulties but add enormous technical complexity.

In addition to Motorola's Iridium system, Bill Gates (Microsoft) and Craig McCaw (McCaw Cellular) propose a system called Teledesic. Loral and Qualcomm are developing Globalstar. Meanwhile, GM-Hughes, the leader in geosynchronous orbit satellites, intends to expand its coverage by adding up to 17 more satellites. Competition among these and other systems will be keen if all parties forge ahead. These firms also face many other difficulties as they work to implement their systems.

Table 7.5 lists some obstacles these systems must overcome.

TABLE 7.5   Obstacles Facing LEO Cellular Systems

Obstacle	Related Factors
Construction financing	$2 billion up to $12 billion
User terminal cost	$750 up to $2000
User operating cost*	Up to $200 per hour
Political	Obtaining electromagnetic spectrum, international regulations
Technical decisions	Spectrum sharing, modulation type
Ground facilities	Facilities sharing agreements

* Predicted operating costs depend very much on traffic volume, traffic type, and other financial and technical factors.

In addition to the technical, political, and financial difficulties inherent in satellite systems development, marketing uncertainties and competition stalk the players. If all current players ultimately launch satellite systems, analysts believe some are destined to fail. Motorola hopes to improve service and reduce Iridium costs by reducing the number of satellites from 77 to 66. In 1997, Motorola announced plans to begin a new 96-satellite network, called Celestri, specializing in data transmission. It uses a combination of geosynchronous and LEO satellites. TRW's 12-satellite system Odyssey and Loral's 48-satellite system both combine satellites, ground stations, and existing networks to provide service.[20] Developers believe that re-tooling some current facilities will help lower development and operating costs. Funding for Iridium is complete, and funding for several others seems assured. Currently Russian and U.S. rockets deploy Iridium satellites. Service is expected to begin in 1998.

Fearing that Europe will lag in the race to develop low-earth satellite communications systems, the European Community Commission (ECC) called for industry and government to coordinate research and technical standards. The ECC intends to stimulate European interest in these systems before companies from other nations capture the field.[21]

One very significant telecommunications development impacts the entire world: the giant network called the Internet. Born in 1973, the Internet evolved from the government's Advanced Research Projects Agency network (ARPANET) and now consists of several hundred thousand connected networks. Millions of computers and tens of millions, perhaps 100 million users worldwide connect to the Internet. (No one knows for sure how many people actually use the Internet.) By extrapolating current growth rates, the Internet will have a staggering 180 million computers attached to its network by the year 2000. Although some claim such extrapolations are unwarranted, many believe the Internet *is* the information superhighway.[22]

Internet users can search databases, share research, transfer files, or chat about almost anything they desire. Thousands of groups use the net to communicate about common interests ranging from solid research to absolute fantasy. Today, as growth is just beginning, the number of e-mail messages exceeds postal messages by a factor of five. Analysts predict the number of e-mail users in the U.S. will triple by 2005 to 170 million. By 2005, users will send five billion messages per day, easily exceeding the sum of regular phone and mail messages.

Electronic commerce on the Internet also grows rapidly—estimates place the number of home Web shoppers at nine million. The more rapid growth of the number of Web pages, however, indicates increasing competition. Although some Internet purists deplore commercial exploitation of the net, many others see electronic commerce as a new, high-potential business environment. But many serious technical obstacles are inherent in commercial expansion including insufficient security, challenges to individual privacy, and communications capacity. Nevertheless, commercial enterprises (.com domains) grow rapidly.

Commercially available online information services attract customers at the rate of about 10,000 per day, according to some reports. Customers of America Online (AOL), Compuserve, Prodigy, Delphi, and others such as AT&T's Interchange and Microsoft's MSN will soon exceed 15 million. Most services seek to upgrade network capabilities to improve response time and offer more data-intensive services. Prodigy, for example, supports ISDN in cooperation with BellSouth, NYNEX, and Pacific Telesis Group. This service intends to open the path to multimedia services and to provide faster data access. As the infrastructure evolves and matures, rapid exploitation of the superhighway will certainly occur.

The Internet takes on new meaning in the corporate world, too, as firms install Web features on internal networks called intranets. These networks permit employees to access directly important company information such as corporate policies and employee benefits data. Within the company, departmental Web pages make essential, real-time business information available to employees. Careful planning and sound security can also make some of this intranet information available to suppliers and customers, thus substantially increasing inter-company information exchange and offering the many benefits of speedy communication.[23]

Although the entire information technology industry is experiencing rapid and profound change, networking advances outpace most others. Created from computer building blocks and fiber-optic and satellite technology, powerful technology dramatically increases individuals' and firms' ability to transport voice, data, and video information globally. To exploit new fiber, wireless, and satellite technology, telecommunications firms invest large sums of money in their race to build a global information infrastructure.

The massive effort of developing the global infrastructure raises extraordinary challenges for private industry and government agencies alike. Businesses rise to the challenge by marshaling capital, people, and technology resources to extend current facilities and to provide new services and capabilities. During the next decade, several billion individuals without phone service will acquire it. Another billion customers, now using wireline and wireless services, will benefit from the business and technological innovations of phone, cable, and content providers. In the U.S. and around the globe, construction of the information infrastructure progresses at breakneck speed.

Individuals and organizations rapidly exploit new telecommunications developments. Small personal computers and large computing systems around the world are rapidly being networked, thus permitting nearly instantaneous, global interchange of important personal or corporate information. As Walter Wriston points out, "Borders are not boundaries."[24] Because networking is the key to exploiting information, today's information systems are increasingly telecommunications based.

Businesses are increasing their interaction with technology providers, not only by building on their routine service, but in more fundamental ways as well. As businesses examine their strategies and delineate core competencies, they frequently conclude that parts of their information system are outside the core. In many cases, they find computer utilities superior to captive operations and their networks better operated and managed by telecommunications firms.

Consequently, organizations forge partnerships with the firms discussed in this chapter and many others, too. These partnerships develop and operate applications, manage system operations, and manage and operate important networks. As organizations' use of information technology matures, the desire for and the tendency toward informations systems self-sufficiency will continue to diminish.

For managers in typical IT organizations, these trends mean change on a grand scale. In effect, IT organizations and their employees are held to increasingly high standards of performance as comparisons are made between them and competing suppliers. For many critical functions, their performance will be unchallenged, particularly if they pay attention to the firm's core mission. For other functions, IT managers seek alliances with outside service providers, negotiate arrangements, manage transitions, and supervise delivery of services. For IT managers and their employees, accomplishing these tasks will be a major achievement.

Today, the IT industry experiences rapid change with profound implications for professional managers everywhere. Wise managers need to understand these trends and their implications and must take advantage of them for their organizations. Unwise managers, trapped in the past, will resist new developments and eventually succumb to forces beyond their control.

Review Questions

1. Using figures from Intel, describe why the semiconductor industry is considered capital intensive compared to other industries with which you are familiar.

2. Where are semiconductor products consumed? Where would one find semiconductors in a private residence?

3. Name as many different parts of the computer industry as you can and identify the significance of each.

4. What are some business ramifications of the rapid growth in mobile computing?

5. Why are alliances and joint ventures so popular in the computer industry?

6. What elements comprise the U.S. information infrastructure?

7. Describe the essential technological differences between telephone companies and cable companies.

8. What factors drove the Modified Final Judgment of 1982 and the 1996 Telecommunications Act?

9. What forms of competition do RBOCs face? How will this change in the future?

10. What is BellSouth's growth strategy? How does this differ from US West's strategy?

11. What is a LATA? What is its significance to the telephone industry? Why will this concept be less important in the future?

12. Who are some key players in the LEO satellite business? Explain why the Iridium project needs about 70 satellites.

13. Describe some of the obstacles that LEO satellite cellular systems must overcome.

14. What is the most important ingredient of US West's strategy?

15. Why is customer satisfaction so important to the telecommunications industry?

16. Describe how industry trends discussed in this chapter will influence IT managers.

Discussion Questions

1. Michael Porter states, "Domestic rivalry not only creates pressures to innovate but to innovate in ways that upgrade the competitive advantages of a nation's firms. The presence of domestic rivals nullifies the types of advantage that come simply from being in the nation, such as factor costs, access to or preference in the home market, a local supplier base, and costs of importing that must be borne by foreign firms."[25] Discuss why Porter's comments are so important in an era of rising global competition?

2. Consider a semiconductor firm that must invest more than $1 billion to produce commodity products that sell for ever-declining prices. Analyze the associated risk factors. What actions can minimize these risks?

3. Discuss the business significance of the fact that many firms supply mobile and desktop computers. How does this relate to strategies of larger computer companies?

4. Many large computer manufacturers produce software for their products, yet growth in independent software developers has been extraordinary. Discuss the reasons for this, noting in particular the reasons for Microsoft's phenomenal rise to prominence.

5. Referring to Figure 7.1, discuss actions that telecommunications firms can take to build the future network.

6. Government agencies must balance the interests of many parties in fashioning regulations. Identify the parties that agencies must deal with in the telecommunications field. Identify these groups' divergent interests.

7. Some analysts believe that regulatory actions always lag behind technology developments. Assuming this is true, do you think this is an advantage or disadvantage for the public? Discuss your reasoning.

8. From 1984 to 1995, the number of local phone suppliers declined but the number of long-distance providers grew. Discuss the reasons for these trends, and explain what you think will happen in the future and why.

9. Discuss the advantages and disadvantages of underseas optical cable as compared with satellites for international communication.

10. In 1994, AT&T purchased McCaw Cellular for $11.5 billion. Discuss the significance of this purchase to the telecommunications industry and its strategic importance to AT&T.

11. Itemize advantages and disadvantages of LEO satellites versus geosynchronous satellites for cellular phone communications.

12. Discuss risks and opportunities in the strategy adopted by US West. How important are future technological developments to this strategy, and what is the importance of the 1996 Telecommunications Act to it?

13. Discuss factors that make the Internet so important to individuals and businesses today. What does rapid Internet growth mean to phone and cable firms?

14. Discuss the meaning and importance of Figure 7.1 to commercial Internet service providers such as America Online and Prodigy.

15. What actions can IT managers take to cope with rapid and tumultuous changes occurring in industry today? In preparing your answers, consider material you learned so far in this text.

Assignments

1. From annual reports or other industry information, sketch the strategy of a major telephone or cable firm. On what technical, business, or regulatory developments is the firm hoping to capitalize? What risks does the strategy involve?

2. Through library research, find agencies and organizations in your state that have telecommunications regulatory authority. Describe several state regulations and the processes involved in establishing and enforcing them.

3. Review the 1996 Act and present a report to the class on the 14-point checklist that an RBOC must pass before it can offer long-distance service within its region.

4. Write a two-page report on telecommunications developments in Europe including privatization actions, regulatory changes, and the development of alliances and joint ventures.

5. By exploring the Web you can obtain considerable information about telecom companies. For example, see the following:

www.sbc.com	for annual report and news items
www.mci.com	for information on merger developments
www.ameritech.com	for product and service information
www.nynex.com	for information on BellAtlantic/NYNEX merger
www.gte.com	for financial and technology information

Which of these sites is most effective in your opinion, and why?

According to the 1996 Telecommunications Act, local access provided by an RBOC to competitors must meet 14 requirements before the region is deemed open to local competition.

1. Nondiscrimination in access and quality of interconnection, rates, terms, conditions, and prices.

2. Unbundled access to the network under nondiscriminatory terms and conditions.

3. Nondiscriminatory access to poles, ducts, conduits, and rights-of-way at just and reasonable rates.

4. Local loop transmission unbundled from loop switching or other services.

5. Local transport from the RBOC switch unbundled from switching or other services.

6. Local switching unbundled from transport local loop transmission or other services.

7. Nondiscriminatory access to 911 services, directory assistance services, and operator call completion services.

8. Directory listings for customers of the competing carrier's service.

9. Access to telephone numbers for assignment to the competing carrier's customers.

10. Nondiscriminatory access to databases and associated signaling necessary for call routing and completion.

11. Provide competing carrier's customers with number portability through several means with little impairment to service.

12. Access to services or information so the competing carrier can implement local dialing parity.

13. Costs must be fairly identified and compensation based on them.

14. Must make telecommunications services available for resale.

As of November 1997, SBC Communications, Ameritech, and BellSouth attempted and failed to pass the FCC checklist.

¹ Southwestern Bell Corporation, 1992 *Annual Report*, 34.

² "The Lure of Distance," *The Economist*, April 6, 1996, 63.

³ According to Paul Strassmann, total worldwide spending on computer technology in 1995 was about $1 trillion, about half of which was spent in the U.S. *The Squandered Computer* (New Canaan, CT: The Information Economics Press), 25-27.

⁴ The contract to build the ENIAC computer, dated June 5, 1943, marks the computer industry's beginning.

⁵ Intel invested about $200,000 per employee and spends about $33,000 per employee annually on research and development. Its wafer manufacturing plants cost more than $1 billion each.

⁶ S & P *Industry Survey*, Vol. 2, December 28, 1995, C82.

⁷ See note 6, C82.

⁸ Walter B. Wriston, *The Twilight of Sovereignty* (New York: Charles Scribner's Sons, 1992), 86.

⁹ This important action is known as the Carterphone decision.

¹⁰ Public Law 104-104, February 8, 1996. "An Act—To promote competition and reduce regulation in order to secure lower prices and higher quality services for American telecommunications consumers and encourage the rapid deployment of new telecommunications technologies."

¹¹ Bryan Gruley and Albert R. Karr, "Telecom Vote Signals Competitive Free-for-All," *The Wall Street Journal*, February 2, 1996, B1.

¹² Many challenges to the Act and its interpretation by the FCC have gone to the courts. Competition in local markets is encountering significant delays.

¹³ Telecommunications Survey, *The Economist*, September 13, 1997, 27.

[14] Smaller than the seven RBOCs and GTE, 15 local companies such as Southern New England Telecom and Pacific Telecom serve 50,000 to five million customers each.

[15] *The Wall Street Journal*, June 9, 1997, B6.

[16] Robert E. Allen, *1995 Annual Report* to shareholders, February 11, 1996, 4.

[17] As reported in *The Wall Street Journal Reports—Telecommunications*, September 11, 1997, R4.

[18] See note 17.

[19] George Gilder, "Telecosm: Ethersphere," *Forbes ASAP*, October 10, 1994, 132.

[20] "Phone Space Race Has Fortune at Stake." *The Wall Street Journal*, January 18, 1993, B1. For a summary of costs, satellite deployment data, and anticipated service start dates, see *The Wall Street Journal Reports—Telecommunications*, September 11, 1997, R4.

[21] Richard L. Hudson, "EC Commission Gives Wake-Up Call For Satellite Phone," *The Wall Street Journal*, April 16, 1993, A5B.

[22] "Just by counting network nodes, the Internet is now doubling in size every year. This growth can't continue for long—at that rate, everyone on earth will be connected in the year 2003. Impossible." Clifford Stoll, *Silicon Snake Oil*, (New York: Doubleday, 1995), 17.

[23] For an outstanding example of the use of Internet technology by an industry leader see www.sun.com/sun-on-net/moneystories.html. Sun Microsystems has over 1000 internal servers and 250,000 Web pages.

[24] Walter B. Wriston, *The Twilight of Sovereignty* (New York: Charles Scribner's Sons, 1992), 129.

[25] Michael E. Porter. *The Competitive Advantage of Nations*. New York: The Free Press, 1990, 119.

Part
Three

Managing Application Portfolio Resources

Application programs and databases are valuable resources for today's organizations. These resources grow more valuable and change character as organizations embrace network technologies. They can provide competitive advantage, but they also incur technical and business obsolescence. IT professionals and skilled users must manage these critical resources carefully. This part addresses management issues associated with application and data resources. Topics include Application Portfolio Management, Managing Application Development, Alternatives to Traditional Development, and Managing Network Applications.

Managing application and data resources is a challenging task. Successful IT managers consider this activity a cornerstone of their mission.

8 *Application Portfolio Management*

Building Big Systems—Managing Huge Risks

The information management industry has created another industry in its midst: rescues. Directed by corporate consultants and information management gurus, specialized shops made millions in the last few years—$10s of millions per year for one accounting firm, according to a *Business Week* article.[1] These shops catch and tame runaways—out-of-control systems development projects.

Many U.S. businesses and government agencies that did not properly manage or understand large-system development risks experienced development disasters. Two of every eight large software systems under development are ultimately canceled; most exceed their schedules by factors of two or more; and most large systems do not function as intended or are never used.[2] Programming remains a cottage-industry craft for many organizations, particularly those whose business is not software development, unlike hardware development, which flourishes under engineering disciplines. Some system development problems are discussed next.

1. Systems must connect with other systems, especially those of businesses that have large databases accessed by many applications and users. Earlier developers built stand-alone systems successfully because systems were smaller and they could envision complete programs. Today, however, developing complex, communication-based systems linked to other networked systems requires different skills and techniques.

   For example, the Department of Motor Vehicles in California intended to let patrons renew licenses at one-stop kiosks conveniently located throughout the state. This meant merging the vehicle and driver registration systems—a seemingly easy task. When the schedule expanded and costs were more than six times the original estimate, the DMV canceled the seven-year project after spending more than $44 million.

2. Because of management inexperience, poor judgment, or excess enthusiasm, developers tend to oversell what they can deliver. Although programmers and analysts building the applications may be overworked and understaffed or lack needed skills, their executives promise unrealistic performance, including low cost, short development time, high reliability, and ease of use. When programming teams can't deliver, organizations typically suffer severe embarrassment, spend large, unplanned amounts of money, and incur other liabilities. In organizations like health care providers where data reliability is especially important, lost or incorrect data can lead to subscriber loss, hardships and danger to subscribers, and business losses for the organization.

3. The more complex the application, the longer development takes and the more it costs. These factors leverage each other; over a period of several years, the basic cost of development grows. Therefore, a project that takes four years to complete instead of two will bill far higher developer rates as the years roll by. Long projects pose higher risk of schedule slips, and delays increase cost. Allstate Insurance, for example, began a project in 1982, scheduled for completion in 1987, with an $8 million projected cost. By 1988 the cost rose to $100 million, and completion date was pushed to 1993.[3] Ultimately, the completed project's price was about 14 times the original cost estimate.

How can these problems be solved? There are no easy answers, but experience indicates that the failure risk rises rapidly as system size increases. Unfortunately, development departments frequently fail to estimate soundly; failing to estimate the project's complexity, they err on the low side. After that things only get worse. But some actions can and should be taken to minimize risks.[4]

First, senior decision makers and those paying the bills must become deeply involved. They must understand the project in detail and stay involved in the process from start to finish. Second, the project status must be reviewed frequently—continuous review and ongoing audits are not unreasonable for large, high-risk projects. Each review must analyze progress reports, budget items, and all performance criteria for variances from the plan. Even small plan changes usually spell big trouble for large systems.

Third, firms must find some way to guarantee the high standards and abilities of people responsible for designing and programming the system, whether they are subcontractors or the firm's own employees. This may require the firm to specify qualified individuals by name in the plan or contract.

Fourth, those who will use the system must be an integral part of the development effort. The project needs a constant series of user trials; waiting until the project's end to bring users on board is not sufficient. Development projects that fail to involve users from the onset are likely to experience transaction problems, cultural problems, and lack of general acceptance. Knowledgeable users can alert developers to possible conflicts in transactions; they can provide information about how data flows and how the organization uses it; and they can champion the system throughout the company during the trial and introduction phases.

But a fundamental shift in software development and management from cottage-industry craftsmanship to professional engineering must occur. Most likely, this cultural change requires decades to develop. Shifts in university computer science curricula, software engineering certification, and the management practice in industrial software development are necessary. Meanwhile, risk aversion in applications development is a prudent course of action.

INTRODUCTION

Part II discussed information delivery systems and the industry that provides them. It concentrated on advances in semiconductor and recording technology, operating systems, systems hardware, and system and network architecture. The strong, dominant trend of rapidly declining unit costs and ever-increasing functional capability is the basis for the rapid technology infusion into nearly every niche of our society.

The inexorable growth of software, such as operating systems, telecommunications programs, and utility programs, continues to support and bring usefulness to advanced hardware. Application programs developed within the firm also grow, adding to the total already in use. One hundred billion lines of COBOL are estimated to be in use today, costing about $2 trillion to produce.[5] In addition, an abundance of commercially developed application programs or business utilities such as payroll, general ledger, billing, and inventory control programs augment programs developed within the firm. Designed for use across a broad spectrum of functional activities, these programs support most of the firm's fundamental activities. The firm's information delivery system consists of hardware, including everything from personal computers to supercomputers, and the software that supports it.

The large databases used in concert with the application portfolio and the networks that deploy this information where it's needed comprise the remaining ingredients of the firm's information system. These vital program and data resources represent enormous investment over a long time. These applications and their associated data are operationally and strategically vital to the firm. Therefore, managing the firm's software and data resources effectively is critically important to the firm's success.

This chapter treats the application portfolio as a valuable asset and describes methods to preserve its value and manage its growth and development optimally. It develops a systematic approach to achieving consensus on prioritized portfolio actions that improve the portfolio's value. It describes which applications should receive new investments.

Chapter 9 describes how selected applications must be managed. It outlines techniques and processes to maximize the organization's returns on investments in application development. Chapter 10 presents tools and techniques important to development and describes useful alternatives to in-house programming. Today, many valuable new systems depend on network technology such as client/servers, the Internet, intranets, and interorganizational systems. Chapter 11 describes how these important systems must be managed.

THE APPLICATION RESOURCE

Depreciation and Obsolescence

For nearly all organizations, application portfolios and related databases grow in size and value over time. To keep these assets vital and useful, organizations must continue to invest in them and apply resources to offset the effects of depreciation and obsolescence. Application program depreciation stems from the accumulation of functional inadequacies due to gradually changing business conditions. For example, the work-in-process inventory system designed for a labor-intensive process requires enhancements as robotic operations gradually replace labor-intensive processes. Obsolescence results from introducing business changes that reduce the current applications' appropriateness or value. Combining labor accounting functions with the work-in-process inventory system on a plant floor, for example, makes the previous labor accounting system obsolete.

Maintenance and Enhancement

In contrast with hardware investments, costs associated with maintaining, enhancing, and improving application programs are mostly personnel related and increase with time. Although new tools to improve programmer's productivity are being developed, no major cost breakthroughs are anticipated for most applications written in third-generation languages. Additionally, and again in contrast with hardware, no easy, low-cost way exists to move from older generations of application software to new, more functional applications.

According to estimates, U.S. businesses spend about $30 billion annually on routine maintenance and enhancement of the 100 billion lines of COBOL currently installed. Peter Keen claims a typical Fortune 1000 company maintains about 35 million lines of code for its business systems.[6] Yet, as the new millennium draws near, major new problems arise because many programs must be reworked to handle the rollover from 1999 to 2000—the "year 2000" problem.

The year 2000 problem stems from the practice of identifying years in two-digit format, i.e., 95 for 1995, 99 for 1999, and 00 for 2000. Because subtracting years to compute elapsed time is common, huge errors will develop from many calculations if dates are not converted to four-digit format soon. This problem differs significantly from many others because it is not a "bug" (a programming error) or the result of changing business or environmental conditions, such as converting to the EU currency in Europe: It stems from failing to recognize application programs' intrinsic long-term value. Most firms would never risk near-term business failure by allowing valuable physical assets to deteriorate; nevertheless, failing to appreciate the enormous, sustained worth of expensed application programs, many firms must now correct this internal deficiency at great cost.

To fix this problem, maintenance (repair) on programs valued at trillions of dollars will cost hundreds of billions.[7] The year 2000 problem is dramatic because it is huge, visible, and occurs everywhere simultaneously. Many other, less dramatic applications problems that occur daily are all repaired under the guise of maintenance. By any measure, maintaining and enhancing computer applications is an expensive proposition.

Organizations' application programs are also valuable because they codify the rules and procedures of the firm's business processes; moreover, in many ways applications describe the organizations' culture because they represent long-term knowledge of the firm's operations. To a considerable extent, functions that application programs provide represent the yardstick for measuring the firm's information system, i.e., the value of the firm's information system is gauged by the functions it performs. In most modern firms, application programs are valuable resources because they enable the organization to meet its objectives. Today, most organizations cannot succeed without critical application programs.

DATA RESOURCES

Databases accompanying application programs are among the firm's essential assets. Representing a continuous accumulation of information, they expand and gain in importance to the firm. Because they collect valuable information, databases also represent an important investment. Generally increasing in volume and detail, the firm's databases support advances in the firm's application technology. These data systems provide the lifeblood that sustains organizations' daily operations. These data resources, their database management systems, and the hardware on which they reside are vital and critical adjuncts to the application portfolio itself.

Because data resources are so closely linked to applications, they may also lose value as the applications themselves depreciate. In addition, the database management system, satisfactory for current applications, may constrain portfolio modernization and make enhancements much more difficult. Also, because most large databases are involved in numerous interactions among and between applications, they link applications together, greatly complicating application program revitalization or enhancement.

Data distribution or dispersion is also an issue for most organizations. Some firm's database resources disperse widely when they expand geographically. Others find that data dispersion accompanies computing dispersion when individual workstations are introduced. As data dispersion develops, the firm's information architecture becomes an important issue for its senior managers. In all cases, however, the databases' intrinsic value demands that organizations install effective protection mechanisms to reduce vulnerability to loss and destruction. In addition, managers must understand systemic forms of obsolescence that attack databases.

APPLICATIONS AS DEPRECIATING ASSETS

In practice, applications pose a serious dilemma for managers. The application portfolio represents a large investment on which important future applications can be developed. At the same time, however, program improvements and enhancements are mandated by ever-changing business conditions, implying application depreciation or even obsolescence. Thus, to capitalize on the stream of opportunities available through their application programs, firms must spend increasing amounts of time and money. To avoid the negative consequences of obsolescence and depreciation as the portfolio enlarges, resource investments must increase, often reducing the resources available for new development. Because the demand for resources to fuel the growing application asset base shows no signs of diminishing, most firms face difficult and important decisions regarding resource allocation.

SPENDING MORE MAY NOT BE THE ANSWER

In most organizations, the applications programming department faces a large workload backlog, ranging from one to four years in length. Firms in which client application development is occurring may also experience backlogs. The backlog stems from organizations' needs to remain competitive via new applications and to keep the current application portfolio functionally modern. In addition, the process for maintaining and enhancing applications is inherently inefficient because programmer productivity in this activity tends to be low. Typical program development organizations struggle to manage the backlog effectively. Attempting to do something for everyone, they perform a less than satisfactory job for the organization as a whole. Well-managed organizations, however, can successfully overcome many of these difficulties.

The Programming Backlog

The following formula defines the application programming backlog.

$$\text{Backlog} = \frac{\text{Work to be accomplished in person-months}}{\text{Number of persons to do the work}}$$

For instance, if the organization has two programming tasks to perform for a total of 36 person-months of work and has three programmers to do the work, then the backlog is 12 months. Some assumptions are made in calculating the backlog this way. One assumption is that no time is wasted due to skill imbalances during the development cycle. This is most likely to be true for large development organizations. Another assumption is that none of the code is reusable, so all code must be uniquely developed for each application. This is less likely to be true in large organizations.

In addition to the identified backlog, an unidentified or "invisible" backlog frequently exists. This usually occurs when the identified backlog is large. An invisible backlog develops because departments needing programming work do not reveal their requirements. They are reluctant to add to an already long list of outstanding work requirements. The true backlog facing application developers is the sum of the identified and unidentified outstanding work. In many firms, the identified backlog varies from two to three years or more. Depending on the firm's dynamics, the invisible backlog may be large and important.

THE NEED TO PRIORITIZE

For many firms, applying more resources to application development is not a reasonable alternative. Because good programmers are costly and difficult to obtain, many firms are reluctant to make long-term investments in programmers. Newly hired programmers are not immediately effective because they need to learn the organization's culture and business aspects. Enlarging the programming staff may actually reduce output because communication needs and overhead increase.[8] Therefore, prioritizing application development resources is a difficult challenge and an important management issue.

Programming prioritization is important for other reasons, too. Because the application portfolio was developed over a long time and will remain important far into the future, it is a strategic resource. In addition, the portfolio is an important source of management expectations. Consequently, expenditures on application development demand high-level consideration. IT managers need thoughtful, long-term development plans and processes for managing these strategic resources and associated expectations.

Development plans must describe the application of available resources, programming talent, and financial support to manage the tasks the enterprise wants to accomplish. Figure 8.1 shows the processes involved and portrays several alternatives the organization can use to accomplish its goals.

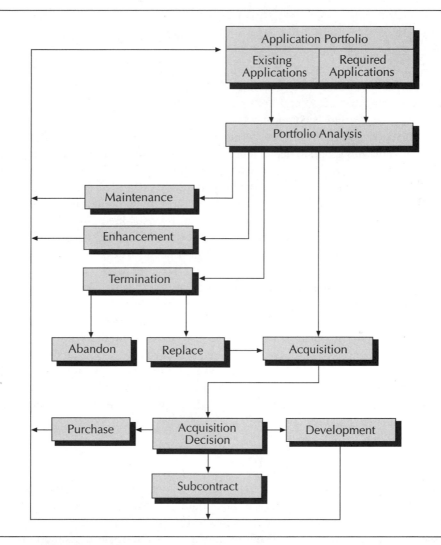

FIGURE 8.1   Portfolio Alternatives

The application portfolio in Figure 8.1 consists of the firm's current applications and some required, but unavailable, applications. Portfolio analysis, described later, provides an organized and businesslike decision-making approach to the alternatives in Figure 8.1. First-level decisions select from maintenance, enhancement, acquisition, or termination choices. The figure also shows decisions between abandonment or replacement, as well as alternatives to acquisition. Processes revealed in Figure 8.1 and discussed below provide a rational approach toward maximizing the portfolio's value. When the processes are complete, resources to optimize the tasks of maintenance, enhancement, or acquisition will be identified.

ENHANCEMENT AND MAINTENANCE CONSIDERATIONS

Many firms commonly deploy programming resources mostly in the maintenance and enhancement of existing applications.[9] Generally, fewer resources are available for new development. New or changing business conditions cause the 100 billion lines of COBOL programs used today to depreciate and become obsolete. To offset the effects of depreciation and obsolescence, this huge program asset base requires continuous resource investments. Managers who cope effectively with program depreciation make important contributions to the firm.

Over the years, vendors produced several generations of hardware; each significantly improved cost and performance compared with their predecessor. Vendors aided the transition from one hardware generation to the next by creating new computing systems with features that let old programs run easily on new hardware. Firms gained cost and performance advantages by upgrading computer hardware; yet, they found enhancing their important business applications much more difficult. For many firms this means that old, possibly obsolete application programs run on new, high-tech hardware.

The IRS, for example, processes thousands of daily transactions on a tape-based system that has been enhanced, patched, and maintained for several decades. During the past five years, the IRS spent several billion dollars on massive, across-the-board modernization efforts with virtually no worthwhile results. The Treasury Department, the White House, and Congress all made recommendations or took action to correct the desperate situation, including hiring an administrator with substantial information systems experience. Businesses have many similar experiences on a smaller scale. Upgrading a CPU, adding new input or output devices, or changing business processes (changing the tax law, for example) are relatively easy; upgrading application software is much more difficult.

The High Cost of Enhancement

Many firms feel trapped by the past. Their applications require significant modernization, yet the resources needed for the job are unavailable. Money spent on hardware brings early returns; however, the costs to correct application program deficiencies remain extremely high and the results slow in coming. What principle factors underlie the situation just described?

Table 8.1 summarizes the reasons why application maintenance and enhancement consume large sums over long periods.

TABLE 8.1   Why Application Maintenance is Costly

1. Obsolete programming techniques were originally used.

2. Documentation is obsolete or absent.

3. Many uncoordinated modifications have been made.

4. Old versions of languages were used.

5. Languages were mixed within the program.

6. Unskilled programmers made enhancements.

7. Architecture changes are required.

8. File structures need major changes.

Many older programs were built using programming techniques now considered unsatisfactory. The design of some programming techniques hindered later program modification. Many old programs are not well structured. They are badly organized, containing strings of code arranged in convoluted and unstructured fashion. Over time, these old programs were continuously modified and enhanced, probably by many different programmers. Repeated modifications using flawed processes generally leads to poor or absent program documentation. In some cases, programmers need more time to understand the program than to fix the programming problem.[10]

Many older programs are written in previous versions of current programming languages, and some may even contain embedded assembly language code. Modifying these applications is an error-prone process that introduces new risks into program operation. In addition, most programmers consider this low productivity work frustrating, undesirable, and less "glamorous" than new program development. As expected, low programmer morale leads to inferior work products.

Many firms assign enhancement and maintenance programming to the organization's least experienced programmers, considering it good training. Inexperienced junior programmers frequently produce less efficient code that contains more errors. Because skilled senior programmers usually prefer new development work, attracting strong people to perform maintenance becomes increasingly difficult. In some firms, individuals reluctantly take responsibility for program maintenance and only under pressure.

Older programs generally require major architectural improvements too. When the firm reengineers its operations, for example, they must convert legacy batch-processing mainframe systems operating sequentially to client/server mode.[11] Sometimes, the original input media in punched-card format must be converted to graphical user interfaces. Or, to capture benefits of modern data analysis (data mining), the firm needs and must implement new database management systems. These new systems impact data currently existing on sequential files and the programs that use it. Such changes require massive amounts of effort because they alter the application's architecture and foundation. Architectural alterations frequently consume resources exceeding those originally invested during initial development.

Sometimes, an application operates with unique files. Changing its function usually dictates large and expensive database changes, especially if the file was originally a sequential tape file copied to disk or if the application must be integrated with others. Enhancements in such cases mean large investments, and, because of the complexities involved, portfolio modifications and enhancements proceed slowly. Program and data interrelationships make maintenance and enhancement processes inherently error-prone and risky.

Trends in Resource Application

The conditions just described result in continued large financial expenditures, increased dissatisfaction with the program development department, and likely deterioration of application quality. Figure 8.2 portrays the deployment of funds over time in maintenance, enhancement, and new development for many organizations in this situation.

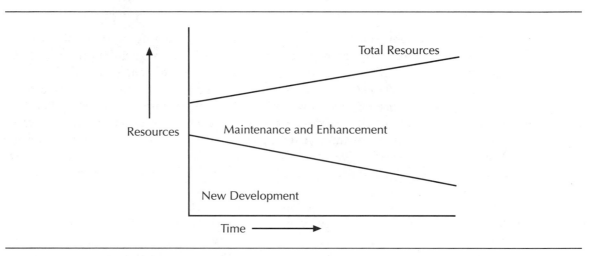

FIGURE 8.2   Typical Expenditures in the Portfolio

Figure 8.2 shows the typical firm increasing its total expenditures on programming. The increase results from escalating costs per programmer, additions to the programming staff, or both. However, resources devoted to new development decline over time because of demands for changes and alterations to current applications. Maintenance and enhancement activity consumes an increasing share of available resources. In some cases, this activity accounts for up to 80 percent of the total planned expense.

Although there are reasons to be optimistic about finding solutions to the applications problem, the current situation in business and industry is far from satisfactory. Most large firms' portfolios hold several thousand applications. Even small- to medium-sized companies typically own several hundred to a thousand or more programs. These applications may be recent portfolio additions, or they may be 10 or more years old. At the extreme, applications designed for obsolete hardware continue to operate on some form of hardware simulator. In this rare situation, the simulator program mimics obsolete hardware, enabling the old program to operate on modern equipment. In other cases, some programs simply can no longer be modified because documentation or relevant skills are no longer available.

Full-time maintenance programmers usually support many applications. For example, one maintenance programmer may be responsible for financial applications and another for marketing applications. Several may support manufacturing applications, and some may respond to work requests from various other departments. As problems arise or when enhancements are required, the maintenance programmer moves from program to program. Because the backlog of work is large, the programmer may take shortcuts and ignore some important aspects of the work. To please the greatest number of clients, maintenance programmers frequently fail to document changes completely. Many totally ignore documentation, further complicating future maintenance.

The term *maintenance* itself acquires several meanings in this context. Some effort is devoted to repairs. These repairs consist of fixing bugs in new programs and finding and repairing bugs in enhancements to old programs. As the portfolio ages and enhancements grow, this effort increases. Frequently, additional effort is required to interface or bridge enhanced programs to old databases. These bridge programs increase the portfolio's size and also require enhancement as other programs or databases change. Bridge programs may also introduce errors, further complicating the situation.

Maintenance may also include minor rewrites. In some cases, rebuilding portions of the application becomes easier than working with old code. Because application maintenance is highly inefficient, firms heavily involved in this activity are on a course depicted in Figure 8.2.

For reasons discussed earlier, most programmers do not consider software maintenance highly desirable work—they usually prefer other assignments. Preventing or deferring application portfolio obsolescence is very important to the firm, however, and poor quality workmanship or untimely maintenance and enhancement can severely reduce the firm's ability to conduct its affairs effectively.

In many firms, the situation is not improving. Maintenance and enhancement consumes more than 50 percent of the programming budget. Costs grow because programmers continually add 10 percent or more per year to the usable code. Proliferation of personal computers and end-user computing may aggravate the situation because small but important programs developed in widely dispersed locations remain largely unknown to any central controlling organization. When business changes or the year 2000 arrives, some organizations may find themselves unable to cope with the maintenance effort.

TYPICAL AD HOC PROCESSES

Many firms lack an organized, disciplined approach to this thorny problem, relying on one or more reactive methods to prioritize the application programming team's work. Some common but unsatisfactory approaches to programming prioritization include:

- Greasing "the squeaky wheel"
- Reacting to perceived threat of failure
- Recovering from embarrassing situations
- Adjusting to threats from competition

Using the "squeaky wheel" approach, managers throughout the firm contend for application development resources, appealing personally to application development managers or directly to programmers. The appeal's success depends on the manager's degree of persuasion, the relative position of the person appealing, or the degree of implied or actual threat. Seasoned managers are more likely to succeed than new managers. Friendship frequently plays a role. A system of scorekeeping (who owes what to whom) may develop. The process is rich with emotion, high on anxiety, and very low in objectivity. Attempting to calm troubled waters, IT managers frequently try to do something for everyone. This usually raises expectations, generates overcommitment, and yields unsatisfactory results. Effective IT managers must use disciplined processes successfully in these situations.

In spite of many difficulties, the squeaky wheel management style is rather popular. Most managers believe they can negotiate a better deal for their organizations than their peers can. Old timers are especially prone to rely on their status and reluctant to relinquish their favored position. They tend to resist a more rational process. In some firms, the squeaky wheel is the corporate culture; in others it is the default management system. Frequent finger pointing and a lack of teamwork characterize firms with this culture or this management system.

In the reactive mode, on the other hand, priorities are readjusted frequently to avert difficulties. When highly visible problems surface within the squeaky wheel environment, corrective action usually follows promptly. Unfortunately, the firm often rushes from one disaster to the next in a futile attempt to keep all systems from collapsing simultaneously. The inevitable result is a terrible waste of resources, generally unsatisfactory long-term performance, and the conclusion by even the most casual observers that management does not know what is happening.

Because reactive processes rarely succeed, embarrassing situations are likely to develop from which the firm must recover. Payroll processing goes astray for all employees to observe firsthand, customers receive incorrect or duplicate billings, or, worse yet, receivables are not collected. Occasionally, collected receivables are improperly recorded. These situations require immediate action. Resources are deployed from around the firm to correct the problems post haste.

Still, when the firm's activities appear to be on a steady course, competitors may marshal their IT forces for offensive action. Detecting these competitive actions always elicits managers' prompt reaction. Again, resources are deployed to contend with these threats and to forestall their consequences, if possible. When the firm's survival is threatened, all other activities have lower priority.

These reactionary actions and out-of-control situations are undesirable, however popular they may be. Is it possible to develop a process which better serves the firm? What management tools and techniques are available to assist in prioritizing the resources? Given most firms' limited and constrained resources, what approach is preferred? Obtaining sound answers to these questions must be a high priority task for business managers because major consequences result from the actions taken or omitted under these circumstances.

The firm's application portfolio represents significant long-term investments; it demands the firm's senior managers' attention as well as IT and user managers'. The portfolio management techniques outlined next let the firm's senior managers combine their strengths with those of the IT and user organizations to yield a preferred course of action.

The methodology focuses on business results. Because resources are always limited, it presents alternatives for managers to consider. Prioritizing alternatives requires intense communication among various players. The methodology causes managers to consider alternatives from the level of the organization's senior people. Focusing on considerations fundamental to the firm, this approach demands a general management perspective. For these and other reasons, this approach is likely to achieve superior results.

Several factors are important in prioritizing the backlog. Among these are the firm's business objectives and the financial and other benefits derived from its applications. Frequently, the applications have important, intangible benefits. Some programs, in fact, may be leading-edge, technically important applications. Usually, relationships exist among and between these factors. For example, an essential business application's failure to generate a positive financial return is not uncommon. Its benefits may be great but intangible or realizable only in the long term. Likewise, an investment to attain technological leadership may be valuable for competitive reasons but may not immediately generate cash.

The questions these considerations pose make resource allocation decisions difficult. IT clients alone do not have enough information to allocate resources. IT and client managers together may lack the vision of top executives. To resolve these issues and prioritize the application development backlog, some firms use a steering committee of top executives.

Results are likely to be mediocre, however, unless the steering committee handles the prioritizing process rigorously and systematically, without political motivations. If the committee incorporates the advice and counsel of senior executives as part of the firm's formal strategy and planning process, superior results are likely. This text strongly favors the planning approach.[12]

How is superior portfolio management accomplished in practice? What steps can the IT organization and other system planners and users take during the strategy and planning process to prioritize the programming work? What is an appropriate management system to carry out these difficult and important tasks? The following analyses present an organized and disciplined methodology to resolve program prioritizing issues.

Satisfaction Analysis

A satisfaction analysis on each portfolio application must be performed as the first step in this disciplined approach. Client organizations and the IT organization each develop satisfaction ratings for each portfolio application. The satisfaction analysis focuses on attributes important to these organizations and quantifies emotional perceptions. Table 8.2 lists typical factors used in the satisfaction analysis for IT and client organizations.

TABLE 8.2 Factors Used in Satisfaction Analysis

Client Factors	IT Factors
Sound function	Good documentation
Easy to use	Modern language
Good client documentation	Ease of operation
Healthy cost/benefit ratio	Trouble free
Sound architecture	Well architected

Although not exhaustive, these lists illustrate most attributes contributing to satisfaction. Specific cases should include some additional firm-dependent attributes.

Each organization ranks portfolio applications from 0 to 10; then the applications are sorted in several different ways. First, rankings are used to reach a consensus, if possible, on which applications will and will not receive additional funding. When the organizations agree that certain applications warrant no additional investments, these applications are not considered further.

Figure 8.3 illustrates how plotting results from this type of analysis usually reveals interesting insights. Figure 8.3 portrays typical analyses conducted by application users and the IT organization. After using the attributes listed in Table 8.2 to evaluate each application, their scores are plotted. Figure 8.3 shows the results for three programs, identified as A, B, and C.

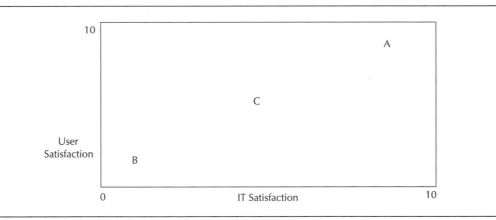

FIGURE 8.3 Graphical View of Some Satisfaction Analysis Results

Newly developed and recently installed programs generally reside in the upper-right quadrant of Figure 8.3, in this case, program A. New programs, such as program A, commonly score lower than 10-10 because of trade-offs made and changes in business conditions during development. Even newly developed applications generally are not completely satisfying because requirements usually become clear only after program implementation.

Many applications, including those currently undergoing enhancement, reside in the lower-left quadrant of Figure 8.3, as does program B. In many instances, most enhancement resources flow to programs with low developer and user satisfaction. Given the information available at this point in our example, managers cannot yet reach final conclusions.

Program C is an average program in the portfolio. Neither clients nor developers find it very satisfactory. Most applications in most firms are like program C. When plotted, all results of this analysis tend to fall along the diagonal as shown in Figure 8.3. Only very unusual applications satisfy client organizations but not IT. The converse is also true. Generally, applications difficulties affect both groups.

When completed, this analysis lets managers deny some applications further consideration and begin the dialogue on prioritizing issues; nevertheless, it provides insufficient information to prioritize the backlog completely. One missing ingredient is the short- versus long-term perspective on each application's value. The next step supplies this information.

Strategic and Operational Factors

To delineate more clearly which applications merit additional resources, managers must obtain insights from a different perspective. They must evaluate the applications' strategic or operational value and understand the applications' short- and long-range importance. The firm's senior executives and its IT and user managers must carefully evaluate each application's value and potential value from the present through the long term. To accomplish these evaluations, managers again rate the applications' strategic and operational importance using the 0 to 10 scale. Figure 8.4 portrays the results of the strategic versus operational analysis for selected programs.

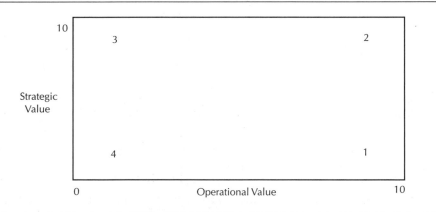

FIGURE 8.4   Strategic vs. Operational Value

Figure 8.4 contains some example applications to illustrate the insights gained in this analysis. The selected examples show how this analysis forces managers to differentiate between various applications the organization uses.

The number 1 represents typical application programs such as payroll, accounts payable, or student accounting. Although very important to the firm's operations, these applications generally provide little or no strategic advantage. The number 2 identifies programs such as the Merrill Lynch Cash Management Account, the Sabre reservation system, or a seismic data processor for an oil exploration company. These programs are operationally and strategically important and vital to the firm operationally and over the long term. They provide strategic competitive advantage in addition to supporting important operational functions.

The number 3 indicates programs such as the CMA in 1977, computer-aided instruction programs for elementary schools, or the next generation vector processor. These applications presently have relatively low operational importance to the organization but offer high potential, long-range strategic advantage. Finally, the Number 4 programs are relatively unimportant strategically and operationally. Typical applications in this category are office furniture inventory programs and one-time statistical analysis programs. Although necessary, these applications are not highly important to the firm's mission.

In these analyses, application owners perform most of the evaluation, taking into account the firm's strategic goals and objectives. IT input should be obtained during the process, especially in relation to anticipated technology improvements. IT expertise is especially useful for emerging applications, or for those applications that are candidates for major enhancements. The list of programs included in this analysis must include new, unfunded requirements in the backlog as well as those the firm currently owns and operates.

Again, the scoring scale is 0 to 10, however, the plotted results, shown in Figure 8.4, are scattered, unlike those on the previous diagram. This result is normal and desirable because it allows managers to discriminate among various programs. Used in conjunction with the previous analysis, this analysis draws a clearer picture of which applications are candidates for resource deployment. Either analysis individually provides some useful information, yet each is insufficient for decision making. Together they build a base for the final crucial step: cost-benefit analysis.

Costs and Benefits

Completing the prioritizing process, the cost and benefits analysis quantifies financially the actions proposed for each application. To be useful for program prioritizing, the analysis must include the cost of failing to do the work and the benefits of intangible results.[13] To reach final conclusions in most cases, highly refined data are not necessary. Most analyses can proceed if the costs and benefits are predictable to within 20 percent or so. Additional refinement may be required to resolve close decisions. This analysis must incorporate the time value of money to ensure financial integrity. Texts on financial accounting provide examples of various analysis methods.[14]

Department managers using the applications provide benefit information; IT managers supply development and implementation costs. Benefits may be purely financial or largely intangible; however, costs are real and tangible. The firm's controller should "book" the benefits and the costs for an application deemed a candidate for resource expenditure. Specifically, the client organization commits expected benefits and the developing organization commits the costs and expenses. The controller tracks both. Because this approach includes commitment, it ensures a higher degree of data integrity than other methods. In addition, it places commitment responsibility on the organizations where it belongs.

An organization's tabulation of all applications upon completing the scoring and cost-benefits analysis reveals much about the portfolio's current state. Table 8.3 illustrates some scores, costs, and benefits of applications that may be in a large portfolio.

TABLE 8.3   Compilation of Program Ratings

Program Name	User Sat.	IT Sat.	Strategic Importance	Tactical Importance	Costs	Benefits
Program 1	8	8	1	8	10K	14K
Program 2	2	3	1	3	30K	40K
Program 3	7	9	8	4	100K	200K
Program 5						
Program 6						
etc.						
"						
"						
Program 2000 ...						

Table 8.3 includes three examples of the analysis results. Program 1, a recently completed application, has fairly high tactical importance. Additional investment in it will yield a modest return. Program 2, an older program, is less satisfactory to both the user community and IT. Although strategically and tactically unimportant, this program will generate modest financial benefits as a result of additional expenditures.

Program 3, a relatively new program, is anticipated to be of great strategic importance. Both the client group and the IT organization are quite satisfied with its current form. Additional investments are required to achieve the relatively high future payoff anticipated for this application. In a large portfolio, many individual differences between applications let decision makers discriminate between candidates for resource allocation.

Sufficient information is now available for making thoughtful decisions. The firm's senior functional managers and their superiors are the decision makers. The appropriate time to conduct this activity is after establishing the firm's strategy but before its planning cycle begins.

Using all factors this analysis reveals, the management team prioritizes the list from high to low.[15] Combining judgment and experience with the data, the team reorders the list until they reach consensus or make a decision. This process puts decision making at the management level where it belongs and reduces overcommitment. The list of work items is closed when the sum of resources required equals that available to the firm. This may mean, for example, that the first 400 of 2000 applications receive development or enhancement resources while the last 1600 remain unfunded. This process indicates what will and will not be done. The firm now has a committed, achievable plan.

Additional Important Factors

Given the analysis outlined in the preceding section, the firm's senior decision makers are well-positioned to allocate critical resources confidently. Although the IT organization directs the analysis process, it must obtain the client's view of the portfolio's applications. IT must also take the lead in developing the discussion preceding the decision to deploy resources. IT managers are responsible for bringing a completed analysis to senior executives.

This process gives each senior member of the review team a clearer understanding of the firm's business needs. If the top individual has established the proper environment, the firm's needs are well served. Rigorously following this process eliminates other, less disciplined approaches such as the squeaky wheel method. This approach puts decision making at the top of the organization where it belongs. It eliminates many small *ad hoc* deals between individuals at lower levels in the organization that can reduce the firm's overall effectiveness. This methodology sheds light on a complex resource allocation problem. It leads to a much better informed organization because it fosters organizational learning. And it prepares the organization to use information technology advantageously.

Frequently, resources are far less available than managers would prefer. As illustrated in Figure 8.1, if money is more available than people, purchasing applications or subcontracting program development may represent attractive alternatives. Although each raises some additional questions, many organizations prefer these alternatives to hiring and training more analysts and programmers.

An additional alternative not shown in Figure 8.1 is using a service bureau to fulfill certain data processing needs. The information in Figure 8.4 is most helpful in deciding whether to buy the service, buy the application, subcontract the development, or develop in-house. Sometimes, a service bureau can best perform routine tasks such as payroll processing. Analyzing the application portfolio systematically usually reveals additional options.

Most firms conclude that strategically valuable applications, or those utilizing proprietary processes or data, should be developed internally. The resident application programming department is most likely to complete local development. In some cases, however, user departments themselves can develop applications. In many cases, clients or users can also perform maintenance or enhancement. Termed *end-user computing*, this activity can be very effective. Chapter 11 discusses it more fully.

When the prioritizing process is complete, IT managers and developers have all the information needed to complete the applications program portion of the IT plan. Management systems for prioritizing and for IT planning are essential portions of strategic application management. The firm's tactical plans contain cost and benefit information developed during this process. This tracking information becomes part of the performance measurement of managers who own or control the applications. In a similar manner, expected applications costs are committed and tracked. IT development managers are measured on this commitment, using the firm's normal cost-accounting system. This process incorporates strategy development, strategic planning, and operational control elements. It ensures a high degree of integrity, secures commitment, and places responsibility where it belongs.

Finally, the process just described offers an additional major advantage. Because the disciplined process involves the firm's senior executives, it provides a splendid opportunity to identify and fund new strategic systems. Using the framework developed in Chapter 2, executives can begin the task of identifying strategic opportunities. They now have a great deal of valuable information on the portfolio, and the environment is favorable. This is the proper time to search for strategic opportunities using, for example, the method Rackoff and colleagues presented.[16]

The prioritizing methodology presented in this chapter provides an ideal background from which to begin the search for strategic information systems opportunities. The right people are present, the timing is right, and a decision-making process has been established. New strategic opportunities can be introduced and resources allocated to their implementation using this prioritizing scheme. All conditions necessary for sound decision making are present.

MANAGING DATA RESOURCES

Data resources accompanying the applications, the database management system, and certain hardware items, must form an integral part of the prioritizing process. In most cases, application enhancements, conversions, or rebuilding activities necessitate resource expenditures on the database. In some instances, database conversions may be important enough in their own right to be considered separate items in the prioritizing methodology. For example, database conversion or reorganization activity or the establishment of a new, comprehensive data warehouse should be evaluated and prioritized the same way that the firm prioritizes applications.

In many cases, database considerations and resources bear on the sequence (and perhaps the priority) of the actions planned for the applications. If the firm does not already have one, an information architecture will begin to emerge from this work. In the final analysis, the interplay among and between the applications and data must be clearly understood. Sources and sinks of data within the firm must be as well defined and understood as the applications themselves for the process to be effective.

Application specialists in the IT organization understand the database system's technical nature and the interactions among applications. IT professionals need to develop the portfolio's architectural considerations for the firm's senior managers. They must conduct this element of executive education thoroughly and skillfully to improve the executive decision-making process. This activity provides another opportunity for the IT executives to obtain agreement on direction and expectations among their peers and with their superiors. It's an opportunity that should not be lost.

THE VALUE OF THIS PROCESS

The methodology developed in this chapter organizes the decision-making process and provides a framework for judging how best to deploy scarce resources. This methodology causes decision makers to focus on issues fundamental to the well-being of the entire organization. It discourages parochialism. The process puts decision making at the senior executive level where it belongs. It serves to integrate the IT function and other developers into the fabric of the business. Executives at many levels in the business will improve their understanding of information technology issues during this process. The improved results are embedded in the firm's strategizing, planning, and control operations.

The management system for dealing with the application portfolio addresses some critical issues facing executives today. It offers the opportunity to advance organizational learning and helps define IT's role and contribution. It focuses on data as a corporate resource. Used in conjunction with strategy and plan development, this management system forces congruence between the firm's strategic plan and the strategic plan for IT regarding application resources. Finally, the process ensures realistic expectations regarding the portfolio. All senior executives know what to expect from IT; and IT knows unambiguously what it must deliver. Many critical issues facing the IT organization and its clients are best addressed by employing these processes effectively.

SUMMARY

The application portfolio and related databases are large and important assets for modern firms. Managing these assets presents significant challenges and unique opportunities. The firm's senior managers must participate in portfolio management. Several IT manager critical success factors pertain to portfolio management, too. This chapter describes methods that successful managers can use to handle application portfolio issues skillfully and effectively.

Many advantages accompany the methodology advocated in this chapter. The process puts decision making in the hands of those most responsible for and most capable of making well-informed decisions about important issues. It makes IT managers jointly responsible for the results and forces proper accountability throughout several management levels. This methodology eliminates many other less-effective prioritization methods and causes decision makers to consider alternatives not likely to be revealed by other, less well-disciplined methods. Additionally, developing unrealistic expectations regarding the applications is less likely with this process.

Managers must be knowledgeable of the fundamental application attributes and their requested enhancements. This knowledge may not be readily available, but informed decisions cannot be made without it. If the firm is not accustomed to handling resource questions in a disciplined manner, this approach requires executive education. This prioritizing process may also be useful in other situations within the firm. Lastly, this process yields substantial educational benefits regarding information technology for the participants. For all these reasons, this approach serves the IT organization and its managers very well.

Review Questions

1. What are the ingredients of a firm's information system?

2. Identify the intangible assets of the firm's information system.

3. What is the difference between depreciation and obsolescence as applied to application programs?

4. Why do the firm's databases grow in magnitude over time? Why is the database management system an important asset?

5. Distinguish between *maintenance* and *enhancement* as these terms apply to application programs.

6. The terms *depreciation* and *obsolescence* are used in connection with programs. How do these terms compare with depreciation and obsolescence of more tangible assets?

7. Define the application backlog.

8. What is the invisible backlog, and under what conditions is it likely to be large?

9. What courses of action for a program can result from applying the process discussed in this chapter?

10. Describe the dilemma the application backlog poses.

11. Why does maintaining and enhancing the application portfolio consume large amounts of resources?

12. Application programs frequently share data files. What additional complications to the enhancement process does this cause?

13. What elements are essential to the methodology presented in this chapter?

14. Why is a plot of the satisfaction analysis (shown in Figure 8.3) likely to look scattered along the diagonal? Why is it unlikely that any application will be rated 10-10?

15. How should the IT organization participate in strategic/operational analysis?

16. Why are costs and benefits important to the process?

17. How does the process outlined in this chapter focus executive attention on a variety of alternatives?

18. Under what conditions may the firm's databases restrict the choice of options or otherwise impact the decision-making process?

Discussion Questions

1. Discuss the causes of the computer runaways in the Business Vignette. What additional causes may be present?

2. Using data presented earlier in this text and appropriate estimates, establish a relative value for the application programs and data resources in a typical firm. Be prepared to discuss your results.

3. Discuss the pros and cons of using application maintenance and enhancement as a training vehicle for junior programmers.

4. What additional attributes might be useful to include in the satisfaction analysis? Under what circumstances might additional attributes be required?

5. Using information from Chapter 2 and Figure 8.4, trace the evolution of Merrill Lynch's CMA program as you believe it occurred.

6. How does the application backlog relate to the issue of expectations discussed in Chapter 1?

7. Discuss problems that will arise if satisfaction analysis alone is used to prioritize the backlog.

8. Discuss issues associated with compiling cost and benefit data for applications. Discuss the effect on decision making of the cost of failure to do the work and of intangible costs and benefits.

9. How can the time value of money be incorporated into the cost-benefit analysis?

10. Discuss how the management process described in this chapter helps the firm identify areas where strategic systems may exist.

11. Some people believe the methodology presented in this chapter is too time-consuming and bureaucratic. What are the alternatives? Compare and contrast them to the methodology discussed here.

Assignments

1. Read the material referenced in endnote 12. Itemize the various costs and benefits Yourdon discusses. Summarize the writer's thoughts on strategic benefits.

2. Study the Rackoff article referenced in endnote 16, and prepare a report detailing the SIS planning process at GTE.

3. What additional factors are important for firms developing application programs for sale? For such firms, how would the processes in this chapter need to be modified?

[1] Jeffrey Rothfeder, "It's Late, Costly, Incompetent—But Try Firing a Computer System," *Business Week*, November 7, 1988, 164.

[2] W. Wayt Gibbs, "Software's Chronic Crisis," *Scientific American*, September 1994, 86.

[3] See note 1.

[4] Subsequent chapters note many additional actions. For examples, see Watts S. Humphrey, *Managing the Software Process*, (Reading, MA: Addison-Wesley Publishing Co., 1990), 17–24.

[5] Peter G. W. Keen, *Every Manager's Guide to Information Technology* (Cambridge, MA: Harvard Business School Press, 1991), 59.

[6] See note 5.

[7] See, for example, Gary Anthes, "Year 2000 Problems Drag On," *Computerworld*, October 7, 1996, and "Feds Garner Failing Grades for Year 2000," *Computerworld*, August 5, 1996, 15. Cost estimates to fix this worldwide problem vary considerably, peaking at about $600 billion.

[8] Frederick P. Brooks, Jr., *The Mythical Man-Month* (Reading, MA: Addison-Wesley Publishing Company, 1975), 16.

[9] The General Accounting Office estimates that the federal government spends 40 to 60 percent of its software budget on maintenance. International Data Corp. found that enhancements consume 50 percent of the maintenance budget, fixing bugs requires 24 percent, and adapting programs to new hardware or system environments consumes 26 percent of federal maintenance expenditures. Scott D. Palmer, "Software Maintenance," *Federal Computer Week*, December 5, 1988, 26. These findings probably pertain to commercial firms also.

[10] Poor programming techniques cause great difficulty in many aged but important systems as organizations struggle to fix the year 2000 problem.

[11] Legacy systems are "systems that have evolved over many years and are considered irreplaceable, either because reimplementing their function is considered to be too expensive, or because they are trusted by users." Edward Yourdon, *Decline & Fall of the American Programmer* (Englewood Cliffs, NJ: Yourdon Press, 1993), 238.

[12] Michael R. Mainelle and David R. Miller, "Strategic Planning For IS At British Rail," *Long Range Planning*, August, 1988, 65. The method used at British Rail parallels that developed in this chapter.

[13] Edward Yourdon, *Modern Structured Analysis* (Englewood Cliffs, NJ: Yourdon Press, 1989). Appendix C of this book contains a thorough discussion of cost/benefits analysis.

[14] For example, see Carl L. Moore, Robert K. Jaedicke, and Lane K. Anderson, *Managerial Accounting*, 6th ed. (Cincinnati, OH: South-Western Publishing Co., 1984), 347. Also see Yourdon, 510.

[15] Eric Clemons says, "Even when it is not possible to compute explicit, precise values associated with embarking on strategic programs, it may be possible to estimate, with enough accuracy, to rank alternatives." "Evaluation of Strategic Investments in Information Technology," *Communications of the ACM*, January 1991, Vol. 34, No. 1.

[16] Nick Rackoff, Charles Wiseman, and Walter A. Ullrich, "Information Systems For Competitive Advantage: Implementation Of A Planning Process," *MIS Quarterly*, December, 1985, 285. The planning process includes instructing executives on competitive strategy and SIS; applying SIS concepts to actual cases; and reviewing the firm's competitive position. This is followed by brainstorming and discussing SIS opportunities; evaluating the opportunities; and developing detail for strategic systems planning.

9 *Managing Application Development*

Building Better Systems[1]

It shouldn't take long for a visitor to Perdue Farms Incorporated to realize that Bob Cook appreciates the people who work for him. If customer-satisfaction scores posted on the walls escape notice, numerous "Associate of the Month" certificates are easy to spot.

"Having satisfied associates (programmers and analysts) is the key to delivering quality products and having satisfied customers," said MIS director Cook. As evidence, he points to his development staff's successes since IS began emphasizing communication and employee development. Since 1989, the department has reduced the average time spent on maintenance from 146 to 52 hours per week. Today, associates spend 94 percent of billable hours building new systems or improving existing ones deemed most valuable to business. These efforts resulted in the elimination of a six-year project backlog. Average program development time fell from 60 hours to 16, while average cost per program dropped from $1950 to $568. These impressive figures add up to a 300 percent increase in associate productivity and an enviable systems development ROI of three dollars of benefit for each dollar invested, according to the company's calculations.

Perdue Farms Inc. is a $1.3 billion integrated poultry producer with 12,500 employees located in Salisbury, MD. In 1988, Perdue, the fourth-largest integrated poultry company in the U.S., responded to a disastrous decentralization effort by reorganizing the data center and rethinking its attitude toward both associates and users. Its new approach emphasizes communication goals up front and publicizes results after the fact. In addition, associates' training increased to five weeks per year. These changes slashed Perdue's IS turnover rate to eight percent, down from 30 percent in 1988, according to Cook. Management and process changes contributed more to the department's accomplishments than new technology, says Cook. "We're interested in proven technologies," he said. "Being a generation behind is acceptable."

IS has certainly come a long way since 1988, the first year Perdue operated at a loss, due in part to decentralization begun in the early 1980s. At that time, Perdue divided its single data center into three separate centers, hoping to push decisions down through corporate ranks and provide more autonomy. When the strategy intended to control costs drove them up instead, company executives responded quickly by reconsolidating and downsizing.

"We needed to focus on basic things, like reacting quickly to business changes and meeting on-time delivery," said George Reiswig, now Perdue's CFO. We were an industry leader in terms of marketing, sales, and logistics. We wanted to create an MIS department which would be consistent with our corporate objectives and culture." IS had to ensure its services' value and relevance to the business units while convincing customers that they had a vested interest in development projects from the start.

Toward these ends, Perdue implemented a monthly chargeback system and incorporated a formalized mission statement and list of Critical Success Factors (CSFs) into its systems-development methodology. One major tenet of the methodology is that automation is never the first step in a project. Rather, it occurs only after eliminating superfluous business processes, and simplifying necessary ones (business process improvement).

CSFs for each project include senior-management sponsorship; limited project size, duration, and scope; precisely defined requirements; and continuous involvement of both the systems staff and the customer. Planning is a collaborative process in which customer/supplier agreements state each project's requirements up front. IS devotes much time to identifying its customers, and determining what they do and how technology can help them.

"MIS spent a lot of time in our area identifying needs," one user said. "They performed a cost-justification and payback analysis and suggested a presentation system for use with our outside customers. Formerly, we did one yearly presentation for the top 20 percent of our customers but now we give presentations to nearly all on a quarterly basis." This is typical of IS and user activity in many Perdue departments.

Keeping users informed about plans, goals, and activities is another important part of IS managers' jobs. "We work closely with senior line managers," Cook said. "We issue monthly status reports which I look on as advertising." As a result, users better understand the traditionally mystery-shrouded IS function. "Knowing about MIS makes you have a buy-in to what they're doing," one user stated.

Perdue's long-range, strategic-planning efforts present one of the IS organization's greatest challenges. "Historically, Perdue wasn't good at strategic planning," says Milton Shupe, Perdue's IS strategic planner. As part of the effort to change and enlarge the IS role, Shupe and his fellow IS managers become educated in strategic planning. "We've always been great executors," Shupe said. "Now we're learning to become great planners."

INTRODUCTION

Chapter Eight concentrated on the most critical application management task—prioritizing scarce resources for application program development, maintenance, or enhancement. It introduced portfolio management techniques building on strategies and planning methods developed in Chapters 3 and 4. Using an organized, logical approach, these techniques prioritize activities to solve critical resource issues. Chapter Eight focuses on doing the right things.

This chapter focuses on doing things right. Emphasizing application project management, it concentrates on developing application programs, one of several alternatives for application acquisition. Development, maintenance, or enhancement activities are performed internally for many applications. Managers prefer this option for large, unique applications or for important, strategic applications because the firm itself controls specialized, proprietary knowledge or owns exclusive, confidential databases. Directing this internal application development option is a critical success factor for managers.

Technical considerations are important in application development; however, management considerations are critical. Weak or ineffective development managers are the most frequent and expensive sources of development difficulties. Concentrating on application project management, this chapter stresses techniques important to application system development. This chapter's purpose is to develop procedures and processes for delivering application systems on time, within budget, and to the client's satisfaction. This discussion uses traditional approaches to application development as a means for exploring these management issues.

THE CHALLENGES OF APPLICATION DEVELOPMENT

Application development is a significant concern for many firms. For some, application development is a very traumatic experience for both developers and clients. For example, a large university introduced a telephone registration system, which subsequently failed, creating major embarrassment for the administration. The failure resulted in an estimated 5700 students waiting in line for up to 10 hours to add classes. "We have just experienced the pain of new technology," commented the embarrassed president, conveniently avoiding the real issues.

The problem seems to spare no one. California's $1.2 billion information technology budget is under legislative review as a result of claimed mismanagement of its largest projects, including the failed DMV system. Amid much finger pointing, the FAA is scrapping portions of its air traffic control system: first estimated at $2.6 billion, costs are now expected to near $7 billion. The IRS efforts to modernize its system cost nearly $4 billion over the past five years and produced virtually no tangible results. The $190 million baggage handling system at Denver's new airport failed; resulting construction delays attracted national attention. In Denver's system, one hundred networked computers control tens of bar-code scanners, hundreds of radio receivers, and thousands of electric eyes that collectively manage 4000 carts delivering luggage to 20 different airlines. Unfortunately, as the airport remained idle, software problems in this massive undertaking cost Denver over $1 million per day in interest and other charges. At one time or another, reliable firms like IBM, Lotus Development, Ashton-Tate, and Microsoft all fell victim to program development problems.

On the other hand, Southern California Edison Company completed a huge new customer service system so successfully that its project manager was promoted to head the organization for which it was built. "Projects that succeed are just about the most satisfying work experience you can have. It's as much fun as you can have and still get paid for," claims Steve McMenamin, former project manager and now vice president of customer service for the company.[2] Many firms and their project managers can experience similar success by adopting proven project management practices.

Reasons for Development Difficulties

Program development is one of IT managers' most difficult tasks. Difficulties seem to fall into one of two categories: those associated with programming itself and those stemming from the firm or its management. Some major difficulties can be attributed to large, complicated programs; greatly increased program complexity; measurement and control weaknesses; and weak theoretical foundations for computer programming.

Over time, programming itself has become more difficult because many of the small, easily written programs were completed years ago, whereas, current applications are larger and more complex. For example, the first accounting programs or early billing systems contained perhaps 10,000 or 20,000 lines of code and cost several $100,000 or less. On the other hand, the Allstate system is a $100 million project, the university student registration system took five years to develop, and the space shuttle project contains 25.6 million lines of code and represents a $1.2 billion investment. The space station project requires an estimated 75 million computer instructions. Management difficulties increase exponentially as programs grow larger.

Many well-known, frequently-used applications are themselves large and complex computer programs. Table 9.1 presents some figures that describe the size, in lines of code (LOC), and cost of some common application systems.[3]

TABLE 9.1   Size and Cost of Common Application Systems

Program	LOC	Cost (Millions)
Lotus 1-2-3, version 3	400,000	$22
Citibank Teller machine	780,000	$13
Supermarket checkout scanner	90,000	$3

Today, many systems rely on complex telecommunication nets. Many users interact frequently with such systems in critical, highly visible ways. Although some are based on older, less complicated, and relatively inflexible systems, many of today's important systems are larger and more complex than older ones by orders of magnitude. While application size and complexity continue to increase rapidly, tools and techniques for building them have not improved in capability and ease of use at the same rate.

Because software development is an intellectual process, software itself is very flexible and easily changed. Changing a line of code is easy. However, completely understanding the change's consequences may be very difficult, particularly if the line of code is part of a very large program. Thus, large programs tend to be rigid, inflexible, and difficult to change. Even small, seemingly innocuous changes can lead to large errors.[4] Advances in software development's theoretical foundations and tools lag behind the demand for new, complex applications. Moreover, in some cases, promising, sophisticated software development techniques have not yet penetrated the entrenched cottage industry of program development.

In many organizations, especially those whose central focus is not software development, programming largely remains an individual craft, lacking solid measurement systems and strong management systems. Software cost-estimating, for instance, is still poorly understood—cost overruns occur in 60 to 80 percent of software projects, according to recent data.[5] Large system development is difficult to understand and predict, and nearly impossible to control. In organizations that have yet to establish or adopt professional software engineering standards and practices, large development projects proceed with great uncertainty and high risk.

In many cases, the firm and its managers are also part of the program development problem. Some firm-based sources of difficulties include 1) environmental factors, 2) inadequate development tools, 3) improperly skilled developers, 4) failure to use improved techniques, and 5) weak management control systems.

Facing with competition and needing to improve productivity, executives hope to shorten development cycles, increase development productivity, and improve application quality. Senior executives who want to capture benefits of new or improved applications sooner naively exert excessive or unrealistic pressure on development managers, hoping to create high-quality, low-cost applications quickly. Under these pressures and others, IT executives continuously search for controllable development processes yielding predictable results. They strive diligently to increase productivity and manage complexities of new programs under development and those receiving maintenance, repair, or enhancement.

Executives' expectations of application developers appear reasonable. Nevertheless, in most cases, large unfortunate gaps exist between executive expectations and program managers' ability to deliver. Gaps result primarily from failure to understand the firm's capability versus its expectations or failure to agree on achievable goals.

In addition to the pressures discussed thus far, developers sometimes are caught in divisive corporate politics arising, for instance, from severe competition among various departments for IT services. With limited resources, IT developers cannot satisfy everyone. For these and other reasons, friction between IT and its clients sometimes reduces the programming team's effectiveness, increasing risks and jeopardizing project success. In other words, organizational environmental conflicts set the stage for application development failures.

Program development tools and techniques are less than state-of-the-art in most firms today. Some look for the one new technique or tool to improve development productivity, failing to understand that progress must be made across a broad front. For example, one of the latest and most important techniques, object-oriented analysis and development, cannot by itself improve the productivity of overworked, unmotivated, and otherwise badly managed programmers. As Yourdon states, "There is no one single bullet. But taken together, perhaps a collection of small silver pellets will help to slay the werewolves of software development quality and productivity."[6] Improved management systems is one silver pellet.

The idea that systems have life cycles underlies most management systems for application development. Embryonic concepts for new applications emerge; ideas develop; and systems are designed and implemented. After implementation, systems are maintained, enhanced, and ultimately replaced. This systems life-cycle concept is the basis for studying management issues and considerations in the traditional approach to application development. Sometimes called the "waterfall" method of systems development, the life-cycle approach is widely used. It accommodates structured development, increased use of development tools, and the influence of prototyping methodologies. Although not the only systems development technique, the use of the popular life-cycle concept is expected to continue. The life cycle model brings order to complex activities and provides the basis for constructive management intervention in systems development. This chapter outlines fundamentals of life-cycle management, beginning with a discussion of the life-cycle approach, continuing with essential aspects of program project management, and concluding with some essential management considerations.

THE TRADITIONAL LIFE-CYCLE APPROACH

The life-cycle approach divides the complex task of system development into phases, each culminating in a management review. Partitioning a project and using a phased approach offers many advantages. Complex activities are more easily understood and controlled in small increments, for example, and, because skill requirements vary considerably over a project's life, both various skill groups and interactions between developers and client functions are much easier to manage incrementally. Finally, managers must evaluate progress and make decisions on an interim basis for maximum effectiveness. Figure 9.1 illustrates the waterfall life cycle and the associated phase reviews.

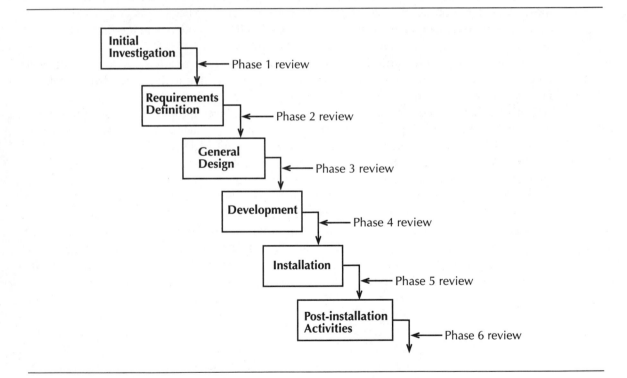

FIGURE 9.1   Waterfall Life Cycle

Each phase has distinct characteristics, and each contains unique development activities. However, some activities within the life cycle span several phases. Included in every phase, management activity tends to be similar from phase to phase.

In the life-cycle approach to systems development, a project usually includes five or more phases. The specific approach selected depends on the project's size and the firm's unique management system.[7] Table 9.2 outlines the phases discussed in this chapter and frequently employed in practice.

TABLE 9.2   Phases In The Development Life Cycle

1. Initial investigation

2. Requirements definition

3. General design

4. Development

5. Installation

6. Post-installation activities

Each phase in Table 9.2 requires both unique and routine management activities, for example, the specific types of management information needed during each phase and at the end of each phase. The remainder of this chapter discusses these activities and information requirements in detail.

Although good texts widely discuss techniques of systems analysis, design, and implementation, much less information is available on *management systems* for systems design, development, and installation. This chapter concentrates extensively on managing analysis, design, and installation activities because, like most information technology endeavors, managing work effectively is the critically important variable.

APPLICATION PROJECT MANAGEMENT

Managing application development projects consists of many elements common to managing other projects and some critical elements unique to programming projects. Application development managers achieve a relatively low success rate, especially for large and complex programs. This occurs because of important and significant differences between application development and other types of projects, and because these differences are critical to the management system development managers employ.

Table 9.3 lists essential elements of the application project management process. Necessary for success in managing application development, development managers use these items as a decision-making framework. Although adopting these notions does not automatically guarantee development managers success, ignoring them greatly increases their chances of failure. Together, these activities and controls form the basis for the application development management system.

TABLE 9.3   Steps In The Application Project Management Process

1. Business case development

2. Phase review process

3. Managing reviews

4. Resource allocation and control

5. Risk analysis

6. Risk reduction actions

BUSINESS CASE DEVELOPMENT

An application programming project's business case itemizes investment resources and estimates investment returns. It illustrates and clarifies the firm's expected costs and itemizes benefits the firm will reap from the completed project. To provide a basis of comparison among alternatives, the business case must address both tangible and intangible costs and benefits. One alternative that the case must consider is the business-as-usual (do-nothing) case. Managers must use one or more traditional financial analysis tools to evaluate the tangible costs and benefits of the selected alternative, remaining alternatives, and business-as-usual case.

Intangible items are harder to quantify than others using traditional financial tools. In many cases, however, these intangible items are very important to the business case and deserve special attention. Some argue that items whose benefits cannot be quantified are perhaps unreal and should be disregarded. Nevertheless, managers should attempt to value items like customer service, product quality, or company reputation.

Department managers who own applications are responsible for providing the business case. The development manager must propose the most cost-effective solution to the business problem. The owner uses this solution to articulate the business case. Subject to revision during the life cycle, the business case is critically important to the phase review outcome. Its primary purpose is to give executives vital information for making important decisions concerning application investments.

The first step in preparing the business case is to establish objectives that the development activity intends to meet. These objectives form the base for all future analyses; they are the foundation for subsequent decision making. Objectives include problems this application intends to solve or opportunities on which the organization intends to capitalize. Issues may be operational, tactical, or strategic, and the system's goals may identify tangible or intangible benefits. In the second step, managers analyze benefits and related costs.

Cost-benefits analysis must be performed for each alternative approach to the problem or opportunity and for the business-as-usual case, too. Managers must clearly and thoroughly understand the current situation. Will the current system be viable five years hence? Will competition render the current approach obsolete in the future, and, if so, what will the consequences be? With this background clearly defined, the decision to proceed with alternative courses of action can be judged objectively.

Analyzing the current system may seem easy, but it can grow quite complicated if it includes costs associated with lost opportunities or costs of unsatisfactory customer service, for example. To make sound decisions, such items must be evaluated in both the current situation and proposed alternatives.

To assess the proposed solutions' operational costs, financial evaluations must cover an extended period. Because the analysis may cover five years or more, the time value of money becomes important. Also, over the extended period the analysis must consider a wide variety of uncertainties and may require a number of judgment calls. Under these circumstances, establishing projected ranges of costs and benefits may be best. In their assessments, decision makers can then consider expected probabilities over various value ranges. In addition, economic assessments must evaluate all tangible and intangible costs and benefits accumulated across *all* departments the application affects, over the appropriate time period.

Development cost estimates include personnel expenses such as salary and benefits, hiring and training, occupancy and overhead costs, and costs associated with development. Development activity costs include hardware or computing costs, travel, and other items directly related to development tasks. These costs arise in IT and client organizations. The business case must summarize them for the period from project inception to application installation.

Operating costs in all functions include personnel costs from implementation to the end of the calculation period. Additionally, operating costs must include all costs associated with hardware and telecommunication systems, purchased services, equipment, and other system operational costs. If system installation is phased, cost estimates need to reflect this accurately.

After completing cost analyses, return on investment can be calculated. Several calculation methods may be used including the payback method (which tells when operating benefits exceed development costs), the net present value, or the internal rate of return.

The net present value (NPV) method recognizes the time value of money over the project's life. NPV accounts for the facts that 1) money received today is worth more than money received at some future time, and 2) costs incurred today are more expensive than future costs. For most development projects, NPV equals the sum of discounted net cash flows over the project's life. Usually, the discount percentage relates directly to the firm's capital costs and may vary with time.

Internal rate of return (IRR) also considers the time value of money and the project's time span. The IRR calculations yield a percentage rate that the firm receives for money spent on the project. Because these analyses are technically detailed, finance and accounting personnel must participate to ensure financial integrity. Their participation also gives financial executives confidence that appropriate financial considerations have been made. This is important because large projects require the controller's concurrence. Most financial management texts explain the details of these financial-analysis techniques.[8]

When considerations are mostly financial and analyses consistently performed, selection among the alternatives, including the business-as-usual case, is straightforward. Managers should select the alternative with the fastest payback, highest NPV, or most positive IRR. Because intangible and other difficult-to-quantify factors complicate most cases, unambiguous results from financial analysis are rare.

Even when purely financial considerations dominate the business case, managers must consider additional factors. The payback method, for example, favors operational systems over tactical systems, and tactical systems over strategic systems. The payback method almost always excludes long-range investments; using this method tends to encourage short-range thinking. On the other hand, development costs incurred over long periods present other difficulties. Technical obsolescence, changing business conditions and management objectives, or other factors increase risk that the system ultimately installed is less desirable than the system originally conceived. Experience and judgment help greatly to balance these conflicting ideas.

Sometimes nonfinancial factors weigh heavily in the decision to develop an application. For example, executive information systems may be good investments because they improve decision making; some systems are necessary to satisfy government or legal requirements; still others may be important for introducing or evaluating new technology or business methods.[9] Each of these cases involves risks in undertaking or scrapping the project. In such cases, risk and benefit evaluation does not rely on primarily financial considerations.

Frequently, large intangible benefits or long-range implications weigh heavily in decision making. Strategic systems and other applications most highly sought by executives usually contain both these conditions.[10] Strategic planning and resource allocation methodologies discussed earlier are especially valuable under these circumstances. These processes, thoughtfulness, and the carefully considered judgments of senior executives over time all increase the probability of success in risky endeavors.

Firms that embark on large, complex application development projects incur large risks. Organizations that proceed with development projects without the benefits of sound management systems and essential disciplines discussed earlier incur needless risk. Executives, their organizations, and the firm itself accept this risk consciously or not. Misunderstanding risks or thoughtlessly accepting them is a principle factor in most runaway application projects.

THE PHASE REVIEW PROCESS

Dividing the work into logical segments best controls important activities in application development. The primary purpose of the phased life-cycle approach is to ensure commitment, understanding, and control of the project's risk elements.[11] Control aspects of each phase have four dimensions: scope, content, resources, and schedules. The detailed phase activities and the information required for phase reviews reveal these dimensions.[12] Because each phase depends on previous phases, project management processes lead naturally from one phase to the next throughout the application's life.

The phase review process is management oriented. It lets managers inspect progress at each step and examine plans for the future.[13] It is a time for decision making. The phase review's end result is a decision to continue the project, continue it with modifications, or terminate it. Frequently, the decision is conditional; i.e., it will be reviewed again before or during the next phase. The possibility always exists that subsequent reviews may uncover facts leading to decisions different than those made during previous reviews.

Phase reviews are typically conducted during meetings but under unusual circumstances may be handled via correspondence. The project's permanent record must contain a report documenting each phase review. Documentation presented during the phase review should evolve smoothly from the project's management system; i.e., phase-review information for well-managed projects is readily available.

Project management systems must be structured, well-defined, and consistent across phases and between projects. They must ensure that phase activities include the items listed in Table 9.4.

TABLE 9.4   Ingredients of a Phase Review

Project description

Well-defined goals, objectives, and benefits

Budgets and staffing plans

Specific tasks planned vs. accomplished

Risk assessment

Process to track plans vs. actual accomplishments

Asset protection and business controls plans

Client concurrence with objectives and plans

Phase reviews should produce documented results useful for implementing subsequent phases. Participants in the next phase must agree that they have all the input necessary for continuing the project. During the review, the application's owner must concur that work completed meets specifications. After a favorable review, work proceeds from one phase to the next.

Although development appears to progress from phase to phase step by step, concurrent activity frequently occurs within and between phases, especially in large, complex applications. Nevertheless, valid checkpoints that determine project status and permit managers to track, measure, or authorize project continuation become even more important when the project involves parallel activity.

Sometimes resolution of all issues is impossible during one review session. Because the judgment to continue, continue with alterations, or terminate is based on current circumstances and available information, application owners may elect to proceed with the project despite unresolved issues. A risk assessment must accompany this option. To ensure project integrity, open issues must be resolved promptly, preferably before the next scheduled review.

Phase Review Objectives

The phase review's goal is to increase the probability of project success to the highest possible level. This means that phase reviews must measure the accomplishment of agreed-upon objectives within the planned time and cost parameters. They must let managers accurately assess activity status and develop alternate action plans, if required. Reviews are a vehicle for analyzing and ratifying plans and objectives for subsequent phases. Everyone affected by the project's design, schedules, costs, data, and operating requirements must have the opportunity to evaluate project status at phase reviews. Finally, the phase review report confirms the decision that all affected managers reach.

Timing of Phase Reviews

Phase reviews should occur when each phase in the system development life cycle ends. A successful review indicates satisfactory completion of the phase. Phases, corresponding schedules, and checkpoints must be identified when applying this process to maintenance or enhancement activities that qualify as significant projects. Interim reviews should be conducted for large projects with phases extending for six months or more. Like normal reviews, interim reviews must concentrate on plans vs. actual accomplishments and expenditures and on estimates for completing the phase.

Large complex systems usually consist of multiple, concurrently developed subsystems. Phase reviews should be conducted for each subsystem. In addition, the entire project should be thoroughly reviewed at least once every six months. The purpose of these comprehensive reviews is to coordinate sub-system reviews and assess overall project status. Each review must establish, alter, or confirm schedules for subsequent phase reviews.

Phase Review Contents

The activities of Phase 1, initial investigation, begin with an idea for a new application or an enhancement to an existing application. These embryonic concepts usually (but not always) arise in a department needing a new application or using an existing application. Strategic systems, for example, may originate from the strategizing or planning processes discussed earlier. Systems analysts conduct preliminary reviews of the existing system and define new system requirements. They develop preliminary system concepts and generate design alternatives. After evaluating feasibility, they develop plans and schedules for the Phase 1 review. Table 9.5 itemizes the management information required for this review.

TABLE 9.5 Phase 1 Review—Management Information Requirements

Statement of need and estimate of benefits

Schedule and cost commitments for Phase 2

Preliminary project schedule

Preliminary total resource requirements

Project dependencies

Risk analysis

Project scope

Plans for Phase 2

Phase 2, requirements definition, consists of modeling the existing physical system and deriving a logical equivalent to which the new system requirements are added. This results in a logical model of the new system, the basis for creating the global technical design. Updated costs and benefits, and system control and auditability requirements are developed and established for the project.[14] System performance criteria are also set at this time. When the Phase 3 plan is complete, the Phase 2 review is scheduled. This chapter's appendix presents management information requirements for Phase 2 and subsequent reviews.

Phase 3 activities consist of developing external and internal general design specifications. During this phase, system software specifications are refined and utility program requirements are specified. At this time, hardware requirements and system architecture definitions become final. During Phase 3, system control and auditability requirements, user documentation and training, and plans for Phase 4 are developed. After these tasks are complete, the Phase 3 review is scheduled.

The primary activities of Phase 4 are program design, build, and unit test. File and data conversion strategies are developed during Phase 4, and program modules are written. Program module testing is completed during this phase and system installation is planned. The development team begins user training and develops user documentation. After plans for Phase 5 are developed, the Phase 4 review is scheduled.

The activities of Phase 5, installation, are critical to the application's success. During Phase 5, user training and user documentation are completed. Acceptance testing is passed during this phase, and file conversions are completed. After the installed system is ready for operation, the Phase 5 review is scheduled. After a satisfactory Phase 5 review, the new system replaces the old according to plan.

Phase 6, post-installation activities, mainly consists of the new system's operation and maintenance. This is the time to evaluate the effectiveness of the life-cycle management system and review the management techniques applied during the development process. Using original specifications and objectives, the system itself is evaluated. Managers conduct an analysis of programming and implementation effectiveness and review requested system enhancements.

The Phase 6 review evaluates results of previous phase reviews and prepares plans for incorporating new knowledge into subsequent projects. Thus, business process improvement techniques are applied to program development. The Phase 6 review provides a great opportunity to review previous cost estimates and incorporate plans to improve estimating techniques.[15] Phase 6 is important to application development managers because it is introspective and reinforces sound management techniques. The development organization learns through experience, just as individuals do. Phase 6 reviews examine strategy, various plans, and actual implementation for consistency and effectiveness. These audits are essential to management control and improved strategizing, planning, and implementation.

These processes require significant amounts of critical information so managers can accurately assess the project's progress and make crucial judgments regarding its future. Without this kind of information and a formal setting in which to evaluate it, managers accept unnecessary risk and face the prospect of operating without control. Phase 6 is mainly a review of the management process quality and an application of business process improvement techniques to application development. This introspective review assists managers in refining processes for the organization's future use.

The Participants

IT managers or their representatives normally direct the review process. The owner of the application system and its data, key client managers, and representatives of all functions influencing or affected by the project must participate in and evaluate the phase review results. Other senior executives should participate when the project's potential benefit is significant to the firm. Usually, gaining executives' attention is easy when large sums of money are involved.

Phases for Large Projects

Exceptionally large or complex projects may require special attention that is best caught by modifying the phase review schedule. The terms "large" and "complex" are relative to the organization's size and skills; i. e., the decision to invoke special treatment must be individually determined for each organization. The proposed project must be specially handled if it is the largest or most complex the firm has ever undertaken. Other projects may qualify for special treatment as well. The phase review schedule for large, complex, major projects includes more events. Table 9.6 lists special additions to normal phased processes.

TABLE 9.6 Expanded Phase Review Process

Phase 1	Initial investigation
Phase 2	Requirements definition
Phase 3	General design
Phase 3A	Detailed external design
Phase 3B	Detailed internal design
Phase 4	Development
Phase 4A	Detailed program design
Phase 4B	Program and test
Phase 5	Installation
Phase 6	Post-installation

The expanded phased approach Table 9.6 presents divides the design and development phases, creating an eight-phase project. The review concluding Phases 3A and 3B evaluates the detailed external and internal designs. Likewise, Phases 4A and 4B evaluate program design separately from programming and testing. This additional perspective greatly helps maintain project control. Certain conditions may warrant further subdivisions. In addition, project managers must define appropriate intermediate checkpoints within each phase to ensure proper controls. The firm's controller may require and large, critical projects may warrant, a continuous audit process.

Phase reviews are essential for controlling programming projects. Directly related to the system development life cycle, they facilitate thoughtful, organized decision making. Therefore, their management is a critical IT activity.

MANAGING THE REVIEW PROCESS

The management-oriented review process leads to sound decision making. Documentation of each phase review must always be unambiguous. Minimally, the documentation must describe the project's scope, content, resources, and schedules. Additionally, a very clear statement of assumptions and dependencies involved in each phase and in the complete plan must be developed. All parties must concur that the documentation is correct. Sometime before successful project completion, assumptions must become facts and dependencies must be explicitly formalized.

Managers must devote special effort to resolving issues and disagreements before or during the phase review. Because reviews emphasize project scope, content, resources, and schedules, any changes to these items must receive special attention. Managers who continually search for the reasons underlying changes and variances are more likely to avoid trouble later.

Managers must develop procedures for issue tracking and reporting. They must assign each issue to an individual who agrees upon a schedule for prompt resolution. Subsequent phase reviews must document that all issues have been resolved satisfactorily. Unresolved issues inevitably lead to an unsatisfactory review and can cause project termination.

Phase review results and conclusions must be documented for the project management file. A summary report should be prepared and promptly distributed to appropriate management team members and all other concerned parties.

RESOURCE ALLOCATION AND CONTROL

Careful project planning describes the ebb and flow of skills from Phase 1 through Phase 6. Analyst and client activity is high when the project begins and just before and during installation. Programming activity is light during project definition, peaks during implementation, and tapers off after installation. Similarly, computer operators, technical writers, database administrators, trainers, and others have unique activity patterns over the life cycle. Managers must track and control these people resources just as they do physical or monetary resources. Life-cycle resource management and control are fundamental to project management success.

Information describing resource deployment by skill type is part of the application development plan. The resource plan must be available during the project's life; additional details and refinements are added at each phase review. During each phase, managers must track resources by skill type vs. the plan. For example, during Phase 3 tracking should reveal analyst effort declining and programmer activity increasing as the application moves toward implementation. Deviations from the plan during Phase 3 in these skill areas signals managers to identify and attack the underlying cause. Analyzing total resources simply does not reveal the detail that managers need.

In an important analysis of system failures, Stephen Keider identifies six tasks most likely to cause failure if mismanaged; additionally, he presents seven early danger signs. Through interviews, he learned what 100 MIS professionals think are the most important causes of system failure.[16] Table 9.7 presents his results.

TABLE 9.7   Major Causes of System Failures

Reason for failure	Number of responses
Lack of project plan	23
Inadequate definition of project scope	22
Lack of communication with end users	14
Insufficient personnel resources and associated training	11
Lack of communication within project team	8
Inaccurate estimate	8
Miscellaneous	14

Although technology or design problems cause some projects to fail, project managers can control most difficulties. Lacking plans or failing to understand or control the project's scope are planning deficiencies that virtually ensure project failure. Because scope changes or expansions occur so frequently, change management is especially critical for project managers. According to the Center for Project Management, San Ramon, CA, 25 percent of projects are canceled due to scope increases, yet fewer than 15 percent of project managers create change management plans.[17]

RISK ANALYSIS

Software projects fail for a variety of reasons; however, inadequate project management systems and techniques are the primary causes. Fortunately, principles developed in this chapter can easily reduce management system failures. Still, one readily available and easily applied technique significantly improves a project's chances for success: risk analysis. Successful project managers include risk analysis as part of every application project.

Application development processes contain analogies to the familiar economic concepts of leading, coincident, and lagging indicators. Some relatively uninformed project managers observe lagging indicators. For example, some development managers discover that the program they developed earlier failed because their clients never embraced it or stopped using it for various reasons. By observing lagging indicators, these managers discover the failure after it occurs.

Better informed programming managers, however, continuously track an exhaustive list of project metrics. These coincident indicators include budget, schedule, function, and myriad other relevant items. Carefully monitoring these indicators, they know the very moment their project fails—in fact, they are the first to know. Using coincident indicators is certainly better than using lagging indicators; nevertheless, it is not nearly good enough. After all, success, not failure, is the goal. Managers need a set of leading indicators that alert them when their projects head for trouble so they can take corrective action. Risk analysis provides these indicators.

What is risk analysis? How does it provide early warning to project managers? What alerts project managers to impending difficulties? To answer these questions, managers must search for sources of programming project difficulties. Next, they must develop quantifiable measures describing the extent of risk to which they are vulnerable in each important area. Finally, managers must track these risk measures to discern trends over the project's life.

Table 9.8 lists the major risk sources in application programming projects. They fall into six categories and carry weight corresponding to the risk each poses.

TABLE 9.8   Sources of Risk

Risk	Weight
1. Client activity	20
2. Programming and management skill	10
3. Application characteristics	30
4. Project importance and commitment	20
5. Hardware requirements	10
6. System software requirements	10

Further subdividing each category into quantifiable and measurable items enables project risk assessment. First, let's describe the ingredients of these six categories and then develop a rationale for quantifying them.

Active client involvement in programming projects is vital for their success. Weak or missing support from the client community jeopardizes project success. We can measure client activity by evaluating the quality and quantity of the requirements definition and the depth and breadth of client involvement in prerequirement activities. Good measures of client activity are 1) client involvement throughout post-requirement activities, and 2) client knowledge of the proposed system and its relationship to other systems. Other valuable measures include training levels and the success of training clients to use the new system.

Project success depends heavily on the knowledge, skill, and experience levels of system implementors and project managers. Managers need to take corrective action when implementors lack satisfactory skills or are understaffed. For instance, if the project requires new telecommunication software and technicians have insufficient experience or training, project managers must be concerned about project risk.

A third, obvious area of concern is the nature of the project itself. Important items include the size or scope of the system under development, its duration and complexity, and anticipated project outputs or deliverables. Project logistics and the size of the geographic area in which development takes place are also significant factors. For instance, application programs developed in several locations present much more risk than those developed at one site. Additional project factors include the sophistication of project control techniques, and the homogeneity of the development team, and their experience working together. This latter item includes the effects of subcontract programming, for example.

The fourth risk area is project importance and commitment. The schedule's aggressiveness, the number of management agreements needed for implementation, and the extent of managers' project commitment are each measurable elements in the project's risk assessment.

System hardware forms a fifth risk area. Project risk increases if the system requires new or unfamiliar hardware or unusual performance. System performance specifications and hardware system capacity planning are also important. Programs developed to run on existing hardware systems incur little or no risk in this area.

The final risk item includes the operating system and other software needed to develop and implement the application. Applications using routine system software and developed in familiar languages minimize risk. Programs developed in new or unfamiliar languages are riskier. Using prereleased or relatively untested software packages is riskier still.

There may be other risks. For example, if part of the application is purchased, risks associated with the vendor may arise. On the other hand, purchasing all or part of the application may reduce the firm's risks if its programming staff is weak or lacks necessary skills. Third-party telecommunications systems are another external risk source. Third-party participation can either increase or decrease risk, depending on their capabilities vis-a-vis the firm's capabilities. Careful analysis can quantify these factors.

Table 9.9 is a useful but somewhat arbitrary guide for project managers to weight or quantify these risk elements. Most managers need to adjust these values to suit their individual situations. Regardless, using a consistent methodology during the project's life is important. Because absolute risk values cannot be measured accurately, value changes are more valid and useful risk indicators. Keeping a history of risk measures is worthwhile to establish trends within the organization and across projects and project managers.

TABLE 9.9   Detailed Risk Items

Risks	Relative weights
1. Client activity	
a. Quality and quantity of requirements definition	6
b. Prerequirements planning activity	3
c. Post-requirements activity	4
d. Knowledge and understanding of proposed system	4
e. Client training	3
2. Programming and management skill	
a. Implementor experience, skill, and ability	5
b. Management experience, skill, and ability	5
3. Application characteristics	
a. System size, scope, and complexity	6
b. Project duration and complexity	5
c. Project deliverables	3
d. Project logistics	5
e. Project control techniques	7
f. Organization considerations	4
4. Project importance and commitment	
a. Aggressiveness of the schedule	4
b. Number of managers	8
c. Extent of management commitment	8
5. Hardware requirements	
a. New hardware	6
b. Stringent performance requirements	4
6. System software requirements	
a. Operating system and utilities software	5
b. Language requirements	5
Total	100

Items in Table 9.9 must be scored and evaluated before each phase review and preferably more often. An item with no risk scores zero; a high-risk item receives the highest score possible for that item. To determine risk trends, weights must be consistent from one review to another. Precision is much preferred to accuracy in this analysis. The absolute risk level is valuable in determining whether to continue the project, continue it with modifications, or terminate it. For example, if the risk totals 75 at Phase 1, many individual items are very risky and most prudent managers would seriously consider terminating the project. On the other hand, if the risk totals 20 or lower at Phase 1 the risk is low or manageable. Risk totals between 20 and 75 require detailed review and analysis before managers decide to proceed.

As the program progresses toward implementation, initial risks should decline. (A successfully installed and operating program has zero development risk.) Programs shed risk as they proceed smoothly from Phase 1 to Phase 6. Increases in total risk over the project's life signal danger and demand more detailed analysis. Yet, even if total risk declines, the project may still be headed for trouble. For example, if an item with zero or low risk suddenly increases in risk value, managers must take note. If in Phase 3 a program found that the value of item 3b, duration and complexity, increased from 1 to 5 because the program became much more complex than predicted, managers must be seriously concerned, even if other items' total risk value declined.

Risk analysis provides leading indicators of project success. Managers commit to deliver a product in the future, at predicted costs, and with stated function. They make these commitments after performing risk analysis, understanding the risks involved. Managers further understand that they must mitigate risks during development to meet their commitments. If subsequent analyses indicate unfavorable risk trends, managers must take corrective action to meet previous commitments. Risk analysis lets managers take action when problems are small before commitments are unmet. Proceeding with projects that face high or rapidly rising risk usually leads to very severe consequences.[18]

Seemingly easy tasks, such as switching to a new computer system, can be extremely troublesome even for firms with lots of experience with computers. For example, in changing to a new management information system, Sun Microsystems lost control of customer orders and inventory, among other things. As a result, Sun failed to pay some bills on time, needed to do its accounting manually, and eventually suffered revenue and profit reductions. Many other firms have displeased their customers because of uncoordinated or failing customer support systems or tardy software product delivery. Application management difficulties seem to spare no one.

RISK REDUCTION

Quantifying project risks clearly benefits prudent managers who can then act to reduce or mitigate them and take advantage of risk trends. Completely eliminating all project risks may be impossible, but identifying risky areas and managing them proactively certainly is possible. Gene Dressler, program manager at GTE, states it well: "Manage the risk. There always will be certain parts more susceptible to going wrong. ....We look at four or five areas with high risk. We develop contingency plans and watch extra closely."[19]

Management actions to control risk may consist of deploying special resources, instituting special control techniques, or using lower risk alternatives. Project managers have many resources at their disposal for managing problems during the project's life. Risk assessment tools alert them to act early to avert trouble later.

For instance, if training lags behind schedule, the project's risk rises. Alert project managers recognize this trend and search for its underlying causes. They know the importance of training users to operate the application, and that poor or late training jeopardizes successful application implementation. Yet, amidst development activities, one might easily assume that training activity will accelerate later when time permits. Aggressive managers resist this temptation and act to eliminate risks promptly, thereby mitigating increased risk. "A slip is a slip is a slip," says Diana Garrett, IS project manager at Intel. "Don't ever count on catching up later, it's not going to happen."[20]

Problem management is a major task for project managers. Risk analysis is an analytical tool that warns of impending difficulties. Alerted to possible problems, managers can act to solve them when they are small and manageable. Risk analysis and reduction actions are indispensable parts of application development management systems.

MORE ON THE LIFE-CYCLE APPROACH

The life-cycle methodology discussed in this chapter helps illustrate management processes and procedures essential to achieving consistent results in application development. Although popular, this method is not the most sophisticated. The life-cycle (or waterfall) approach has disadvantages that other methods strive to overcome:

1. Tangible client results come late in the cycle.
2. It depends upon stable initial requirements.
3. It tends to be paper intensive and bureaucratic.
4. Parallel activities are permitted but not encouraged.

During the first three phases in the life-cycle approach, developers and clients deal mostly in paper, analyzing current procedures, proposing new procedures, defining new system requirements, and developing new system designs. During these phases, the results are words, symbols, diagrams, and resource statements about time, money, and people. At this point, no operational results are available for inspection.

In some cases, clients have difficulty precisely stating system requirements while developers struggle to translate clients' statements into workable designs. Although needed to compose a complete and workable requirements statement, clear but intense communication between developers and clients is often difficult to achieve. Marketing managers and IT analysts may use the same words but not share a common meaning. During requirements definition, "Build what I mean, not what I say," is a frequently felt sentiment.

Alternatives to the traditional life-cycle method attempt to circumvent these communication difficulties by introducing more parallelism into the process and involving clients in different ways. In addition, the results of the waterfall approach and its alternatives improve significantly with the use of sophisticated development tools. Computer-aided software design (CASE) tools provide an automated formalism to assist both clients and developers in producing satisfactory systems.

Using sophisticated tools during development produces observable results earlier that clients can study and analysts can refine. These tools permit testing of some system parts before others are fully defined. They give everyone the opportunity to develop ideas based on tangible results from earlier study. They also remove the daunting challenge of defining all requirements at the outset.

Today, the many variations of methodologies used for software development include prototyping, object development, incremental waterfalls, structured techniques, and information engineering, as well as variants of these. Some approaches are superior for operational systems; others, such as prototyping, work better for data-driven systems like decision support or client-server applications. Subsequent chapters discuss alternatives to the life-cycle approach.

Whether the conservative life-cycle approach or the free-flowing prototyping method is used, managers must understand their commitments and have methods for evaluating progress toward meeting them. To this end, managers need checkpoints to understand scope, content, resources, and schedules. Regardless of the methods employed, managers must quantify risk and respond promptly to contain or eliminate it. Managers who cannot predict or respond to risks lead uncontrolled projects that generate unpredictable and usually disastrous results.

SUCCESSFUL APPLICATION MANAGEMENT

Successful project management flows from well-designed and smoothly functioning management systems. Successful application development thrives on controlled processes, yielding predictable results and dealing with increased complexity. Its surprise-free products meet all technical and functional specifications. They satisfy schedule and budget agreements and conditions expressed in the business case regarding operating costs and realizable benefits. The review process and efforts to complete it ensure attainment of these goals. Successful application management yields predictable products that contribute important assets to the organization.

Processes defined in this chapter apply not only to programs but also to documentation. Programs, program documentation, and operating documentation must all be managed similarly. Documentation products must be as critically scrutinized at phase reviews as the emerging application is.

Business managers in most firms today are intensely interested in improving productivity and enhancing performance. Application project managers must also concentrate on productivity. They must be able to demonstrate productivity improvements. The thoughtful, organized, business-like approach to the development process this chapter details is a necessary condition for productivity improvements but is not sufficient by itself. Tools and techniques discussed in subsequent chapters must be combined with superior management systems to achieve productivity gains.

SUMMARY

Managing application development projects is a difficult task. Today, application development involves more resources, occurs over a longer time, and is inherently riskier than it was a decade ago. *Ad hoc* management techniques and routine project management methodologies must be supplanted by disciplined processes specifically designed to cope with application development issues found in today's larger, riskier projects.

Phased development divides the project into manageable tasks culminating in reviews focusing on fundamental project issues. Project validity is the foremost issue. Business case analysis and periodic review ensures continued project viability. Without sound business cases, all efforts to develop successful applications ultimately fail, regardless of methodological subtleties.

Managers need control points during the project's life to carefully re-evaluate the business case and focus on the application's scope, content, resources, and schedules. Carefully designed phase reviews illuminate important issues relating to the project's continuation and let managers resolve critical items. Risk analysis gives managers leading indicators of project difficulties and provides opportunities for managers to resolve small, easily handled problems. Failure to heed these early warnings generally leads to major problems and to extreme distress for project managers and those who depend on application development.

Successful project managers use management systems fostering rigorous, thoughtful processes and candid, open communication among all participants. Antecedent activities like strategy development, strategic and tactical planning, technology assessment, and application portfolio asset management must precede project management tools, techniques, and processes.

Thus, application management systems build on previous management processes and contribute to business goals in a positive, well-understood manner. The firm's managers know that IT's development activities match their objectives and that strategic development, planning, and resource prioritization efforts are successful. Executives' expectations of IT are likely to be realistic when projects are well managed. Realistic expectations and successful projects are critical success factors for IT managers.

Review Questions

1. What is the connection between critical success factors and application development management?

2. What are the goals of successful application development? Why is this area a critical success factor for the IT manager?

3. Why is local development the only reasonable alternative for many applications?

4. Why does the life-cycle approach divide the project into phases? What are the advantages of this phased approach?

5. What are the phases in a typical life cycle?

6. What activities take place in each phase? What information is required?

7. Why is Phase 6 important?

8. Do system development life cycles and phase reviews apply to large application maintenance? Under what conditions are these activities most important?

9. What are six essential elements of a sound project management system for application development?

10. What are the similarities and differences between managing a computer system application and managing construction of an office building?

11. What are the ingredients of a computer system application business case?

12. Why is developing and assessing application business cases becoming more difficult?

13. Why are phase reviews an indispensable part of the management system for application development?

14. What are the objectives of a phase review? How often should phase reviews be held?

15. How should the phase review process be modified for very large applications?

16. What role does documentation play in the review process?

17. Describe the risk analysis process. What elements are reviewed in risk analysis?

18. Why is risk analysis considered a leading indicator?

19. What are the disadvantages of the waterfall methodology?

Discussion Questions

1. List all factors you think contribute to Perdue's success in application development. In your opinion, which factor is most important, and why?

2. Who should participate in systems analysis and design? How does the degree of involvement for analysts and programmers change from concept to system implementation?

3. Briefly discuss what systems analysts do in each of the six phases. What does the client manager do at the completion of each phase?

4. Discuss the phase review discipline for very large or complex applications.

5. Describe the management system that ensures project integrity from phase to phase.

6. Describe some intangible issues in application business cases, and discuss their importance.

7. What intangible issues are likely to be present when developing a new online order entry system?

8. What major risk factors are present when replacing the current payroll program with a new one incorporating the latest tax changes? How would these risk factors vary over the development cycle?

9. What risks are likely when developing a totally new strategic information system? How do the risk areas differ from those of the payroll program in Question 8?

10. What management actions are central to successful project management systems used in application development?

11. What management system elements developed in this chapter are most valuable to firms discussed in the Chapter 8 Business Vignette?

12. Describe the firm's antecedent activities needed for application development projects' success.

Assignments

1. Outline the agenda for the Phase 3 review of an internal application program. Name the management positions that should be represented and identify the order in which they present their findings and opinions. If the firm intends to sell the program being reviewed and Phase 3 just precedes product announcement, what additional considerations are important?

2. Read an article on systems development in a scholarly journal like the one note 5 references, and summarize the management techniques it describes in a report for your class.

APPENDIX

The tables in this appendix list the minimum management information requirements for phase 2, 3, 4, and 5 reviews. Additional application or organization-specific information is needed in most cases.

TABLE A.1   Phase 2 Review—Management Information Requirements

Documented statement of requirements

Refined benefits commitment

Schedule and cost commitments for Phase 3

Refined project schedule

Refined total resource requirements

Updated analysis of dependencies

Risk analysis

Project requirements and scope

Plans for Phase 3

TABLE A.2   Phase 3 Review—Management Information Requirements

Final general design

Final benefits commitment

Schedule and cost commitments for Phase 4

Committed project schedules through Phase 5

Committed costs through Phase 5

Resolution of remaining dependencies

Risk analysis

Test plans

Preliminary user documentation

Preliminary installation plan

Plans for Phase 4

TABLE A.3   Phase 4 Review—Management Information Requirements

Final installation plan

Satisfactory completion of program test

Schedule and cost commitments for Phase 5

Committed project schedules through Phase 5

Reaffirmed commitment to system benefits

Commitment to system operational costs

Risk analysis

Final installation plan

Plans for Phase 5

TABLE A.4   Phase 5 Review—Management Information Requirements

Satisfactory completion of system test

Final user documentation

User acceptance document signed

Final application business case

Reaffirmed commitment to system benefits

Commitment to system operational costs

Risk analysis

Plans for Phase 6

[1] Megan Santosus, "Perdue's New Pecking Orders," *CIO*, March 1993, 60. Reprinted through the courtesy of *CIO*. ©1993 CIO Communications Inc.

[2] Kathleen Melymyuka, "Project Management Top Guns," *Computerworld*, October 20, 1997, 108.

[3] Brenton Schlender, "How to Break the Software Logjam," *Fortune*, September 25, 1989, 100.

[4] The three most expensive known software errors cost the firms involved $1.6 billion, $900 million, and $245 million. Each resulted from changing a single line of code in an existing program. Peter G. W. Keen, *Every Manager's Guide To Information Technology* (Boston, MA: Harvard Business School Press, 1992), 46.

[5] Fiona Walkerden and Ross Jeffery, "Software Cost Estimation: A Review of Models, Process, and Practice," in *Advances in Computers*, Vol. 44 (San Diego: Academic Press, 1997), 62.

[6] Edward Yourdon, *Decline* and *Fall of the American Programmer* (Englewood Cliffs, NJ: Yourdon Press, 1993), 37.

[7] Little agreement exists on the number of phases into which the systems life cycle should be divided; however, the management principles apply regardless of the number of phases.

[8] For example, see Paul M. Fischer and Werner G. Frank, *Cost Accounting* (Cincinnati, OH: South-Western Publishing Co., 1985), 219-226.

[9] C. James Bacon, "The Use of Decision Criteria in Selecting Information Systems/Technology Investments," *MIS Quarterly*, September 1992, 335.

[10] For more on this topic, see Eric K. Clemons, "Evaluation of Strategic Investments in Information Technology," *Communications of the ACM*, January 1991, Vol. 34, No. 1.

[11] Joel D. Aron, *The Program Development Process: The Programming Team* (Reading, MA: Addison-Wesley, 1983), 340-348.

[12] For another perspective on the need and content of phase reviews, see Watts S. Humphrey, *Managing the Software Process*, (Reading, MA: Addison-Wesley Publishing Company, 1990), 78-80.

[13] A management axiom is, "You get what you inspect, not what you expect."

[14] System control and auditability features are critically important to applications programs. Chapter 18 discusses them extensively.

[15] See note 5, 122. The authors claim evaluating "an estimating process with improvement as a goal may be the most effective way to improve."

[16] Stephen P. Keider, "Managing Systems Development Projects," *Journal of Information Systems Management*, Summer, 1984, 33.

[17] Alice LaPlante, "Scope Grope," *Computerworld*, March 20, 1995, 81.

[18] Aron (note 11, 343-348), presents risk assessment somewhat differently.

[19] See note 2, 109.

[20] See note 18, 81.

10 *Alternatives to Traditional Development*

Edward Yourdan, a prominent authority on programming, is so concerned with the demise of the American programmer that he wrote a book on the subject.[1] Critics of the American craft and firms that outsource programming offshore offer considerable evidence supporting his view. According to Yourdan, unless U.S. firms adopt key software technologies, American businesses will continue to outsource software development overseas.

Many firms experience critical skills shortage in new technologies such as object-oriented design, C programming, and graphical user interface (GUI). According to Forrester Research, 86 percent of firms surveyed reported that current programmer skills presently limit or will limit development plans.[2] Today, programming teams from Eastern Europe, Russia, India, and mainland China are rapidly developing many key skills. For example, the Shanghai Software Company advertises programming services ranging from ADA to C++ to PASCAL for eight development environments and five operating systems.

But, while offshore programming trends develop, events in the U.S. conspire to provide American programmers job security for many years. First, the urgent need to rework millions of legacy systems to handle properly the date change at the turn of the century—the year 2000 problem—has opened opportunities for and increased salaries of many COBOL programmers and others knowledgeable of these critical applications. So urgent is this task that firms are recruiting retired employees to help. Some IT organizations have put all else on hold as this work rushes ahead.

Second, some tasks are just too difficult to perform overseas because they require company-specific skills. For example, responding to the 1996 Act requiring open markets, Baby Bells must now let competitors access billing programs developed over many years for monopoly providers. This massive overhaul requires more resources than the year 2000 problem. "This is the largest development program ever performed in the history of the company by far," said Lee Bauman, vice president of Pacific Bell.[3] In addition to supporting new hardware and building new strategic applications, such demands have corporations scrambling for programming talent.

Third, the number of computer science graduates declined 43 percent between 1986 and 1994, just as the demand for computer skills exploded. Business schools, however, still experiencing strong enrollment.[4] Demand for these business school graduates with computer knowledge remains high.

Left with few alternatives, today's corporations increase salaries, give signing bonuses, and recruit and train as never before. People who had not previously considered a career in computer programming now enter the field in college or from other job markets and enroll in training programs offered by U.S. corporations. For example, NYNEX invested $50 million to train 1000 new technical employees to help with urgent IS activities. Given the dynamics of the programmer labor market, forecasters predict that demand will exceed supply for the next decade.

Cost is another factor that complicates the situation long term. In addition to offering bright workers with key skills, offshore developers' labor rates are very attractive. For example, software writers in the former Soviet Bloc countries earn about one-fifth of their U.S. counterparts' wages.[5] Even after adding incremental management and overhead costs, offshore development is a relative bargain. Citing statistics that coding accounts for only about 10 to 15 percent of total project costs, some critics contend that software development is not a cost-based business. Others argue that the government look hard before allowing offshore programming activity to siphon corporations' money and diminish the number of U.S. jobs.

Whether offshore outsourcing is a permanent trend depends in part on the U.S. programming industry itself. Eventually, thousands of COBOL programmers must be retrained in C++, HTML, Java, networking technologies, and software quality assurance methods. On the other hand, offshore programmers are not burdened with old habits nor large mainframe experience: they assume that ISO9000 quality standards, graphical user interfaces, and client/server implementations are routine.

Some believe that challenges from overseas developers will motivate U.S. programmers and firms to adopt technologies that increase quality and productivity and lower costs. Others seek comfort from the rapidly exploding need for programmers, arguing that everyone will always have enough work. Recent developments in the U.S. have strengthened their arguments; nevertheless, the long-term market for programmers is global. For now, offshore development is relatively inconsequential. Only time will tell whether it remains a small factor.

INTRODUCTION

Chapters 8 and 9 developed the foundation for managing the critically important application portfolio. Building on this foundation, Chapter 10 explores additional development tools and techniques and acquisition alternatives. It considers prototyping and object-oriented programming, and the merits of purchased applications. It explores subcontract development, service-bureau organizations, and joint development activity through alliance formation.

Focusing on management systems, this chapter develops important concepts for handling local development alternatives. Its approach builds on traditional alternatives and exposes readers to additional strategies valuable for managing the complete portfolio asset. Chapter 11 explains additional network alternatives like client/server and Internet applications.

Fourth-Generation Languages

During the past 40 years, computer programming has advanced through several stages or generations of technology, normally distinguished by the language types used to solve programming problems. Early computers were programmed in machine language, sometimes called the first generation. Machine language used the binary number system—the hardware's language. Because using machine language was difficult and cumbersome, assembly language quickly replaced it. This second generation language replaced computer operation codes and memory addresses with easier-to-understand terms or mnemonics similar to natural language. In assembly language, one line of code generates one computer instruction.

With the introduction of third-generation languages, computer programming leapt forward. More like natural languages, these languages are easier to use; because each language statement generates several machine instructions, programmers using them are more productive. Third-generation languages represent a major step forward in computer programming. Costing more than $1 trillion, 100 billion or more lines of code created in these languages run on today's computers. Programmers continue to write programs in these popular languages.[6]

New features in some enhanced languages, such as COBOL and Pascal, make them suitable for today's new mainframe, client/server, and Internet applications. In addition, the development of many new language tools supports new hardware and improves programmer productivity.

We desperately need programmer productivity beyond that attainable with third-generation tools. We also need new productivity aids that let more people create programs more effectively. In addition, programmers need tools that are easier to learn and use. Fourth-generation languages (4GLs) promise to partially satisfy these needs. Used with development tools supporting documentation and library functions, they satisfy many developers' needs and also improve productivity. Indeed, one authority, James Martin, defines fourth-generation languages by their ability to improve productivity: "A language should not be called fourth-generation unless its users obtain results in one-tenth of the time with COBOL, or less".[7]

Table 10.1 lists many language types qualifying as fourth-generation.

TABLE 10.1   Types of Fourth-Generation Languages

Database query and update

Report generators

Screen and graphics design

Application generators

Application languages

General-purpose languages

The past traps many firms. They must maintain and repair millions of lines of code running valuable operational systems to solve the year 2000 problem, convert to European Union currency, or handle new billing strategies. Often, programmers no longer with these firms wrote the code in earlier-generation languages. Converting these applications to more sophisticated languages, for different architectures such as client/server, or for other types of processing is a formidable if not impossible task. However, emerging tools can assist firms engaged in these difficult tasks.

Table 10.1 lists categories that include many products and languages. Database query languages such as SQL or QUERY-BY-EXAMPLE, information-retrieval and analysis languages such as STAIRS or SAS, report generators like NOMAD or RPG, and application generators like MAPPER, FOCUS, and ADF are examples of fourth-generation languages. Many more language tools (perhaps 100 or so) have fourth-generation characteristics. Most of these languages or programming tools, however, lack the general capabilities that conventional third-generation tools offer. Powerful and easier to use, they are more specialized and less flexible. Hence, programming departments generally use several 4GLs to satisfy their needs. Properly used by trained programmers, 4GLs improve programming productivity and quality. Table 10.2 summarizes characteristics of fourth-generation languages.

TABLE 10.2   Characteristics of Fourth-Generation Languages

Advantages	Disadvantages
Easily used	Low performance in large systems
Reduce programming time	Possible slow response
Improve productivity	Inefficient computer memory use
Improve program quality	Restricted capabilities
Reduce maintenance effort	
Problem-oriented	

In operation, programs written in fourth-generation languages usually consume more machine resources than others. This means their use must be reviewed and matched to the hardware and application tasks for which they are most suited. Very large programs supporting many simultaneous, online users are perhaps best written in conventional languages. Fourth-generation languages optimize programming talent—a costly resource in short supply, at the expense of hardware—an abundant resource rapidly declining in cost. Thus their popularity grows. Still, transaction processing systems associated with very large databases may perform poorly if inefficiently programmed in fourth-generation languages.

To capture the best of both worlds, tools exist that translate programs written in fourth-generation languages into lower-level languages. Their purpose is to take advantage of both language systems and improve program development *and* operation.

4GLs improve programmer productivity considerably because for many applications they are easy to use and more powerful than their predecessors. Today, most applications in most firms can be developed using fourth-generation languages. Some firms improved productivity and quality by converting totally to one or more fourth-generation language. Fourth-generation languages can increase programmer productivity by a ratio of 20 to 1 and significantly reduce expense. For example, Kawasaki suspended COBOL programming entirely; its programmers now write all new functions and programs, including all new online applications, in Pro-IV. Arco Coal eliminated COBOL programs from its portfolio and replaced them with programs written in Focus. The firm accomplished these changes while reducing programming staff.

Many seasoned programmers resist new languages because new technology makes skills developed over a long time obsolete—4GLs requires them to retrain. Some programmers regard fourth-generation languages as technologically unsophisticated, suitable only for end users or beginners. Managers frequently feel more comfortable with third-generation languages because they, too, fear introduction of another language. They are unsure about resolution of the compatibility issues; they dread the expense and frustration associated with the transition period. Some programming departments and their managers, fixated on the year 2000 and other problems, remain firmly anchored in the past, surrendering to the lure of the familiar.

As networking technologies such as client/servers, the Internet, and Web technology introduce more languages into the computing arena, matters become even more complicated.

First proposed in 1965 by Ted Nelson at Xerox PARC, Web technology relies on the hypertext concept. The construction of hypertext documents lets readers study them linearly like a conventional book or jump around from one appealing topic to another. When the reader clicks on highlighted key concepts (hyperlinks), the screen refreshes and displays a document or page that discusses the concept more fully. The HyperText Markup Language (HTML) accomplishes this.[8] HTML is a means for coding documents for online publication with embedded hyperlinks and other features.

HTML and the increasingly popular Java language make change a way of life for leading-edge programmers and their managers. In the increasingly networked world, maintaining the status quo while drifting in a sea of change is a fatal strategy for managers and their technical staff.

In some cases, programmers use advanced languages supported by appropriate tools as direct replacements for earlier languages. Tools supporting development processes in these languages are sometimes considered a separate technology. In the future, however, the distinction between languages and tools will disappear. Performing their tasks at programmer workbenches, developers will be unable to distinguish clearly between various support mechanisms. This melding of tasks, tools, and languages will characterize the computer-automated software engineering era.

CASE Methodology

New and improved programming languages will be helpful only if the entire support environment, including management systems and people management practices, is tuned for success. Among other things, tools that remove drudgery and manual record keeping from programming tasks promise to improve morale and productivity. However, tools must be wisely implemented and used by fully trained programmers. In supportive environments, Computer-Aided Software Engineering (CASE) tools (individual workstations with extensive software supporting the programmer) are valuable assets to program developers.

Advanced languages applied through effective programmer-support tools increase potential for enhancing productivity. These automated functions result from the effort to use computer technology to benefit computer professionals, programmers, analysts, programmer technicians, programmer librarians, and others. Some tools assist parts of the development process; others support the process from requirements definition through maintenance. Support includes diagramming and modeling, code generation, test case development, and many forms of documentation. Typical CASE tools are implemented through online, interconnected workstations containing functions and features designed to help developers, clients, and managers produce quality products.

Several hundred firms, including CASE manufacturers, hardware and database vendors, consultants, and educators market CASE tools. Not all provide full CASE function; the type of business they serve biases many. For example, firms supplying database software tend to provide tools emphasizing data management. Consulting firms tend to stress methodology. CASE tool buyers must thoroughly understand products before committing to them. CASE program prices vary widely from a few $100 to $100,000 or so, depending on their functions.

Sophisticated workbench technology depends on networked workstation devices to assist in the complex communication task among development team members. The network supports timely and accurate information exchange. The most successful CASE systems support the entire project team, including its managers. They provide controlled processes to manage emerging product versions and releases, including their documentation. CASE provides automated tools to manage test case libraries and assist program validation. Some CASE tools feature code-generation or code-rebuilding capability to rebuild or refurbish old programs rapidly and productively. To accomplish these tasks in a user-friendly manner, CASE technology uses graphics extensively.

Some tools are specifically designed to rebuild current programs. File descriptions, database configurations, and source code placed in the system design database are enhanced and updated, or migrated to new languages or data management systems. The process includes both reverse- and forward-engineering.

Several types of CASE tools have evolved. Some support requirements-planning, analysis, and design. They create a model of requirements, check for consistency, and document the results. These tools feature rapid production of design graphics and system documentation. Those that sustain the development life cycle's early phases are termed upper CASE or front-end tools. Lower CASE or back-end tools support code generation, test-case construction, and database development. They help produce the programmed application's documentation. Most tools support program specification and implementation standards. Some development tools also assist in design, code, or test case reuse.

General-purpose products supporting the complete systems development life cycle are called Integrated or I-CASE tools. Generally used in conjunction with some form of systems development life cycle, they may be used with other methodologies. Supporting many common languages, I-CASE tools are useful on a wide variety of applications. Small CASE systems operate on individual or interconnected workstations, but others require mainframe or server support for maximum performance.

Modern CASE tools also assist project managers by collecting development statistics, preparing status reports, and communicating among and between team members and project managers. These tools support life-cycle administration by developing and communicating project metrics and other project management information. To obtain maximum benefit, programming managers and program developers must be thoroughly familiar with the tools and firmly committed to their use. Half-hearted or partial commitment to CASE tools causes confusion, frustration, and ineffective program development.

The U.S. Department of Defense recognizes the importance of controlled programming methodologies. In defining and analyzing its software requirements, DoD standards direct developers to use systematic, well-documented methods. Many CASE tools help developers satisfy this requirement.

CASE technology is important for several reasons. Users believe that CASE improves design quality and greatly assists in developing system documentation. CASE strengthens communication among developers while enriching communication between developers and clients. Although some developers resist new tools or methodologies, user-friendly tools reduce resistance and encourage programmers to adopt proven new methodologies. Management commitment to CASE involves significant expenditures for the tools themselves and requires investments in programmer and manager education and training.

Controls are essential to achieving predictable program development processes. A controlled environment is also indispensable to high-quality development, which, in itself, is an essential condition for high productivity. CASE tools, proven project management systems, and improved programming methodologies, such as the object paradigm, have the potential to improve program quality and maintainability substantially. Phase reviews, sound business case analysis, and risk analysis focus on management processes. Metrics developed from these processes augmented by development metrics from programming tools provide critical information. Managers who use them can work under schedule and budget constraints throughout their projects' life.

Automated tools and techniques supporting highly productive languages will grow and develop significantly in the future. Their deployment in application development will also expand greatly. Organizations demand significant programming quality and productivity improvements, and invest in tools and advanced languages to help attain these goals. Con Edison, for example, uses a CASE application generator to transform design specifications into COBOL code. Automating code generation increased Con Edison programmers' productivity from 65 to 400 lines of code per day in two years. Because quality also increased, future maintenance costs are expected to decline. BDM International, a large systems integrator, reduced costs on a fixed price Air Force contract by nearly $5 million. Error rates also declined by 75 percent. As a result of saving $1.2 million on its first four projects using a CASE tool, Souvran Financial Corporation has adopted the tool as a company standard. Because CASE-developed programs contain fewer lines of code, Souvran also anticipates maintenance savings.

Improvements like these stem from automated programming tasks, better management techniques, improved training, and superior development methodologies. Introducing new languages, advanced tools, and superior techniques creates some technical problems; nevertheless, most challenges can be met by sensitively managing people under changing conditions. Application development departments have little choice but to embrace new technology and approaches. Astute IT managers use their influence and management skills to encourage early adoption and use of promising new approaches to application development.

The Object Paradigm

Today, object-oriented programming gains popularity in academia, business, and industry. The technology originated in the 1960s with Simula, a language that scientists in Norway developed. A Xerox research team advanced the approach, developing an object-oriented language called Smalltalk. In 1981, Bell Labs developed the C++ language, which promptly gained early acceptance at academic institutions and other organizations interested in object development. Many vendors enhanced C++ for use in client/server and Internet applications.

Object technology consists of object-oriented design, development, and databases. It also includes object-oriented programming and object-oriented knowledge representation.

Object programming approaches a problem from a different level of abstraction than conventional programming. Conventional languages separate code and data, but object languages bring them together in a self-contained entity called an *object*. Each object includes code appropriate for use with its data. These code modules are called *methods*. Objects sharing common methods and attributes are called *classes*. For example, a file defined as an object may include appropriate methods such as copy, display, edit, and delete. The object and its common methods form a single entity. A second file, for instance, containing methods identical to those in the first file belongs to the same class as the first file. One class can have many members.

One of object technology's most powerful concepts is *inheritance*. New object classes can be defined as descendants of previously defined classes. New classes inherit their ancestors' methods, but these methods can be altered by adding new methods or redefining previous methods. For example, one can define a new object in the file class called *output*. Output inherits the methods of copy, display, edit, and delete. Object programmers may remove the copy and edit methods for a particular application and then add a method called print. Methods defined with the output class will then consist of display, print, and delete. Generally, object-oriented programming offers numerous possibilities.

Object technology is especially useful for user interfaces; screen applications such as menus, displays, and windows; and text, video, and voice databases. It isolates changes and permits modular expansion of features. It greatly facilitates program reuse, thus significantly increasing programming productivity. Object technology deals with complexity through abstraction. Its many uses include operating systems, programming languages, databases, Internet applications, and others.

The object paradigm is widely used in application development today. Object-oriented versions of COBOL and Pascal are becoming popular, and many developers use C++, a version of C with object-oriented capabilities, now available on a wide variety of platforms.

Some more popular versions of C++ are Optima++ from PowerSoft, VisualAge C++ from IBM, Visual C++ by Microsoft, and Borland C++. Some of these tools are excellent for client/server development. Many new tools are rapidly emerging as computer manufacturers, application developers, government agencies, and others strive to incorporate object technology into all kinds of applications.

Java Programming

According to some experts, object technology will be to the Internet what structured techniques and 4GLs were to traditional applications. Designed to optimize Internet applications, Sun Microsystem's object-oriented programming language, Java, is especially interesting.

According to George Gilder, "Fueled by the efforts of some 400,000 developers who continue to report as much as five-fold increases in productivity, Java has the power to break the Microsoft lock-in of applications profits and lockout of rival operating systems."[9] Java's importance to Internet applications lies in its great cross-platform strengths and ability to deal successfully with the Net's many separate networks and millions of computers. It does not optimize the computer desktop—it optimizes the programmer. It does not standardize the Internet world around Windows 98, NT, or Unix—it lets individuals operate in an open, platform-independent environment. It is the purest form of the object-oriented model now available. Time will reveal Java's full importance to the Internet world.

Forward-looking program developers use CASE tools, advanced languages, and object-oriented techniques, including versions of C++ and Java, to improve programmer productivity and product quality for mainframe, client/server, desktop, and networked applications. Recently created development teams, such as overseas contract programming shops, depend on advanced tools and techniques and global networks to maximize their competitive advantage. To improve the development process, all programming teams must embrace advanced tools and techniques while maintaining close contacts with their clients.

Prototyping

Because program specifications are so difficult to develop, many systems begin their life cycle with serious management and technical problems. Problems in early design stages are difficult to manage and expensive to correct later. For many applications, in fact, specifying the problem is the hardest part of system design. New languages, better tools, and more training for developers and clients cannot solve these problems. Most individuals facing specification difficulties prefer to experiment somewhat with the problem to obtain a more realistic sense of its solution. This process is called *prototyping*.

In many cases, expecting the development team, analysts, and clients to agree on the precise, final specifications early in the project's life cycle is unrealistic. On one hand, developers want to freeze specifications because they know that changing them is a leading cause of slipped schedules and cost overruns. On the other hand, clients unsure of their ultimate requirements and reluctant to adopt final specifications prematurely want to reserve some flexibility. Prototyping offers a compromise to these conflicting objectives.

Such naturally occurring conflicts result in severe consequences if not quickly and satisfactorily resolved. According to a recent survey, 80 percent of development projects do not follow schedule and budget plans because "frozen" requirements change. Cost overruns or schedule slips of 10 to 50 percent occur in 68 percent of these cases; overruns exceeding 50 percent occur in nine percent. Changing requirements stem from poor initial requirements, unfamiliar applications, prolonged project development, and changing business conditions.[10] Avoiding scope changes and creeping requirements serves everyone's best interests.

Design experiments and prototyping solutions offer compelling compromises to the conflicting demands of developers and their clients regarding design specifications. In prototyping, developers and clients work closely together and communicate intensely as they build and evaluate incremental system functions. Successful prototyping demands rapid implementation of many small system alterations and equally rapid evaluation of their merits.[11] Prototyping is best performed in highly automated environments using tools supporting these needs. Third-generation languages operating in batch mode are unsuitable for prototyping. CASE workbenches supporting advanced languages make prototyping feasible. Figure 10.1 illustrates the prototyping process.

When prototyping, system developers and users generate ideas for the new program, then devise an elementary model for evaluation. Developers focus on hardware, software, architecture, and communication technology, while users define human/machine interfaces, functions, and usability criteria. They integrate their ideas, evaluate the model, generate new ideas, and repeat the process until they specify the problem and most of its solution parameters. Experimentation ends when both parties are comfortable with the results. When the system is refined, fully developed, and satisfies user needs, it becomes operational and production-ready. CASE development tools generate most documentation and other supporting material.

Prototyping is advantageous because developers and users can change system specifications if early experience with a running model is unsatisfactory. As a result, the final system closely meets client needs and desires. On the other hand, prototyping processes are difficult to manage. One tough question is: When is the prototype finished? Intelligent users can always find still more ways to improve the application. Another disadvantage is that users may adopt the system before it is completed or integrated with other systems or databases. Finally, obtaining complete documentation on a prototyped system may be difficult as well.

As an alternative to traditional development, combining prototyping and the life-cycle approach can capture many benefits of each method. Developing specifications through prototyping removes some uncertainty during the system's initial development stages. Prototyping techniques refine and solidify user interfaces and architectural details. Following these crucial tasks, the life-cycle methodology beginning at Phase 2 is used to complete the application through Phase 6. This approach is particularly attractive for large programs with complex human interfaces. This alternative approach uses prototyping to develop critical initial specifications; the life-cycle methodology disciplines the final development stages. The combined advantages of each approach form a superior process.

One large, popular computer printer was developed combining prototyping and life-cycle methods. The user interface on the printer consists of back-lighted panels for operator messages and touch panels for operator response. The printer uses five microprocessors to manage the interface and control printer operation. During development, the printer interfaces were prototyped in software and hardware. When the specifications were developed and approved, formal development processes continued for both hardware and programming. The product was introduced on an accelerated schedule, at minimum development cost, and with high confidence that its human factors were satisfactory.

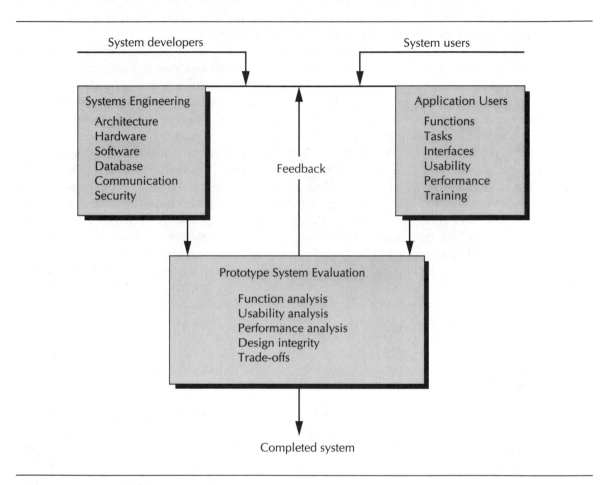

Figure 10.1   Prototyping Process

The prototyping methodology or its variants are especially valuable to some applications. An application to extract detailed information from a very large database is one example. In a large retail operation, for example, programs to obtain information about customer preferences from the large database of cash-register-generated data are important strategic systems. After the first information is obtained from the data warehouse more refined questions are posed and further query program development occurs. The process continues, as the data warehouse is for increasingly detailed information. The query program can be defined only after using and refining its early forms. Defining the final program at the outset is simply not possible.

This example demonstrates prototyping's superiority in certain situations. In some respects, prototyping runs the system development life cycle in reverse. An elementary program is developed and enhanced repeatedly until it takes its final form.[12] Prototyping is an important development method when rapid response to changing conditions is critical or when requirements cannot be known early in the development cycle.

Several other well-known techniques involve system users at the outset and provide early operational models. Rapid application development (RAD) and joint application development (JAD) techniques require intensive developer/user interaction in the early design stages. Combined with CASE tools and prototyping, these interactive methods resolve specification uncertainty before extensive coding and testing occurs. Furthermore, combining these techniques with traditional life-cycle approaches captures some benefits of each.

Managers have several available alternatives for developing application portfolio additions. Traditional life-cycle methods are useful for easily specified applications. These methods are valid for applications that demand careful control because of business case or application risk factors. Prototyping, perhaps combined with RAD or JAD, offers many advantages and gains increasing popularity. When appropriate, prototyping solves several important and vexing problems. Combining prototyping with other methods extends its usefulness. In all cases, advanced languages and sophisticated support tools enhance the process.

Choosing among alternative methodologies must include consideration of development time and cost, and operational and maintenance expenses, programmer productivity, program quality, and user satisfaction. The decision to use one method or to combine methods requires careful analysis. Steadily declining computer hardware costs and rising programming costs increasingly favor prototyping methodologies.

PROGRAMMING PROCESSES IMPROVEMENTS

Many organizations have difficulty meeting programming projects' cost, schedule, or quality targets because they lack basic techniques for understanding their programming processes. Many formal classes for software engineers or application programmers teach analysis, design, and various languages but fail to focus on programming processes or methods.[13] In many organizations, programmers work as individual artisans who developed their craft (their individual methods) through experience. Exceptional programmers, sometimes called "eagles," are 10 times more productive than average programmers. But, in most cases, the organization does not know why this disparity exists and cannot teach the high performer's techniques to the entire group. Although several factors are important in achieving success, high-performance programmers or programming groups study their processes, know their defect injection and removal rates, and utilize techniques that greatly improve their product's quality.

Program quality is a serious problem. Programs delivered to customers typically contain several defects per thousand lines of code (KLOC), whereas the standard for high quality products is three defects per million lines.[14] Poor quality causes product delays, raises development and service costs, and lowers customer satisfaction. With defect insertion rates of nearly 100 per KLOC from undisciplined programmers, organizations find defect removal expensive, time-consuming, and never ending. Techniques exist to find, correct, and reduce defect injection rates early in the development cycle when removal costs are low. Developers who use these techniques receive rewards of shortened schedules, lower costs, and higher product quality.

Consulting firms, private research groups, and, most notably, Watts Humphrey of the Software Engineering Institute at Carnegie Mellon University[15] have developed formal concepts for process assessment and improvement. Humphrey's maturity model describes five levels of programming process attainment, from the Initial or *ad hoc* level to the Optimized level.

At the Initial level, little formalization exists as individuals ply their craft. (Many application development departments operate at or near this level.) At level 2, the Repeatable level, organizations use statistical measures and exert statistical control. Organizations at the third or Defined level establish quality and cost parameters and maintain process databases to gather and contain statistics. At this level, the basis for sustained quality improvement exists. At the Managed level, level 4, organizations analyze process databases, scrutinize programming processes, and initiate process improvements. At the Optimized level, the 5th and final level, organizations improve process measures based on prior experience and optimize processes further. Students and organizations interested in this important topic should study Humphrey's work referenced in the endnotes and other current writings.

Programming process improvement methods, like Humphrey's maturity model, are critically important to professional managers. First, managing application development successfully depends on using predictable, repeatable development processes. Credible schedule, cost, quality, and performance commitments demand controllable design and development processes. Without them, schedule and cost predictions are simply guesses unworthy of professional managers. Second, advanced languages, CASE tools, and object technology are more effective in disciplined environments. In uncontrolled environments, sophisticated languages and tools produce more defects more quickly. Uncontrolled automated processes increase defect removal costs, lengthen schedules, negate the tool's productivity improvements, and lower product quality unnecessarily.

Poor programmer productivity and low program quality are management problems rather than technology problems. Statistical process controls, noted for greatly improving manufacturing operations, apply equally well to programming activities. Today's high performance organizations embed these techniques into disciplined processes to control and improve programming activities. Organizations that fail to adopt sound management techniques and instill discipline on development processes become candidates for outsourcing and other application acquisition alternatives.

SUBCONTRACT DEVELOPMENT

Not all needed programs must be developed by the firm. Frequently, some application development can be off-loaded to specialized programming firms. Called *subcontract development* or application development *outsourcing*, this activity reduces in-house programming by paying subcontractors to develop applications externally.

Because subcontract development trades money for application developers, it lets firms balance resources advantageously. This path to program acquisition also reduces the firm's personnel requirements, compared with local development, and may help resolve skill imbalances within the firm. Subcontracting lets firms manage their long-term staffing more effectively because subcontracted tasks are workload buffers for the firm's permanent staff.

Subcontracting also lets the firm temporarily obtain skills for specialized tasks, again reducing permanent staffing needs. Because subcontractors may have both skills and experience in the general problem area, their work products may be superior to those from other sources. Together, these factors may lead to lower application acquisition costs.

Subcontracting also has some disadvantages. Involving contracts, reviews, reports, and more formal supervision than local development, subcontracting is bureaucratic. Thus, it is much more management intensive than in-house development. Analysis and requirements definition for subcontractors must be especially thorough because contract terms and conditions are formally documented. After work commences, altering the work scope or content usually results in financial penalties and schedule changes. Unless the contract provides for prototyping, the product delivered meets the contract's specifications.

Subcontracting also diffuses or disseminates company information. If system development involves confidential information, nondisclosure agreements must be negotiated. Ultimately, exposure risks or confidentially breaches may be greater with subcontractors than with the firm's permanent employees.

Both parties must also understand and resolve conflicting objectives. For example, the client firm wants the job completed on schedule, to specifications, and within budget. The contractor wants maximum profits. A cost-plus contract encourages the contractor to increase the job's size or scope and lengthen its schedule. A fixed-fee contract motivates the contractor to perform as little work as possible to meet contract terms. Contract terms and conditions must explicitly recognize this conflict. In addition, the firm's management system must enforce the contract in fairness to both parties.

A risk analysis according to the discussion in Chapter 9 is mandatory before entering development contracts. Periodic risk analyses must continue during the contract's life as well. Subcontract development burdens the firm's managers and the IT management system considerably.

PURCHASED APPLICATIONS

Increasingly, businesses turn to external sources for some application programs. Since the mid-1990s, the trend toward purchasing applications and the decline of in-house development have been particularly dramatic. According to a survey of 1100 companies worldwide by Meta Group, Stamford, CT, applications built in-house rose about six percent annually during the early 1990s but dropped 10 percent from 1994 to 1995 and 50 percent from 1995 to 1996.[16] Of the several reasons advanced to explain this phenomena, the availability of high-function, commercially developed applications is certainly an important contributing factor.

Indeed, widely available, highly useful applications have been the major force driving the explosive growth of personal computing. Although widely associated with PCs, purchased applications are becoming more commonly used in minicomputer and mainframe installations too. For instance, a portfolio of several thousand applications is available for the IBM AS/400. Most were developed in collaboration with IBM customers and can be purchased with the hardware.

Software is big business. IBM is the largest software supplier, but thousands of firms produce program products for sale to others. Microsoft, Computer Associates, and Oracle are major suppliers in an industry generating nearly $100 billion in yearly revenue. Growing 12 percent annually, worldwide packaged software revenues are expected to reach $153 billion by 2000. These firms and many others provide a smorgasbord of applications for organizations worldwide. They form the backbone of the software development industry.[17]

Advantages of Purchased Applications

In most large organizations, the resident programming staff must develop and maintain many applications. Organizations can readily purchase some applications, however, from software vendors. Still, the build-or-buy decision is many faceted, depending heavily on the advantages and disadvantages of purchased applications for the particular firm. Table 10.3 outlines the advantages of purchased applications.

TABLE 10.3   Advantages of Purchased Software

Early availability

Well-known function

Known and verifiable quality

Inspectable documentation

Lower total cost

Availability of maintenance

Periodic updates

Education and training

In contrast with extended in-house development cycles, installation and use of purchased applications is generally relatively prompt. Because programming is such a large part of application development, purchased applications can save considerable time and substantial expense. Because benefits also accrue earlier, the cost/benefits ratio is improved. Thus, in many cases, purchased applications are more financially attractive than other alternatives.

Purchased applications usually include well-defined or easily determined functions. Compared with locally developed programs, their functional reliability reduces risks: purchasers can be relatively certain of the program's functional capability before making commitments. Purchasers can objectively determine the application's value by reading the vendor's documentation and trade reviews, running program tests, and listening to other purchasers' experiences. These outside information sources also reduce some biases associated with in-house development.

Through discussions with application users, industry experience, and pre-purchase application tests, purchasers can obtain a good understanding of the program and documentation quality. Again, this reduces risk. Program and user documentation, usually a major part of the application product, is frequently inadequate for locally-developed programs. Good programmers are not necessarily good writers; they may produce low quality documentation, available much later than needed. Preparing good documentation is difficult and tends to be deferred in favor of other more interesting or seemingly more important activities. Thus, if anything in the schedule slips, it's likely to be the documentation. Purchasing applications eliminates these risks because both the product and documentation are available simultaneously.

In addition, purchased applications tend to lower total costs, thus improving the business case. Application developers spread development costs over many customers, thus reducing costs for all. In contrast with hardware or other physical products, per unit software manufacturing costs are very low. Most costs are in document reproduction, packaging, and distribution. Therefore, purchased applications include the benefits of large-scale economies.[18]

When many widely available applications are enhanced or updated they are available to users at reduced prices. These updated versions add functional improvements to the application that support new hardware features, offer new software functions, or enable integration with applications from the same or other vendors. Designed to improve or broaden the product's function, enhancements increase the product's value to customers and tie them more closely to the vendor. Usually these incremental releases or updates are available at relatively modest prices. Perhaps each customer cannot use all the added functions, but the useable enhancements typically offer a favorable cost/benefit ratio.

Lastly, purchased applications simplify user training. Outstanding training materials support many popular applications: training manuals, reference manuals, user aids, help lines, user groups, and, in some cases, vendor or third-party classes. Some applications are so popular that vendors can offer modestly priced training. And, in contrast to most locally developed training material, many vendors supply high-quality training aids to support their popular applications.

Disadvantages of Purchased Applications

Because purchased applications offer an impressive array of advantages, their popularity and rapidly growing markets are easy to understand. However, we must consider some important disadvantages of purchased applications. Managers must analyze and thoroughly understand advantages and disadvantages so that they can reach intelligent decisions. Table 10.4 lists some disadvantages of purchased applications.

The principal disadvantage of purchased programs relates to functionality. In some instances, purchased applications cannot perform functions the firm desires. For example, strategic systems usually cannot be purchased. Because of unique business characteristics, confidentiality needs, or unique and proprietary information or processes, purchased applications may not offer viable alternatives to local development. The resident programming staff must develop most strategic and other unique systems in-house.

TABLE 10.4   Disadvantages of Purchased Applications

They may have functional deficiencies.

Program and database interactions may cause difficulties.

They may be difficult to customize.

They may contain unnecessary function and unuseable code.

Unique management styles may not be supported.

Some application program groups are so interrelated and highly dependent on common databases that integrating a purchased application may be practically impossible. For instance, a cluster of financial programs may depend so heavily on locally defined databases that integrating a purchased general ledger program is difficult or impossible. Integration cost exceeds development savings in some cases. For the same reason, replacing the entire financial and accounting application set may be infeasible. The difficulty of inserting a commercially available application increases as interactions among and between application program sets increase.

Several factors make modifying purchased programs difficult. These include source code availability, the source language itself, and the availability of program logic manuals or other program documentation. Test-case development may be another detraction. Unavailable source code or incomplete or unavailable programming documentation may make program modification impossible or impractical. In any event, the firm must conduct a cost/benefit analysis to determine the financial reasonableness of proposed modifications.

Surrounding the purchased application with customized programs that bridge to present applications may be necessary or appropriate in some instances. Doing so may be preferable to modifying or customizing code within the purchased application. In addition, preserving compatibility with future software releases is important. Modifications may need to be replicated when the next release arrives. Vendors usually feel no responsibility to migrate your custom code to their next release! Financial considerations involved in this contingency must be factored into the proposed application's business case.

Still, the application's function may be deficient, incomplete, or implemented in a manner foreign to the firm's business operation. Application programs generally implement the firm's management system. Sometimes they include functions to satisfy management style considerations. To a considerable degree, customized application programs embed culture in code and represent "how we do things around here." Inclusion of these nuances in a popular, widely distributed application is unlikely. Purchased applications may require modifications to the applications themselves, the management system, or both. Devoting the resources needed to perform these modifications detracts from the application's business case.

Most firms never evaluate costs to modify the management system to accommodate an attractive application program. Many functional managers believe that programs are easily changed but management systems representing the corporate culture are not. Managers who believe that programmers are employed to support managers or administrators remain completely unwilling to consider alternatives. This myopic thinking causes many organizations to spend large sums developing customized systems whose primary functions are common in the industry.

Examples of this phenomenon are widespread. Tens of thousands of programmers develop and maintain unique ledger systems, payroll programs, manufacturing applications, and inventory control programs. These unique programs maintain the culture, offer no competitive advantage, and cost dearly. Some of these applications must now be reworked to solve the year 2000 problem. For many reasons, the functionality issue looms large when considering purchased applications.

Documentation, support, quality, and maintenance of purchased applications varies widely. Some purchased applications contain bugs or glitches that can embarrass or cost their purchasers.[19] And other purchased applications come without training or educational support. Thoroughly investigating all aspects of a product prior to its purchase has no substitute. The vendor's reputation is important, too. A product with a reputable brand name may be worth its additional cost. Good applications tend to live long. A mutually beneficial association between your firm and its vendor will endure for years.

Varied and complex issues relate to purchasing applications for the firm's portfolio. This is especially true when an application replaces or supplements a current application and uses the firm's traditional databases as information sources and sinks. The issues are somewhat simpler if the proposed application is a "stand-alone" product or is introduced along with a new automation area for the organization. Given all these considerations, however, purchased applications' popularity increases because they tend to be financially attractive and also offer a viable way to reduce the application backlog through contained financial investments.

From a business and financial perspective, microcomputers, applications and the hardware on which they run are generally considered together and frequently purchased together.[20] Purchased applications supporting microcomputer strategies is usually a given condition. As micros' capabilities grow and as they penetrate business processes more deeply, purchased applications will be extremely important. Globally, new businesses and many not highly automated businesses will center their strategies on purchased applications. Some firms will choose programmerless environments; many that do will succeed. For the successful, managing program development will not be an issue.

ADDITIONAL ALTERNATIVES

Other alternatives to in-house application development are evolving. One of the most promising is the formation of alliances. For example, Kidder, Peabody Inc.'s outdated systems presented two prospects: spending six years and $100 million, or acquiring technology and systems from its rival, First Boston. Ultimately, Kidder worked a deal with First Boston; by sharing resources, both firms benefited. As major systems' costs increase, many firms defray these costs by sharing technology and resources, even with competitors when doing so makes sense. Each firm separately develops the system's proprietary portions—usually small in comparison with the whole—for their own use. Some computer manufacturers form partnerships with customers to market customer-developed programs along with their hardware platforms. These marketing arrangements benefit manufacturers and their customers.

Large and small businesses are also forming alliances to advance their strategic interests. Useful for many endeavors, alliances are becoming increasingly popular in computer technology. This relatively new approach to development can be very effective under favorable circumstances. When a firm can find a satisfactory alliance partner, each partner can potentially benefit from reduced costs, improved schedules, and increased function. Successful product development alliances offer both partners the chance to increase revenue and profit.

Northwestern National Life and Infodata Systems Inc., Avon Beauty Group and IMI Systems Inc., and Hilton Canada and Control Key Corporation, among others, have successful joint development and marketing efforts. Some firms engage in joint ventures to solve mutual problems without intending to sell the resulting product. For example, Security Pacific and five other banks are cooperating on an imaging technology development project. If successful, the project will eliminate millions of pieces of paper and reduce data processing costs.

In another case, Baxter Healthcare and IBM formed a joint venture to sell computer hardware, software, and services to the health-care industry. Baxter, with strong connections to the health-care industry, and IBM, with solid image-processing, networking, and workstation technology, hope to change the way hospitals work. Patient records, X-rays, and CAT scan reports can be rapidly routed wherever they're needed. Electronic records can be stored at reduced cost and with improved efficiency. Arrangements such as these and others drive the trend toward alliances and joint ventures.

Firms selling IS services offer another alternative to reduce expenses and off-loading program development work. These service bureaus provide systems and operating environments capable of processing routine applications for their clients. Payroll processing, for example, is an application service bureaus frequently run. The firm supplies payroll input, the service bureau processes it, delivers the checks or makes direct deposits, and completes and returns the payroll register along with other reports to the client.

The advantages of service bureaus are that they off-load work and responsibility, and reduce in-house computer requirements. In some cases such as payroll, the service bureau makes program changes required by law or regulation and keeps the payroll system functionally modern. It updates the payroll program to account for new federal withholding tax changes, for example, and to keep the program compliant with state or local regulations. The bureau spreads its maintenance costs over many clients, reducing expenses for all.[21]

Service bureaus eliminate application development and maintenance costs for some applications, while also reducing hardware capacity requirements. Using service bureaus is appropriate for many operational applications that offer little strategic advantage. Identifying these applications and considering service bureau processing frequently occurs during the prioritization process discussed in Chapter 8. Service bureaus present viable means to trade money for people and computer capacity and to reduce costs too. Using service bureaus is the first step toward outsourcing, an important current trend discussed in Chapter 11.

Chapter 8 examined the resource prioritization problem inherent in maintaining an application portfolio. This chapter and Chapter 9 present alternatives available for portfolio acquisition. Given the range of alternatives, what methodology can be employed in the selection process? How can managers choose among the alternatives?

Many firms' prioritization methodology reveals that programming talent is the most constrained resource and that money is a lesser constraint. This conclusion is frequently drawn during final prioritization discussions: it becomes obvious that the firm's programming staff cannot respond to all work requests.

Answers to five questions lead to optimum selection among alternatives:

1. Which applications can be processed at a service bureau, saving people and computer resources?

2. Which applications needing development or replacement can be purchased to save programming resources and time?

3. In which ways can the firm enter into agreements with others, through contracts or joint development, to optimize the firm's resources?

4. Can new development tools and techniques improve development productivity?

5. What alternatives can increase human resources available for application development? (The next chapter discusses end-user computing.)

Given the array of alternatives and the thought processes involved in answering these questions, the firm has the tools to balance its approach to prioritization. The firm can most likely off-load programs with low strategic but high operational value to service bureaus. Highly valuable strategic applications or those with potential strategic value are most appropriate for in-house development. The firm must manage these highly valuable assets carefully. Usually they involve proprietary information and reside at or near the top of the priority list.

The firm must make choices for the remaining applications, applying available resources to optimize its goals and objectives. The strategizing and planning processes discussed earlier are critically important foundations for critical decisions. IT managers must ensure that the environment supports and encourages high productivity by using productive tools, advanced techniques, and disciplined management systems.

SUMMARY

Application acquisition offers IT and user managers a variety of opportunities to optimize the firm's resources. It also provides opportunities to develop the firm's important strengths. Managers must respond to these opportunities in a disciplined manner. This response begins with a clear vision of the application portfolio's contribution to the firm's success. Careful strategic planning develops and enhances this vision. Tactical and operational planning fine-tunes resource allocation to strengthen the portfolio.

Traditional life-cycle methodologies or alternative approaches augment the application portfolio. Although all alternatives offer opportunities for gain, all involve risk. Managers must thoroughly understand and mitigate these risks in some manner. Managers must also use technical and people management skills to implement and use the available advanced tools and system development techniques. Advanced tools shorten development cycles and greatly improve productivity, product quality, and user satisfaction. Achieving these goals is a high priority for IT managers. Indeed, these objectives are critical success factors for IT managers and client managers alike.

Exploring alternative approaches to system maintenance, enhancement, and acquisition causes firms to reassess their programming development and computer operation functions. Commercial application development grows rapidly as purchased applications, attractive for both large and small systems, gain popularity. Recognizing the alternatives' economic value, firms more willingly consider them. Many firms, reluctant or unable to accept the long-term commitments associated with a permanent programming staff, welcome alternatives. Many others, after reconsidering their IT self-sufficiency strategies, actively pursue alternatives. In many cases, CIOs lead firms in these dramatic new directions.

Review Questions

1. What is the year 2000 problem? What bearing does it have on the supply/demand for U.S. programmers?
2. Where are the leverage points in using fourth-generation languages in conjunction with CASE tools?
3. What is Java, who developed it, and what is its potential importance to the industry?
4. Under what circumstances can prototyping be highly effective when combined with normal life-cycle development?
5. What are the advantages of using prototyping with CASE tools?
6. What factors accelerate the trend toward purchased applications? Do you think these factors will increase or decrease in importance, and why?
7. What are the disadvantages of purchasing application programs? Are these disadvantages more or less important for well-established information technology departments?
8. What are the risks in using purchased applications? How can these risks be quantified and minimized?
9. What is the relationship of purchased applications to the explosive growth of personal computers?
10. In what ways does product quality enter into the purchasing decision?
11. What are the advantages and disadvantages of subcontracting applications development?
12. How does subcontracting application development assist in balancing resources?
13. What are the advantages and disadvantages of using a service bureau for some of the firm's applications?

Discussion Questions

1. Discuss the advantages and disadvantages of using offshore programmers. For what kinds of systems, and under what circumstances, would this approach be least risky?

2. Global competition in the programming development business is increasing. What are its implications for American programmers, their managers, and their clients?

3. Using the risk analysis techniques discussed earlier as a starting point, identify the risk elements inherent in subcontract development. Prioritize your list of risk elements, and discuss your rationale.

4. What trends in information processing favor service bureau firms? What trends are unfavorable?

5. Draw a flow chart of the questioning process discussed in the section on managing alternatives.

6. Discuss the relationship between alternatives to traditional development and the critical success factors discussed in Chapter 1.

7. Why is programmer productivity such an important issue?

8. Discuss the significance of the object paradigm. What is the importance of Java to programming in general? Discuss the importance of Java in the struggle for industry leadership.

9. How does the analysis leading up to Figure 8.4, Strategic vs. Operational Value, assist in the task of managing alternatives?

Assignments

1. Using library resources, analyze two firms in the commercial applications industry. Compare and contrast these firms' products and services. You may want to visit www.cai.com for information on Computer Associates International, a large developer of commercial application programs.

2. Obtain descriptive material on two fourth-generation languages. Compare and contrast their capabilities and limitations. For which class of problems is each language most suitable?

3. Read the first chapter in Edward Yourdon's book, *Decline & Fall of the American Programmer*, and summarize its main points. Present an argument that supports or contradicts Yourdon's thesis.

[1] Edward Yourdan, *The Decline And Fall Of The American Programmer* (Englewood Cliffs, NJ: Prentice Hall, Inc., 1993).

[2] Martin LaMonica and Elizabeth Heichler, "Operation Offshore," *Computerworld*, August 8, 1994, 73.

[3] Kim Girard and Robert L. Sheier, "Bell Legacy Systems Plague Deregulation," *Computerworld*, April 28, 1997, 1.

[4] Julia King, "IS Labor Drought Will Last Past 2003," *Computerworld*, June 30, 1997, 1.

[5] G. Pascal Zachery, "U.S. Software: Now It May Be Made in Bulgaria," *The Wall Street Journal*, February 21, 1995, B1.

[6] Companies worldwide use many languages. Today, 44 percent use C, 42 percent Cobol, 27 percent C++, 24 percent assembler, 18 percent PowerBuilder, 18 percent Visual Basic, and 95 percent use other languages too. *Computerworld*, July 14, 1997, 53.

[7] James Martin, *Application Development Without Programmers* (Englewood Cliffs, NJ: Prentice-Hall, Inc., 1982), 28. There is no common definition of fourth-generation languages, but hundreds of vendors claim to have superior offerings.

[8] Unlike most other languages, HTML is not compiled into machine language but is interpreted after reception at the client by an interpreter or a browser like Netscape's Navigator.

[9] George Gilder, "Will Java Break Windows?", *Forbes ASAP*, August 25, 1997, 123.

[10] Computerworld/First Market Research Corp. survey of 160 development professionals as reported by Gary H. Anthes, "No More Creeps!" *Computerworld*, May 2, 1994, 107.

[11] Edward Yourdon, *Modern Structured Analysis* (Englewood Cliffs, NJ: Prentice-Hall, Inc., 1989), 97-100, presents a more complete discussion of the prototyping life cycle.

[12] W. H. Inmon, *Developing Client/Server Applications*, rev. ed. (New York, NY: John Wiley & Sons, Inc., 1993), 40.

[13] Watts Humphrey summarizes the problem: "Currently, software engineers learn software development by practicing on toy problems. They develop their own processes for these toy problems. These toy processes are typically not a suitable foundation for large-scale software development processes."

[14] Quality improvement processes leading to the "6 sigma" norm are equivalent to three defects per million lines of delivered code.

[15] Watts S. Humphrey, *Managing The Software Process* (Reading, MA: Addison-Wesley, 1989), is the definitive work on this subject. See also *A Discipline For Software Engineering* by Humphrey (Reading, MA: Addison-Wesley, 1995).

[16] Sharon Gaudin, "Pace of Change Stymies Coders," *Computerworld*, May 5, 1997, 47.

[17] *S&P Industry Surveys*, Computers: Software, October 10, 1996, 7.

[18] Lotus 1-2-3, Version 3, cost $7 million to develop and $15 million to ensure quality. It can be purchased for less than one-one hundredth of one percent of the development cost alone. This is an extreme case, but it illustrates how economies of scale act to the purchaser's advantage.

[19] Joan E. Rigdon, "Buggy PC Software Is Botching Tax Returns," *Wall Street Journal*, March 3, 1995, B1. Some popular tax-preparation programs contain errors causing the IRS to assess back taxes and interest charges, according to this article.

[20] Government agencies purchased thousands of PCs that included preinstalled game programs. They are now attempting to deal with the fact that many employees are wasting valuable time playing them.

[21] Payroll processing typically costs about 50 cents per person per payroll period. Most firms cannot process payroll for anything near this cost, yet many consider payroll processing part of their culture.

11 *Managing Network Applications*

Merrill Lynch Develops Client/Server Applications

Even before Merrill Lynch's CMA program was fully developed, the firm embarked on a massive project, the Professional Information System (PRISM), to support retail brokers with a completely automated and fully networked information system. PRISM is an advanced workstation tool that moves brokers from manual order-entry to a completely automated platform.[1] It was designed to support future growth in business volume and to capitalize on advancing workstation and network technology.

PRISM's primary objective is to enable brokers to scan client information and stock market data simultaneously using multiple windows. Through PRISM, brokers can retrieve client data, stock market information, and research opinions from the company's mainframe systems in New York.

High-tech PRISM workstations in each retail office connect to LANs linked to Merrill's backbone network. All 500 domestic branch offices and international branches in Tokyo, London, Singapore, Sydney, Bern, and Toronto link to its New York and New Jersey headquarters' facilities with a $200 million backbone network supporting PRISM and other applications.

More than 17,000 workstations were installed jointly by IBM and Automatic Data Processing supervised by Merrill Lynch personnel over a period of 21 months. But, before the installations were complete, plans for more extensive capabilities were already being made.

PRISM's capabilities are stunning. Its latest version lets Merrill's financial consultants view customer securities portfolios and trading positions; scan news services such as Dow Jones, Knight-Ritter, or Reuters; and watch internally generated video programs.[2] Consultants can analyze a customer's portfolio with current prices and instantaneously access profit-and-loss statements of both realized and unrealized gains with just three keystrokes. With a few more keystrokes, they can generate pie charts of the portfolio mix and compare them with the desired mix. The system completes hours of manual calculations in seconds. Customers receive much better service, and consultants' performance improves as well.

Merrill is considering methods that will link its customers more fully to account and market data so routine service requests and transactions can be handled without Merrill's intervention. Access to some CMA information is now available through touch-tone phone service, but Merrill plans to considerably expand customer access to its central systems. Merrill's consultants will not be involved in the routine transactions of knowledgeable customers.

But Merrill looks beyond systems excellence to internal efficiency. Howard Sorgen, Merrill Lynch senior VP and managing director of global systems and technology, standardized software across operations and combined 11 data centers into two to increase IS productivity.[3] Consolidating systems into two mainframe sites reduced infrastructure costs by $100 million and its host-based processing workforce by 40 percent. Large business volumes make elimination of the company's mainframes impossible. Running in batch mode, they continually update seven million accounts and process statements. In addition, Merrill supports its popular stock market data Web site on an Amdahl mainframe—customers accessed this site 18 million times last year.[4]

Programmers responsible for host-based production systems work in one central group. Sorgen states that Merrill Lynch benefited from other centralization moves, too, for example, by combining personnel who develop business unit applications. Merrill reduced its mainframe computing costs by 45 percent, reinvesting half the savings in client/server applications.

Merrill spends more than $1 billion annually on information technology and presently allocates $200 million to eliminate year 2000 problems. To prepare for the millennium, Merrill employs 100 permanent employees and 150 consultants who are fixing 170 million lines of code in 1450 applications. This work follows a massive reprogramming effort to change the transaction settlement standard from five to three days. At the time, "We joked, 'What could they possibly give us that would stress us out more than T+3?'" says Susan Luechinger, Merrill's coordinator of its worldwide year 2000 effort.[5] Meanwhile, stock exchange volumes continue to increase steadily.

Merrill Lynch ranked 29th on ComputerWorld's Premier 100. Merrill employs 42,650 people—more than 2500 in information systems. Merrill dedicates 70 percent of its software investment to client/server applications.

INTRODUCTION

Earlier chapters discussed challenges associated with managing the firm's application portfolio and related databases and presented several promising methods for prioritizing the development backlog and managing the portfolio. Chapter 10 focused on promising alternatives to traditional development and explored the popular option of purchasing commercially developed applications. The material presented these alternatives' advantages and disadvantages and reviewed situations in which one or more alternatives would be attractive to the firm. This chapter describes another popular and appealing alternative that has the potential to improve business processes and operating excellence.

Capitalizing on telecommunication technology and new, low-cost workstations, most organizations installed some form of distributed computing. By introducing capability at employees' workplaces, organizations let end users operate business applications and even encourage them to develop applications. Commercially developed applications tuned for operational departments and designed for installation on networked hardware ease the transition to networked computing and also reduce pressures on the firm's programming group. Advanced hardware and new application programs leveraging the potential benefits of distributed computing encourage firms to invest in telecommunication-based computing.

Firms are increasing their funding for distributed computing and devoting a larger share of their IT resources to end-user support. Studies show that leading IT organizations spend approximately 20 percent of their budgets on end users, while client organizations spend a significant amount of their own money on information systems and services. In some firms, more than half the information technology expenditures are spent by or for end users.[6] Resource deployment of this magnitude dramatically influences organizations and their people.

Distributed computing places special demands on employees, managers, and organizations, and its implementation raises many important issues. Critical organizational and political barriers must be removed before firms can benefit significantly from distributed processing. Firms must understand its role, form, the situations in which it is appropriate, and how to manage its introduction and use.

DISTRIBUTED COMPUTING

Distributed computing is the alternative to centralized or mainframe computing. It links user workstations to each other and to a controlling computer or server through a network. Because it rearranges workflow, distributed computing involves not only hardware, but software, applications, processes, and people as well.

In its earliest and simplest form, a central processor or server connects to terminals at the client's workplace where employees enter data, initiate programs, or retrieve information. In this configuration, shown on the left side of Figure 11.1 (see page 289), most processing capability resides at the host computer. For many applications, such as travel agency terminals accessing an airline reservation system, this arrangement is satisfactory because the workstation application itself requires little desktop processing power.

Advances in workstation hardware and telecommunication systems have enabled various forms of distributed processing to gain popularity. Today, refined architectures of networked systems include client/server systems, the Internet and intranets, and cooperative or peer-to-peer network processing. In peer-to-peer and client/server processing, individual workstations on a local area network connect to a server (a CPU) that controls some processing operations and manages data stores. In cooperative processing, any node workstation can initiate an application while other workstations may supply data or computing power in a manner transparent to the initiator. This mode of operation, sometimes called *network computing*, requires sophisticated network operating systems.

Client/server operations divide the application into two parts: one resides on the server and the other on the client workstation. Clients initiate transactions and rely on the server for some processing and data management services. To complete the client's tasks, servers may seek information from databases on other servers or from processors at higher levels in the network. The center and the right side of Figure 11.1 show these arrangements. Client/servers are the most popular form of distributed computing.

The more general term, *end-user computing*, refers to stand-alone or networked PCs operated and sometimes programmed by end users. In some cases, programs that previously ran on centralized systems are loaded on user PCs and become user departments' responsibility. Generally, these programs are called *downsized mainframe applications*. Client/server systems of networked PCs loaded with applications customized to specific department tasks are rapidly displacing stand-alone PCs. Most organizations find that workstations networked to each other and to servers, and sometimes to the firm's mainframe, offer advantages unattainable with individual processors.

Internetworking can extend network client/server LANs to many other larger networks. For example, if the server connects to the firm's central computer, which connects to the Internet, phone system, and other nets, then global e-mail and even electronic commerce is possible. Many other possibilities also exist. Figure 11.1 shows the client/server physical and logical architecture.

Client/server implementations are becoming ubiquitous. They deliver e-mail services and support multimedia applications, program development, group or collaborative work systems, and many other services and applications. This infusion and diffusion of information technology requires large investments, detailed planning, and careful implementation.

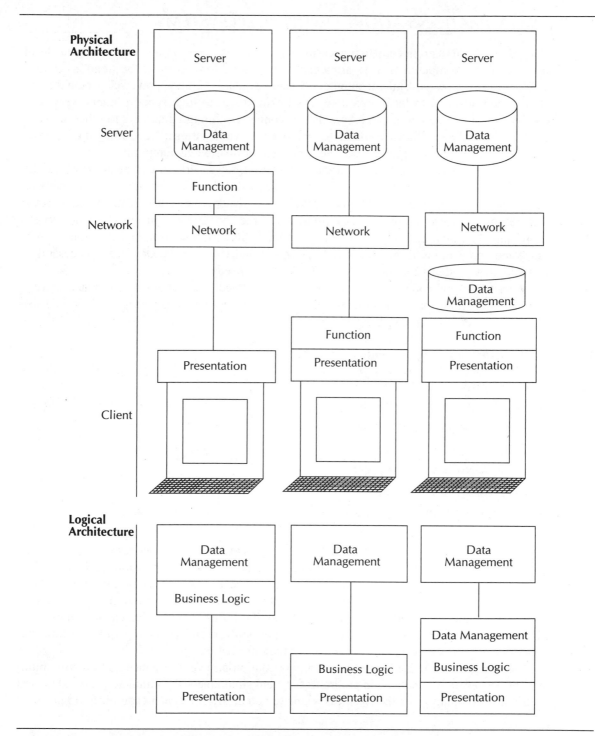

FIGURE 11.1   Client/Server Physical and Logical Architecture

The trend toward distributed data processing is not new; it started in the 1960s with the development of minicomputers and departmental computing. More recently, the trend accelerated as increasingly powerful microcomputers became available. Today, powerful personal workstations are attached to local area networks linking them to high-speed printers, very large databases, and huge central processors. LANs connected through gateways to other networks make additional capabilities available to employees at workstations. New, emerging hardware and software technology greatly favor decentralized or distributed computing.

Business executives in manufacturing, distribution, sales, service, and other areas find distributed computing attractive because it offers operational flexibility and increased responsiveness to business pressures. In addition, it gives them greater control over costs and expenses as they respond to changing market conditions. Given the choice, most business managers prefer to manage their own information systems. Accomplishing local control, however, means moving some applications from centralized systems to individual or departmental processors or installing new systems carefully tuned to departments' needs.

Because several compelling features make them attractive, variations of distributed processing are widely implemented in business, industry, and government organizations. Table 11.1 lists important factors encouraging this trend.

TABLE 11.1   Factors Favoring Distributed Processing

Helps optimize business processes

Empowers employees with new tools

Communicates important business information

Availability of low-cost workstations

Supporting telecommunications systems

Growing base of skilled employees

Availability of workstation applications

The major factor driving distributed computing is organizations' desires to empower their employees with powerful tools, more information, and increased responsibility, thus optimizing their operations. Blending technologies with skilled people creates entirely new types of business solutions built on communication, information generation and sharing, and new forms of collaboration. Decentralizing and dispersing information so that capable employees can productively interact with it is rapidly augmenting and replacing older, traditional processes. These transformations characterize the information age.

Very significant reductions in personal workstation price/performance and growing numbers of cost-effective application programs for them simultaneously stimulate the trend toward distributed computing. As the Business Vignette noted, Merrill Lynch capitalized on personal workstation hardware and highly functional networks to bring its skilled employees powerful new applications. Although retaining central processing for customer account and statement processing, Merrill Lynch's move toward decentralization is obvious. Inexpensive, reliable, and widely available workstation hardware and software makes Merrill's client/server architecture feasible and productive. Today, technology advances and economic considerations foster widespread adoption of personal computing.

Advanced local area networks and sophisticated communications support for centralized systems and servers encourage program and data sharing. Workstation users benefit significantly from these improved capabilities. Networking increases workstation utility, bringing the power of servers and mainframes and their database resources to the employee. When all the firm's employees work together simultaneously toward common goals through networked individual workstations, parallel processing occurs at the firm level. This important phenomenon is rapidly becoming the norm in modern organizations.

Large backlogs of work facing application development groups and increasing programmer and program development expenses also drive distributed computing. Today, many high-function applications for most common tasks are available from independent programming firms. These cost-effective purchased applications reduce dependence on in-house programming and greatly accelerate the trend toward departmental systems serving individual workstations.

The number of skilled individual users of personal computing applications grows rapidly. Professional, technical, and office workers and managers are all learning workstation and application skills. The ranks of skilled users grow quickly as information technology penetrates the firm's operations ever more deeply. Firms have established numerous precedents for distributed computing by implementing technology in many innovative ways. Altogether, these factors encourage those who contemplate using distributed computing technology.

Adopting distributed computing is very significant for the firm, its departments, and its employees and managers. Distributed technology is a microcosm of the larger phenomenon of electronic data processing. Growth stages, change management, business controls, and many other topics are as important to distributed computing as to centralized processing. Distributed computing is important not only because of its high visibility in today's business world, but because it enables business process innovation, leading to operating excellence, resource optimization, and improved responsiveness.

Client/server and Internet issues top the list of topics of most interest and concern to IT managers. Today, the Internet and the World Wide Web engage nearly everyone in business and most of the general public as well. Experts forecast an ever-increasing, global use of electronic communication. Are the Internet and World Wide Web high-tech fads or technologies offering real advantages for individuals and organizations today?[7] Do businesses benefit from client/server computing and Internet applications, or does hype exceed reality? Do these network technologies offer most organizations the right approach, and, if so, when and how should they be implemented? The remainder of this chapter addresses these and other questions.

To understand whether to adopt one or several forms of distributed computing, firms must answer some fundamental questions about their businesses and structures. For example, are the firm's business practices and business operations optimally tuned to its business environment? Can they be improved in light of changing conditions?

To help answer these important questions, operations analysts develop process models of current business systems. Using these models, they analyze work and information flow from inbound processes, through operational activities, to outbound processes, and to sales and services activities. Analysts review the firm's value chain and all supporting activities as they strive to improve operations by restructuring and applying information technology. Although this activity in not new, its most popular name is *reengineering*.[8] The terms "business process improvement," Strassmann's third point, or "business process innovation" more accurately reflect this activity's nature.

When analyzing current business processes, the first goal is to identify and terminate activities the firm should not be doing. The second goal is to identify essential processes that someone outside the firm can accomplish more efficiently. The firm should consider consigning or *outsourcing* essential but routine processes beyond its core business to firms specializing in these activities. For sound business reasons, outsourcing non-core and perhaps even some IT business processes is common practice today.

The General Services Administration, for example, is assessing the benefits of outsourcing many IT facilities. These activities include data networking and local telecommunications facilities, computing centers, the Federal Information Center, and the Federal Procurement Data Center. The commissioner of GSA's IT Services believes that data services are a good starting point to determine the value of privatizing these activities. The consulting firm Arthur Andersen & Co. is doing much of the analysis for the GSA.[9] Chapter 19 more fully discusses outsourcing.

To improve operation, the organization should devise innovative new processes for its remaining activities. Usually information technology supports these new processes. "Information and information technology are powerful tools for enabling and implementing process innovation. Although it is theoretically possible to bring about widespread process innovation without the use of computers or communications, we know of no such examples."[10] Tom Davenport's statement minces no words in describing the relationship between information technology and improved business performance. In today's business world, networked applications are the most popular means for linking technology and process improvements.

Firms today perform much work sequentially, not because they need to (although they may in some cases), but because the means for parallel operations are unavailable or presumed unavailable. Information technology, particularly distributed processing, lets individuals perform parallel processing at the firm level. Powerful workstations, robust networks, and large, well-structured databases permit many employees to attack problems at the same time, in parallel. And, with current networking technologies, one firm's processes can link in parallel to those of supplier or customer firms. The tremendous potential of these concepts drove much of the reengineering effort to date.

In performing reengineering, more accurately called *business process innovation*, the correct sequence of events is to 1) outline current processes and workflow, 2) search for superior process models, and 3) then find resources, including information technology, to implement innovative new processes. Deoptimized business processes create opportunity; information technology helps capture benefits. By themselves, however, client/server architectures, Internet applications, and distributed computing are solutions looking for problems. Thus, adopting ill-considered network technologies may not solve problems.

For example, about $1 trillion worth of proprietary systems developed over decades run on mainframe computers worldwide. These legacy systems have been expensed and the computers running them are fully depreciated; nevertheless, some firms spend large sums rushing to convert to distributed applications. In some cases, the firm reasons that it can save money because PC MIPS cost less than mainframe MIPS. But what's important is not the cost of MIPS, but the cost effectiveness of the organizations using the applications to run the business. Sometimes, firms rushing to embrace new technology overlook some critical issues. Legacy systems converted to client/server applications may be more effective but, in most cases, only if they improve business process.

INTRODUCING INFORMATION TECHNOLOGY INTO NEW PROCESSES

When new processes or procedures are identified and outlined, affected analysts and employees must determine how to best implement them. Usually, but not always, this means that technology will be used to assist the processes and empower employees to conduct them. In many cases, computing power is positioned in the workplace so that important information and process data is available to, and can be collected by employees.

Many factors must be considered in the decision to install distributed systems and move applications to the workplace. Table 11.2 lists some of them.

TABLE 11.2   Installing Applications at the Workplace

Advantages	Disadvantages
Improves responsiveness	Requires management attention
Increases flexibility	Increases user skill demands
Empowers employees	Distracts managers
Increases user control	Weakens central control
Decentralizes costs	Usually increases costs
Fosters purchased systems	Encourages unique systems
Reduces IT workload	Disperses databases
Encourages innovation	Encourages parochialism

In addition to improving responsiveness and flexibility, decentralization asserts local control over important systems and encourages employee productivity. Most departments prefer to purchase rather than build their applications. Some organizations, however, report significantly lower costs and shorter schedules when users themselves develop applications. Still, whether they build or buy applications, operating managers gain the important advantage of experiencing IT costs directly, rather than being charged by a centralized organization.

Distributed computing encourages user innovation because there is little historical or cultural background to overcome. For example, traditional IT organizations tend to develop applications locally because that is what programmers do; however, because user organizations purchase so many other necessary items, they are inclined to purchase commercial application programs, too.

Downsizing, or off-loading mainframe tasks to user systems, increases user workload and demands specialized skills from employees and managers. The complexities of managing an application portfolio and running a small client/server network require skills and talent not usually found in operating departments. These responsibilities also divert attention from the department's primary function, so some managers prefer to leave them to others.

Some highly integrated databases are difficult to separate into distributable units. Downsizing increases data management difficulties and may lead to redundant data elements, increased storage costs, and, possibly, asynchronous conditions. Firms with highly developed information architectures suffer less from these difficulties than others.

Sometimes reverse economies of scale and redundant development accompany decentralized applications. Corporate goals may be subordinate to unit goals because parochialism tends to favor unit performance over firm performance. Firms with highly disciplined planning processes can avoid these difficulties and are much better prepared to capitalize on decentralization opportunities. As always, disciplined management processes pay dividends in many areas.

Some applications are much more suitable for distributed systems than others. Small, single department applications with isolated databases are ideal, particularly if the programs are stable. These low-risk applications are preferred to large, complex applications. Enterprise applications with large integrated databases are probably best centralized. Large evolving systems interfacing with many employees are best left undisturbed. Decentralizing customized applications or technically advanced programs may be risky. Technically complex or other high-risk systems are also poor candidates.

The management system developed in Chapter 8 to handle the firm's applications portfolio provides valuable insights for decentralization strategies. The portfolio management system described there focuses on business objectives for the firm, individual functions, and departments. It puts decision making at the proper level in the firm, thus reducing the emotionalism and parochialism that usually surrounds these decisions. In making these decisions, business goals and objectives must dominate politics or emotions.

Decisions to decentralize information systems, however well informed, must recognize the management issues they raise. "Most of our distributed systems are badly managed and many are unmanageable. It's apparent that the daunting task of keeping myriad distributed and home-grown client/server applications up and running just hasn't been given a high priority," states Patricia Seybold.[11] Because client/servers, Internet technology, and other distributed computing forms are so valuable, they must be managed in ways that defuse the issues accompanying them.

THE ISSUES OF DISTRIBUTED COMPUTING

Issues surrounding distributed computing fall into the eight general categories listed in Table 11.3. Management practices and specially designed organizational changes can contain these issues. Strategic, tactical, and operational reasoning and analysis can optimally resolve each problem area. Corporate culture, political considerations, and company policy also enter into analysis of these issues. These are governance issues in Strassmann's Information Management Superiority model discussed in Chapter 1.

TABLE 11.3   Distributed Computing Issues

Software and applications issues

Hardware compatibility and maintenance

Telecommunications concerns

Data and database issues

Business controls

Financial concerns

Political, cultural, and policy issues

Staffing and personnel

Compatibility among commonly used applications significantly enhances distributed computing's effectiveness. Application software compatibility simplifies networking, training, and hardware installation. Achieving software compatibility simplifies program installation and use and makes benefits obvious sooner. Standards and guidelines for user-developed application programs are necessary to achieve program compatibility. Application compatibility also simplifies and increases the effectiveness of user training programs.

Workstation compatibility policies greatly improve cost-effective hardware installation, maintenance, and upgrade. Hardware compatibility also simplifies networking, software installation, and application portability. Hardware maintenance can be performed on compatible client equipment with minimum disruption of activities. Also, hardware can be upgraded conveniently and inexpensively when units are physically compatible. Finally, policies for workstation ownership must be established.

Communications issues include the network's physical architecture, software, and policy questions. Important policies for using outside databases, linking through Electronic Data Interchange (EDI) or the Internet, and dial-in capability must be established. The firm's senior executives must help answer these critically important policy questions. Network management and maintenance must be transparent to client activities. IT must lead the management of these technical and policy items so that client operations function smoothly and reliably.

Databases and their uses raise many questions in distributed computing. Certain questions must be answered unambiguously: Where do the databases reside? Who owns them? Who controls them and how? Personal computer access to large databases greatly increases the organization's risk. Serious damage or loss results from improper or inoperative backup procedures. In networked environments, data integrity demands well-conceived controls for uploading and downloading. Properly resolving these issues is vital because networked systems in distributed environments thrive on available and secure information assets.

Like information assets, application programs must also be protected. Programs that maintain important records or control valuable assets require special attention. The firm's managers and application owners must carefully consider security risks before deciding whether to distribute these programs to individual workstations. Programming processes and procedures must attain acceptable levels of programming quality.[12] Managers must manage and enforce appropriate standards for program documentation. End users writing applications are more likely to omit documentation than professional programmers. Because future maintenance is a certainty, managers must insist that user-programmers document their work. Network security, however, is an IT responsibility.

As technology dispersion brings powerful systems and applications to user departments, some former IT responsibilities also migrate to user managers. Disaster recovery planning and management is one of these. Because department managers own and operate major information processing resources, they must take precautions against various potential problems. Frequently, user managers do not understand the nature of, and need for, recovery planning. The central IT organization must assume responsibility for providing training and assistance in this important area.

As departmental computing develops, maintaining positive financial returns demands special attention. Conditions most likely to achieve satisfactory results include an aggressive attitude toward benefits accounting and specific attention to cost containment. Often intangible benefits figure prominently in cost justification. Both tangible and intangible benefits should be evaluated and recorded. Unwarranted expenditures for trendy hardware and software upgrades, failure to account for all people-related expenses, and incomplete benefits recording leads to serious difficulties later. Most studies show that annual costs per user rise significantly when migrating from mainframe systems to client/server implementations. Thus, ensuring value received for client/server investments is critically important.[13]

Benefits of decentralizing can be large, but costs are significant, too. The business case partly depends on whether reductions in centralized computing can be achieved. For example, Merrill Lynch closed nine mainframe centers and consolidated or distributed systems. Costs were shifted from centralized processing to distributed systems that closely support consultants. Investments in client support systems yielded superior customer service and increased revenue and profit. Increased organizational effectiveness generated important benefits for Merrill Lynch.

Political issues are important because decentralized computing shifts the firm's power structure. Power shifts raise concerns among individual managers, frequently causing emotions to interfere with rational decision making. Strategy development, long- and short-range planning, and application portfolio management are effective tools for dealing with political issues. Management systems and techniques like these put decision making on a business basis and help diffuse political concerns.

Although all the issues Table 11.3 lists are important, staffing and personnel are most critical. End users require considerable training to become productive and skillful programmer-users. To add information technology to their traditional professional repertoire, employees must significantly increase their skills. Not all individuals make this transition smoothly. Some take the lead quickly and make the transition easily, setting the example for others.[14]

Managers must handle employee transitions skillfully. For some employees, transition is traumatic. Because transitions are so serious for some organizations, one writer proposed that an "organizational impact statement" accompany them, an analysis of system and organizational changes that explains their effects.[15] Managers should be trained to be especially sensitive to employee reactions during stressful transitions.

Not all issues are equally important at all times during transition. User support is highly important in the early installation phases. It encourages and smooths the transition for hesitant users. When adoption progresses smoothly, financial considerations, data management, and business controls become increasingly important. During implementation, the organization undergoes substantial and permanent change. At all levels, the firm's leaders must manage and shape transition in accordance with long-term goals and objectives.

Given the issue's importance and pervasiveness, how can the management team cope? What actions can they take to deal with potential problems? What must the firm do to capitalize on potential benefits? The following sections help answer these and other key questions.

ORGANIZATIONAL CHANGES

IT organizations have many opportunities to resolve the challenging issues of change. IT must define support levels, preferably through service-level agreements, and assign support tasks. User support includes defining management processes and allocating resources to support services. Many firms establish two new entities to do this most effectively: the workstation store and the information center. These organizational units must receive IT assistance, but their staffs generally include non-IT members as well.

The Workstation Store

The workstation store begins supporting users by centralizing the firm's workstation and software purchasing. By purchasing workstations for client managers in large quantities at discount prices, the store saves the firm time and money. Discounts may amount to 30 or 35 percent of the unit price. In addition to reducing costs, this approach helps ensure hardware compatibility for the firm even though securing compatible hardware from more than one vendor may be necessary to satisfy the firm's requirements.

The store also purchases and distributes licensed applications with the hardware. It ensures favorable software prices, satisfies legal requirements by securing site licenses, and achieves software compatibility. Today, intranet and other network techniques offer the store several attractive options for distributing and upgrading user software. In firms planning to use intranets, the workstation store should be a first adopter.

The store also provides central workstation maintenance and supplies substitute equipment during maintenance. Although maintenance is critical, clients prefer to focus on their jobs. Their workstation should be handled like their telephone—if it fails, someone must correct the problem promptly. The store does this.

The store helps manage upgrades and hardware and software migration by obtaining approved hardware and software upgrades and making them available to clients with justified needs. Clients and the firm value these important services—the store provides them efficiently and effectively.

The store also performs other important functions. For example, it gathers important information from its sales to users and from inquiries regarding client requirements. IT can monitor the pace of end-user adoption and obtain trend information through sales analysis. This information is valuable as IT develops further plans for distributed computing. The workstation store also gathers benefits information from clients at the time of purchase. It requests clients to submit this information when they accept delivery of equipment or software. It analyzes benefits data along with other cost and benefits information and uses this information to provide a continuing financial justification for distributed computing.

As workstations proliferate, demands for hardware and software improvements arise. The store can satisfy these demands in much the same way it did when obtaining and distributing the initial products. As a condition for obtaining upgrades, customers must provide an approved benefits statement that reconciles the cost of additional equipment or programs.

In addition to implementing sound procurement policies, the store is an effective internal control mechanism for physical assets, too. Its record-keeping function is an important business control. When the store distributes units to users, its records document the equipment's location, its configuration, and the owner's name. For several reasons, this task is critical.

For instance, when taking inventory to prepare for year 2000 work, Basin Electric Power Cooperative in Bismarck, ND, discovered "hardware we never would have known about," according to lead analyst Dave Anderson. At Fortis, Inc., a New York insurance company, IS "used to have to ask purchasing what we owned," states Joe Hays, PC/LAN project analyst, but then "we couldn't tell where it all went—into another cubicle or out the door."[16] Well-managed workstation stores eliminate these serious problems.

In addition, routine client information is valuable to the organization for understanding trends and establishing client preferences. Combined with other IT planning data, this information serves as a superior leading indicator of future demand.

The workstation store usually reports within the IT organization and has a strong mission to support end users. The store is not optional—it is essential to maintaining effective distributed computing operations.

The Information Center

Other activities also demand careful attention to ensure the smooth progress of distributed computing. An organization called the information center usually performs these activities. In some organizations, this entity is known as the help desk. Usually reporting within the IT department, this important unit maintains close contact with client groups, supporting their use of information systems. Table 11.4 lists information center activities.

TABLE 11.4   Information Center Functions

Conduct or provide user training

Provide development assistance

Evaluate new applications

Distribute customer information

Collect trend information

Determine problems

Gather planning information

The info center trains employees on workstation hardware, software, and procedures, or coordinates training activity provided by others. Most firms' formal training programs for end users mix internal trainers, contract trainers, and outside classes coordinated by the center. The info center also assists clients in program development and trains them in business controls and recovery management techniques.

The center evaluates new application programs and other software; it distributes information on client-developed programs. The info center serves as a first-level clearing house for meeting software requirements of distributed computing. Determining trends in software requirements is also an important center function. It values trend information for establishing future computing requirements and for shaping strategies and plans.

The center provides guides and assists clients with software and hardware migration and upgrades and helps prevent duplicate development activity. An effective center encourages clients to seek advice before taking significant action and coordinates computing activities among client departments.

In addition, the center maintains a telephone helpline and performs initial problem determination. The center answers user questions, even those that seem trivial to experts. It is a place to contact when things go astray, as they occasionally do. When other sources of help are not available, the info center is the client's safety net. Adopting a helpful attitude toward clients is this group's key to success.

The info center plays a central role in distributed computing. It complements the workstation store. It solves or averts problems that the firm is likely to encounter during transitions. In addition to the important task of employee training, the center must train managers on development processes and control issues. Management skills and employee proficiency with new tools are learned traits. The importance of training looms large.

Intranet and client/server technology itself offers many opportunities for the info center to provide effective services. Using network technology for the center's purposes, it can enroll users in training sessions, distribute important information to clients, post solutions to typical user problems, answer frequently asked questions, seek user input, gather planning information, and accomplish many other tasks to help users. In its role as an information provider, the center should not overlook the technology it's promoting.

Successful distributed computing requires skilled management actions and adoption of structural changes to reduce risk and capture benefits. Most firms benefit substantially from network technology and experience significant change in the process. Not only do the firm's operational units become more flexible and more responsive, but the firm's employees become better equipped to assume more responsibility and are more productive. Therein lies the payoff.

POLICY CONSIDERATIONS

Governance, the art of achieving consensus on policy matters, is very important in distributed computing. Organizations must make several important policy decisions when implementing distributed computing. Some policies smooth the way for technology adoption and encourage client computing; others affect computing costs or fundamental concerns such as business controls, security, or employee well-being. Because of their long-term importance to the firm, these policy matters warrant the attention of senior IT executives and others.

Hardware and software compatibility is the first and most important policy issue. For example, the firm must decide whether to adopt one or more of several popular spreadsheet applications currently available. Adopting one reduces costs and simplifies training but may reduce overall capability. Should the firm adopt one word-processing application, or should it permit employees to choose from several? Limiting the choice to one ensures easy document interchange and reduces training and cross-training. Because each application offers many choices, developing application policies is often complex. Letting the issue resolve itself by default, however, is an irresponsible alternative that leads to problems later.

Selecting an application policy paves the way for a hardware compatibility policy. Again, restricting hardware options to clients reduces costs, improves maintenance, and makes migration to future systems easier. Most firms limit hardware options to one or two popular brands but may purchase compatible models or clones from several manufacturers. Some firms maintain a small advanced technology group to evaluate new hardware and software, thus ensuring that new technology developments are known and well understood.

Firms must develop policies on ownership and control of distributed hardware, applications, and data. Individual or local ownership of these items improves security and reduces the risk of loss or damage. Local or individual control may discourage sharing of these assets and may result in negative organizational consequences. Executive policy must choose among conflicting factors.

BUILDING CLIENT/SERVER APPLICATIONS

Planning for client/server applications is a continuous process: The task of introducing and implementing new tools to empower knowledge workers seems never-ending. Because client/sever applications are associated with or drive business process innovations, they must be part of the firm's strategy and planning process, just as restructuring and reengineering are.

Many planning questions address issues such as what technology to introduce next, where to begin its introduction, and when and at what speed changes should occur. Sound planning answers many of these questions.

To succeed, managers responsible for installing client/server, intranet, or other networked architectures must use several planning methodologies, recalling basic concepts such as Nolan's stages of growth, critical success factors, and business system planning methods. Because information technology already deeply penetrates many firms now implementing networked architectures, eclectic planning methods are appropriate. These firms operate in a complex environment. Their planning must be overt, systematic, and well developed; their distributed-system planners must be proactive and have strong strategic perspectives.

Client/server programmers must have strong skills in new tools and techniques, too, because client/server infrastructures change application development in many important ways. For example, object-oriented techniques are superseding structured-programming techniques in client/server development; PowerBuilder, Delphi, and variations of C++ replace COBOL, formerly the language of choice; and knowledge of unique database systems such as Informix, Oracle, and Sybase is critical for database administrators. After developing client/server applications, critical implementation tasks must begin.

IMPLEMENTATION CONSIDERATIONS

Implementing client/server, Internet technology, or other network systems poses many challenges for IT and client organization managers. Some concerns are familiar to centralized operations' managers; however, others are new to the firm. Generally considerations can be grouped into four categories—organizational factors, information infrastructure, systems management, and management issues.

Organizational Factors

The most critical factor in introducing networked systems is a clear perspective on why the firm is changing system strategy. Senior executives, line managers, and IT managers must share a unified perspective that emerges from carefully examining the firm's business practices. Managers' desires to make operations more effective and efficient drive this thoughtful, introspective examination. Its results should be incorporated into business plans.

After identifying new operational processes, managers should define information technology in detail and introduce it to streamline process effectiveness. This task is iterative because new business processes are constantly redefined consistent with available IT capabilities. IT and business managers must cooperate to develop optimum plans that consider technology and business conditions. This collaboration results in new business methods and technology applications.

Today, most new business processes lead to new, flatter organizational structures because technology empowers employees with more responsibility and better tools. Thus, the plan to introduce client/servers, Web technology, or interorganizational systems must include transition plans for the new structure and changed responsibilities. Senior executives must be attentive to this rebalancing because ignoring its effects on people usually leads to severe problems later. Reengineering, restructuring, and advanced technology is a powerful mixture that must be handled carefully to achieve the best results.

Client/server and other types of distributed computing also distribute information technology activities and expand IT's scope. IT must provide training in system development and operation, develop distributed system architectures, and provide development tools and other support. Although distributed programming physically delivers applications or parts of them to using departments in client/server environments, program maintenance and enhancement frequently remain IT's responsibility.

Information Infrastructure

Implementing client/server or other networked operations implies alterations to the firm's information infrastructure that involve hardware, software, and databases. As Figure 11.1 shows, the hardware architecture includes individual workstations networked to servers that may themselves be linked to more extensive systems and/or to external networks. Selecting and connecting these components is a complex systems engineering task requiring detailed knowledge of proposed applications, estimates of possible future applications, load factors, response times, scalability or expansion capability, and other technical details.

Because the firm already owns or leases hardware and network components, additional equipment and services must fit those already in place. Numerous vendors provide equipment, software, and services to support client/servers. Firms must carefully select from among many options because they will live with the consequences for a long time. Claims of open systems must be fully investigated because interoperability must be a reality, not only for the present but for the future too.

Obviously, the operating system managing the client and the server hardware must support the firm's applications. The hardware and system software selected must support database management systems that the applications need. And the database architecture depends on the application, the defined infrastructure, and the firm's future needs. As a general rule, the tendency to underestimate system data-storage requirements is high because, as users gain power, they find advantages in implementing new tasks. Wise managers plan sufficient capacity for now and allow for future expansion.

Systems Management

Many distributed system management tasks are like those associated with centralized systems and networks. Managing problems, changes, capacity, performance, and developing emergency plans and recovery actions are equally important to departmental systems and centralized operations. Later chapters discuss these topics, as well as network management. However, some tasks are unique to client/server and similar networked operations. Most involve system user actions. These important tasks include software introduction and control, workstation security, password management, license requirements, and software distribution.

Some tasks require employee training and others attentive manager supervision, most of which the information center and workstation store can support. These are new, unfamiliar tasks for employees whose primary responsibilities lie elsewhere. Consequently, successful client/server operations require strong IT support. Regardless of hardware and network ownership, successful operations require shared responsibility. IT and operating department managers' results should be jointly measured.

MANAGEMENT ISSUES

Understanding needed investments and evaluating their returns is one of the most difficult management issues in distributed computing because the value of network computing systems lies mostly in increased organizational effectiveness.[17] For reasons noted earlier, benefits are difficult to quantify. Returns are highly judgmental, intangible, and generally not measurable with typical financial tools. Unfortunately, quantifying the investment itself is equally difficult for most organizations; thus, the tendency to underestimate costs is high.

Research indicates some difficulties in tying client/server implementation to cost savings. The question posed in surveys of 500 or more large companies was, "Will migrating to client/server save you money?" In 1992, 53 percent responded no; in 1993, 76 percent said no; and in 1994, 81 percent said no. Some surveyed admitted that the technology had been oversold and that their firms had unreasonably high expectations.[18] Most firms believe client/server technology ultimately delivers benefits, but migrating to it does not immediately save costs.

Total client/server costs over five years can exceed $48,000 per client, according to a Gartner Group study. In a large, complex network, the cost per end user can range from $50,000-$60,000, a large increase from the $2,000-$4,000 usually associated with desktop systems. The study revealed that labor costs constitute more than 70 percent of client/server expenses.[19] Table 11.5 lists the costs this study identified.

TABLE 11.5   Client/Server Cost Elements

Cost Element	Percent
End user labor	41
End user support labor	15
Application development labor	8
Enterprise server operation and other labor	8
Education, training, and professional services	7
Purchased applications software	3
Wiring and communications	3
Client and server hardware and software	15

The report advised that reducing system complexity, installing development tools such as client/server CASE, and developing and implementing architectural guidelines can reduce costs somewhat. Many other studies validate these conclusions and show that hidden user costs total about 50 percent.[20]

End user spending on information technology is large and growing in the U.S. and other countries. BIS Strategic Decisions, a Norwell, MA, firm, expects end user spending in the U.S. to reach $53.2 billion in 1999, up from $37.3 billion in 1994.[21] Much of this spending is in addition to that of the central IT organization, which spends increasing amounts on client-department systems. Financial considerations, important in distributed computing today, grow as firms attempt to capitalize on intranets, extranets, and other interorganizational computing.

People Considerations

Installing client/servers or other types of distributed systems generally alters people's behavior and their attitudes toward their jobs. Individual roles and responsibilities and the organizational reporting structure usually change too. For example, some firms that installed distributed computing systems reassigned IT analysts and programmers to the client manager so their IT skills could directly assist that business unit's implementation. The opposite occurred at Merrill Lynch. Although IT people continued to work in the business units, they reported to the centralized organization so implementation would be uniform across various business units.

Client/server computing directly affects traditional reporting relationships because empowered employees are more like the symphony orchestra musicians Drucker mentions in Chapter 1. Loyalties shift somewhat from the immediate manager to the function and to peers. In procurement, for example, loyalty to the function may become stronger as loyalty to the procurement manager declines somewhat. This is a direct consequence of empowering employees and increasing the managerial control span.

Today's integrated organizations are far more complex than those of 20 years ago. They are more difficult to manage because, like the symphony conductor, the manager must be coach, counselor, and teacher to skilled and empowered employees. Technology changes, administrative reorganizations, and the nature of the work itself greatly impacts the new organizational environment. Accordingly, firms must address human concerns of distributed computing. Their plans should focus on employee communication. Managers must inform employees about system changes, obtain and use employee input on planned changes, explain changes in policies and procedures, and respond to employee grievances promptly. In a rapidly changing work environment, overcommunicating is almost impossible.

Managing Expectations

When first deciding to install client/server computing, organizations should select an operation that can serve as a prototype for future implementations. In the initial effort, predicting detailed results is sometimes difficult and can alter preliminary conclusions. Feasibility studies provide insights for future installations. For instance, if client/servers are planned to support shipping and receiving at distribution centers, cautious managers test the plan at one location, thus improving plans at subsequent locations. The prototypical installation is a testbed for the technology, reinforces planning details, and serves as a valuable training site for subsequent installations. Installing distributed computing usually results in difficult-to-foresee consequences. Feasibility studies or prototyping provides confidence for later implementations.

Planning and implementation teams must manage expectations carefully.[22] Prototyping efforts must proceed with senior managers' support and understanding. Their support indicates confidence that the results will improve the firm's operations. Individuals involved in the installation must be trained and prepared to adopt change and cope with disruptions in routine workflow. Managers must tolerate infrequent difficulties and permit technology to be implemented according to plan without undue pressure or overconfidence.

Prototype installations help managers solidify plans to introduce technology throughout selected functional areas. Feasibility studies and prototyping also focus on current applications and assist in developing cost and benefit information about new technology applications. Prototypes also confirm software and hardware choices, physical installation preparation, and final practices and procedures accompanying new technology implementation.

Human factors are highly important to successfully introducing distributed computing. Software must be easy to understand and use. Smooth interactive capability and well-constructed option menus characterize easy-to-use software. Well-designed systems let users alternate easily between products or functions with the keyboard, mouse, and a graphics screen. Some current windowing systems' splendid design makes them easy to use. One window can run tutorials and help sessions while others implement function; this lets users easily learn the software by using it.

Other environmental factors are also important. The physical environment, including lighting, seating, and dimensioning, must be carefully reviewed to reduce physical stress. Facilities planning must include sufficient space and address other human factor considerations such as noise levels, temperature, humidity, and office appearance. Careful attention to important human factor issues reduces resistance to change, speeds implementation, and improves morale.

Although sophisticated information technology in the workplace entails many risks, the information center and workstation store can avert most. Business controls, data security, and other control issues must be resolved. In its staff role, the IT organization provides guidance on these issues to ensure their consistent treatment throughout the firm. In many of these matters, IT managers have staff responsibility for client/server operations just as for other distributed computing activities. Successful mutual interactions between client/server implementors and IT staff members assist considerably in ensuring systems success.

Change Management

During client/server or other distributed installations, change management can be divided into three phases. During the first phase, when prototype installation is underway and innovative employees are being trained, managers should publish progress bulletins and conduct information meetings for other employees. When managers carefully explain system operations and their impacts, they can establish realistic expectations about later events. Information center personnel should answer employee questions, and managers must respond candidly to employee concerns. In particular, they must promptly address issues that may impact morale.

In the second phase, when employees begin to use the new tools, experienced employees should assist beginners. Additional staff must be available to offset the reduced output of employees who are learning new operations. Training activities must explain each employee's place in the system and system's place in the plan for the restructured environment. Some users will experience frustrations with the system. Managers must be especially sensitive to their concerns and take steps to reduce or eliminate their problems. Managers should provide frequent, positive feedback to employees as they progress.

During the final phase, employees should be competent in using the new technology and should demonstrate confidence in it. They should be able to recommend or initiate improvements to the applications and their use. Managers must always encourage and try to adopt innovative ideas.

Some employees will experience difficulty adapting to changing work patterns and to the structural changes that usually accompany them. These employees can achieve success if managers clearly explain the reasons for change, help them plan for the personal effects of change, and let them appreciate the benefits of change. Manager and employee flexibility lets the organization respond to new opportunities. Successful change management strives to overcome personal anxieties and reduce apprehension. Ideally, resolving these initial problems leads to genuine enthusiasm about the changing environment.

Managers should recognize that changes can occur in technology, organizational structure, management style or technique, and organizational culture. Almost all technological and organizational changes shift employee attitudes or behavior patterns. When technology creates jobs with higher skill requirements, traditional occupations are upgraded, modified, or eliminated, creating employee stress.

When managers make changes affecting employee attitudes or behavior, success or failure of the changes may well depend on whether workers perceive personal rewards or benefits. It may also depend on whether employees can acquire new skills and want to learn them. The scope of communication and the quality of orientation and training play important roles in managing change successfully.

Several points merit emphasis. All individuals will not enthusiastically welcome the impact of change—change threatens some people. Challenge comes from the need to learn new skills and begin again. Attentive managers will cope with this difficulty through effective people management practices, such as one-on-one communication, coaching, counseling, and training. Successful managers deal with problems individually. Recognizing that each individual struggles with change in his or her own way is the manager's most important task.

THE INTERNET AND INTRANETS

Applications of Internet technology grow rapidly in today's organizations.[23] Many firms use the Internet for electronic commerce, and many more use Web pages for advertising and many other purposes. Studies project that corporate spending on Internet commerce software will exceed $1 billion in 1999. Although most firms and many individuals believe they must maintain a presence on the Web, not all claim financial gain from the Internet. However, its popularity evidences that for business firms, especially large ones, the Internet is an enormous success.[24]

The World Wide Web creates a revolution in client/server computing because businesses are installing local Web sites in secure client/server infrastructures for internal purposes. Known as intranets, these local Webs are becoming powerful communication systems. Stressing their importance, David Linthicum writes, "I think this is the single most significant change in the way we are building client/server systems, and it's important that client/server developers learn how to leverage the power of Web-enabled technology."[25] Today, about two-thirds of Fortune 500 companies use intranet technology, according to International Data Corporation.

Many organizations find using Internet technology internally is increasingly valuable for delivering information to the desktops of people who need it. For example, firms load widely used company documents such as corporate standards, operating procedures, benefit plans, and internal phone or address books on internal Web sites, making them accessible to employees. Intranets make other current information such as company news, corporate newsletters, training information, and much more easily available. Only imagination limits their use.

Intranets are also useful departmental or functional communication aids. Remembering that over-communicating with employees during stressful times is almost impossible, managers can use intranets to discuss planned workplace changes and receive employee feedback. This important tool helps managers learn about employee concerns and respond with appropriate information. Additionally, routine information such as sales results, performance figures, and production goals can alert employees to the firm's performance, promoting loyalty and improving morale.

However, intranets are not exclusively for internal use. Firms can share some important internal information with selected outside concerns, too. For example, procurement managers can advertise bids to appropriate vendors; manufacturing managers can inform subcontract firms of parts requirements; and subcontractors can keep manufacturing informed of delivery schedules. With care, firms can perform all these communications without compromising the security of internal operations. This kind of communication promotes good vendor and supplier relations.

The Internet is an important, rapidly evolving technology in today's business world. Like most new technologies, it comes with considerable hype—things are not always as advertised. Managers must analyze the many possible Internet applications for their potential value. As with all technology, it pays to experiment some to better understand possibilities and limitations and to anticipate opportunities and pitfalls. Undoubtedly, the Internet's value and importance to global businesses will grow substantially in the future.

SUMMARY

Networked application systems are one of the most important developments in information technology's history. Professional employees in modern firms have immediate and convenient access to large digital computers. There are many reasons to take advantage of these developments.

To capitalize on distributed computing, the firm's managers must select and implement software and hardware systems. They must engineer telecommunication systems to support the new environment. In addition, managers must deal successfully with financial issues; they must resolve business controls problems, data management and data ownership issues, and, most importantly, they must handle people's concerns skillfully.

Healthy, active information centers and workstation stores provide organizational resources to deal with many important issues. These organizations are the IT manager's support staff, helping to manage distributed computing and its many business, technical, and organizational transitions.

When firms adopt distributed computing, IT organizations themselves experience significant change. Information technology managers' jobs encompass greater staff responsibilities. IT's role in the firm expands, translating into more responsibility for IT people as well. The workstation store and information center create several new and exciting jobs. Employees in client and IT organizations should be given the opportunity to fill these jobs. With an active plan for job rotation, many employees can benefit from these opportunities.

Distributed computing offers high potential payoff but poses many significant risks. When managed properly, it is a win-win situation for everyone. The firm gains substantial benefits, managers gain greatly improved resources for dealing with problems and opportunities, and employees gain new, more valuable skills and increased productivity. When managed properly, its benefits far outweigh the costs and efforts expended.

Review Questions

1. What changes in the computing environment are taking place at Merrill Lynch, according to the Business Vignette?

2. What did Merrill Lynch do for its consultants to make the new system appealing and easy to use?

3. What factors encourage the trend toward distributed computing?

4. What does distributed computing mean, and what role do advances in telecommunications play in it?

5. Define the terms reengineering, outsourcing, and downsizing.

6. What are the advantages of downsizing mainframe applications, and what are its disadvantages?

7. What issues or problems must the firm solve for distributed computing to succeed?

8. Why is compatibility such an important consideration in implementing distributed computing?

9. What questions arise about databases during the installation of client/server computing?

10. Distributed computing shifts some problems normally faced by mainframe operators to client departments. What are these new problems for client managers?

11. Describe how the importance of distributed computing issues varies over time.

12. What functions does the workstation store perform? How does the workstation store help maintain sound business controls?

13. What are the information center's responsibilities?

14. To what department does the workstation store and information center report in most firms? What departments provide staffing for these organizations?

15. What is the relationship between the store, the information center, and the firm's planning process?

16. What are some policy issues accompanying the thrust toward distributed computing?

17. What are the connections between client/server computing and information infrastructure?

18. What percentage of client/server computing costs are labor related? Why does this make cost control difficult?

19. Why is it important that managers deal with employees on an individual basis when implementing workplace changes?

Discussion Questions

1. What steps can be taken to ensure realistic expectations when implementing distributed computing?

2. The text presents a number of factors motivating the trend toward distributed computing. Which factors most relate to the firm's competitive posture and why?

3. Earlier chapters discussed advances in hardware and telecommunications. How do these advances increase or decrease risks associated with distributed computing?

4. Firms need to establish policies concerning the use of purchased applications and user-developed applications. What policy questions do these programs raise?

5. Client/server operations increase management responsibilities in using departments, especially regarding data and business controls issues. What additional considerations do managers need to address in these areas?

6. What alternatives are available regarding who should own workstation hardware and purchased software? Discuss the issues involved.

7. Discuss the pros and cons of funding workstation hardware and purchased software at the corporate level vs. the individual department level.

8. What financial advantages can an effective workstation store capture? Can you make some estimates to quantify these benefits?

9. Elaborate on the notion that the workstation store and the information center provide a "leading indicator" function for the IT organization. How does this relate to IT planning?

10. What benefits can the organization and its people achieve by the staffing opportunities presented in the workstation store and information center?

11. Why is securing senior management involvement during reengineering activities important?

12. Discuss the potential organizational implications of reengineering, outsourcing, and introducing client/server operations.

13. Discuss the difficulties in establishing investments and returns in client/server implementations. How can executive managers help in this task?

14. How can managers ease the transition for employees into the client/server environment? What approaches might succeed most with those who tend to resist?

15. Upon successful implementation of client/server computing, IT manager responsibilities include an expanded staff role. What skills are useful in discharging these new responsibilities? What measures of success can the manager employ in these endeavors?

16. Effective client/server computing empowers employees and increases management's control span. Discuss implications of these changes for managers and employees.

17. Wireless networks and portable workstations promote the "logical office" concept, i.e., the office is where the workstation is. What opportunities does this concept present?

18. Discuss opportunities for firms giving suppliers intranet access. What new considerations does this scenario introduce to the firm's application development department?

Assignments

1. Client/server applications are now built with CASE tools and object-oriented techniques. Research this topic and develop your own thoughts on what this means for employees in operational departments and IT organizations. How might your conclusions relate to the IT organization's structure?

2. Visit a firm using distributed computing. Interview an operating department manager, and develop an opinion on applications of distributed computing for colleges and universities. Discuss how your school could further employ intranet technology.

3. To compare your school with others, you may want to visit the Web sites for Cornell, Michigan, Mississippi, or Stanford at cornell.edu, umich.edu, olemiss.edu, or stanford.edu.

ENDNOTES

[1] Jennifer L. Janson, "PRISM Produces Productivity Spurt for Merrill Lynch," *PC Week*, October 1, 1990.

[2] Alice LaPlante, "Merrill's Wired Stampede," *Forbes ASAP*, June 6, 1994, 76.

[3] Craig Stedman, "Bullish on Technology," *Computerworld's Premier 100*, September 19, 1994, 33.

[4] Jaikumar Vijayan and Tim Ouellette, "Big Iron Gets a Case of Web Fever," *Computerworld*, January 26, 1998, 1.

[5] Thomas Hoffman, "Merrill Lynch Fights Off Recruiters," *Computerworld*, January 12, 1998, 26.

[6] *Computerworld's Premier 100*, September 19, 1994, 46-53. Client/server development expense varies widely from zero to 100 percent of software development expense among these top 100 firms.

[7] Clifford Stoll presents an interesting view on the Internet's value in *Silicon Snake Oil* (New York: Doubleday, 1995).

[8] Someone once said that you can't reengineer what wasn't engineered in the first place, i.e., most business processes did not result from engineering activities.

[9] Brad Bass, "GSA Mulls Outsourcing of Business Activities," *Federal Computer Week*, January 23, 1995, 3.

[10] Tom Davenport, *Process Innovation: Reengineering Work Through Information Technology* (Boston, MA: Harvard Business School Press, 1993), 300.

[11] Patricia Seybold, "The Sorry State of Systems Management," *Computerworld*, March 4, 1996, 37.

[12] Today, many CASE development tools are available and immensely helpful to sound client/server application development.

[13] Estimated annual user costs rise from $5600 to $9640 when moving from mainframe to client/server processing according to the article, "Why Mainframes Aren't Dead Yet," *Fortune*, April 17, 1995, 19.

[14] For more on this important topic, see James C. Brancheau and James C. Wetherbe, "Understanding Innovation Diffusion Helps Boost Acceptance Rates of New Technology," *Chief Information Officer Journal*, Fall, 1989, 23.

[15] Richard Walton, *Up and Running: Integrating Information Technology and the Organization* (Boston: Harvard Business School Press, 1989).

[16] Patrick Dryden, "Cover Your Assets," *Computerworld*, May 26, 1997, 6.

[17] J. William Semich, "Can You Orchestrate Client/Server Computing?" *Datamation*, August 15, 1994, 36.

[18] Source: Sentry Market Research, Westboro, MA, *Computerworld*, November 7, 1994, 71.

[19] Rick Whiting, "C/S Labor Costs Exceed Technology Costs," *Client/Server Today*, July 1994, 20.

[20] B. Caldwell, "Client-Server Report," *InformationWeek*, January 3, 1994. Caldwell reports total five-year expense per user to be $40,400, 18.7 percent in direct costs, 31.7 percent in indirect costs, and 49.6 percent in hidden user costs.

[21] Source: BIS Strategic Decisions, Norwell, MA, *Computerworld*, August 8, 1994, 90.

[22] Julia King, "Keeping a Grip on User Expectations," *Client/Server Journal*, November 1993, 46-50.

[23] Internet households are also growing and will double to 29.4 million from 1996 to 2000; 27 percent will have some form of high-speed access. Joshua Levine, "Good-bye, Wing Tips, Hello, High-top Sneakers," *Forbes*, September 8, 1997, 42.

[24] Large businesses have universally adopted Web technology. However, according to the National Federations of Independent Business, 70 percent of small businesses have computers, but only 40 percent are online. *The Wall Street Journal*, August 21, 1997, A1.

[25] David Linthicum, *Client/Server and Intranet Development* (New York: John Wiley & Sons, Inc., 1997), xii.

Part
Four

Tactical and Operational Considerations

Business operations critically depend on effective and efficient operation of information and telecommunication systems. Disciplined management processes for routine business system operations are key success factors for IT managers. Systematic techniques to deal effectively with numerous operational issues and potential problems are needed in this complex arena. Part Four begins with Developing and Managing Customer Expectations and continues with Problem, Change, and Recovery Management, Managing Systems, and Network Management.

This part develops systematic techniques for handling operational issues. These techniques enable IT managers to attain high levels of customer satisfaction from their information systems.

12 Developing and Managing Customer Expectations

John Singleton began experimenting with service-level agreements and defining his Management by Results program when he was vice president for operations and data processing at Maryland National Bank in Baltimore. Upon moving to Security Pacific as IS Chief, he refined and implemented these concepts with outstanding results. Under Singleton's leadership, Security Pacific Corporation reduced IS expenses, expanded its business, and improved IS performance. His accomplishments paved the way for his promotion to vice chairman of Security Pacific and chairman and CEO of Security Pacific Automation Company.[2]

Upon joining Security Pacific, Singleton was assigned to control IS spending, which was then growing nearly 30 percent annually. Within a year, he reduced annual spending growth to about 10 percent, bringing it more in line with average rates for IS organizations of its size. He reduced spending growth by implementing his Management by Results program and by centralizing the IS organization, contrary to current trends toward downsizing and dispersing IS activities.

Singleton believes that centralized IS organizations have significant cost advantages over downsized operations and can better maintain control over IS activities. In addition, centralized IS operations offer bigger jobs to IS professionals and let firms retain critical IS management talent. Decentralized operations fragment the job and create isolated talent pools. These conditions encourage aggressive managers to seek advancement opportunities outside the firm. To succeed in large companies, however, centralized IS operations must respond to diverse and remote business units' needs. This is where the Management by Results program flourishes.

The program has four related components: IS strategic planning, service-level agreements between IS and users, IS performance tracking and reporting, and personal performance appraisal and salary programs. The goals of Management by Results are to link IS business plans to users' plans, understand customer needs exactly, contract with users to fulfill their needs, report accomplishments and results to users and IS people, and tie individual accomplishments to salary compensation. The program ensures that IS people and their organization are committed to achieving company and user objectives.

Security Pacific Automation prepares an annual strategic plan tuned to user goals. The company uses a steering committee composed of board members from the parent organization to oversee IS activities. For example, Singleton reviews budget requests with the steering committee before going to the full board. To do this, he uses IS achievements obtained during the measurement process. This approach ensures that top level managers agree with IS strategies and plans and virtually guarantees a strong link between IS strategies and overall firm strategies.

Service-level agreements (contracts between IS service providers and users) are central to the Management by Results system. IS works with each user group to establish exactly what they need and how they measure delivered performance. IS and users agree on performance standards, performance calculations, and methods for reporting unsatisfactory results. The agreement also links IS costs and performance because Singleton believes cost effectiveness is a common goal. He believes that a balance exists between performance, quality, and customer relations. Agreements must reflect the balance among these critical performance areas. IS managers and users sign the final document, attesting to their mutual agreement. Security Pacific has more than 500 active service-level agreements.

Security Pacific Automation achieves high credibility with users by measuring and reporting all IS activities. The program sets standards and permits users to evaluate IS performance regularly and frequently—sometimes daily. User evaluations of IS performance help improve objectivity. Ninety-four percent of Singleton's users rated his performance excellent or above average, a tribute to the program's effectiveness.

Individual performance appraisal and salary increases are tied to IS performance, thus ensuring personal commitment to achieving results. Measurement techniques similar to those in the service agreements let managers track individual performance. Employees are measured on managing service levels, budgets, personnel, and new business. Singleton believes this feature helps employees take responsibility for directly managing their careers.

IS employees generate proposals that Security Pacific Automation uses to bid for customer business. This in-house IS division competes with outside vendors for client services or equipment. Largely because it is cost competitive and committed to results through service-level agreements, the Automation Corp. captures business from outside vendors. Additionally, when IS underspends its budget or profits from outside sales, it returns the excess funds to its internal customers. This strategy reinforces and cements relations between IS and its customers.

The Management by Results system yielded high returns at Security Pacific. It substantially reduced in-house staff and cut expenses five percent over two years, even as the company itself grew through acquisition. During the past several years, Security Pacific purchased The Oregon Bank and The Arizona Bank. Cost containment at Security Pacific Automation partially financed these acquisitions. Electronic transactions grew 20 percent in three years; nevertheless, Security Pacific capitalized on its IS effectiveness again by absorbing the data processing departments of three out-of-state banks with combined budgets of $100 million.

When Singleton joined Security Pacific, unhappy users were his most serious concern. At that time, the company believed that IS incurred unreasonably high costs, performed inadequately, and finished projects late. In short, IS had low credibility with users, who wanted improvements quickly. If Singleton had been unable to correct the situation within six months to a year, users were prepared to dismantle the IS operation, divide the pieces among themselves, and take charge of their own destiny.

Singleton moved rapidly but IS managers and users alike resisted his program. Some managers believed their responsibilities were immeasurable, and users resisted centralization efforts until benefits become apparent. IS people unable to adapt left the company voluntarily or were terminated. Still, Singleton demonstrated that centralized IS organizations can be flexible, responsive, and effective. He believes that IS must lead in reengineering the business and restructuring old ways of doing things.

For his accomplishments, Singleton received an Information Systems Award for Executive Leadership from the John E. Anderson Graduate School of Management at UCLA. The accounting firm of Peat Marwick Mitchell & Company found Security Pacific's IS expenses to be about 10 percent lower than those of comparable operations. John Singleton's pioneering work at Security Pacific set an example that many firms now follow as they seek to measure and manage IS strategic resources effectively and manage expectations within the firm.[3]

INTRODUCTION

Application program acquisition, development, and enhancement and managing important databases involve many long-term, strategic issues. Although long-term concerns are critical, applications also raise many important short-term issues. Because the operation and use of applications and their databases define the firm's *modus operandi* in many instances, they demand serious management attention in the near term and beyond.

This chapter focuses on tactical and operational concerns arising in the IT organization's production operation. Production operation is the routine implementation and execution of application programs at centralized mainframe sites or at locations widely dispersed throughout the organization. Some of these applications are nearly continuously online, supporting the firm's mission in many ways. Operating interactively, they serve hundreds, perhaps thousands, of users. Many other systems operate periodically—daily, weekly, or at month's end according to predetermined schedules. Scheduled production work is a critical operational activity that also has important longer-range implications. Because the firm's essential business activities depend so heavily on these production activities, these applications must operate reliably, delivering high quality results. One of IT's critical success factors is to conduct this activity competently.

The first and most important step in building a disciplined approach for managing production activities successfully is to establish expected information service performance levels. This chapter formulates important service-level concepts and describes how IT and its customers can reach agreement on service levels. It describes components of comprehensive service-level agreements and outlines methods for acquiring customer satisfaction measures. This chapter's central themes are developing, managing, and satisfying customers' operational expectations.

TACTICAL AND OPERATIONAL CONCERNS

Adopting disciplined techniques considerably strengthens the management of computer and network operations, whether centralized mainframes, department minis, client/server systems, or Internet operations. Organizations achieve success managing computer and network operations by carefully and systematically managing several important operational processes.

Computer operations are highly visible to many employees who can quickly observe failures when they occur. For example, when server failure in a distributed processing environment causes failures throughout the network, many notice. When the central processing unit supporting the office system fails, everyone promptly notices from executives to warehouse employees. A failing Internet application is perhaps even more widely visible than expected. The organization's production operations are vital to its business and also represent its image to many within, and perhaps to others outside.

In contrast to routine computer operations, a strategic planning error or omission, although perhaps more devastating to the firm long term, may not become apparent for months or even years. However, what happens today and what will happen next week or next month dominates production operations managers' lives. More than others, computer system managers must rely on procedures and disciplines to cope with the wide variety of challenges that come their way. To succeed, operations managers must use management systems that provide tools, techniques, and procedures to maintain stable, reliable, and trouble-free operations.

EXPECTATIONS

This text's early chapters stressed the need for IT managers to meet expectations that the firm's senior executives hold. Fulfilling executives' expectations, whether they are realistic or not, is the principal standard for measuring managers. But, expectations have many origins and are established through various means. Some evolve from sources external to the firm, from meetings with vendors, articles in the trade press, and business association meetings. Numerous internal factors, such as information technology budgets or IT's prominence and visibility within the firm, reinforce these externally developed expectations in executives' minds. In many cases, IT organizations themselves help set expectations, sometimes even encouraging unreasonable ones. Expectations may not always be soundly based; nevertheless, given their important role, organizations must use rational management techniques to help set and meet realistic expectations.

Part Three discussed tools, techniques, and processes designed to handle expectations surrounding the application portfolio. Although these processes occur in the intermediate timeframe when tactical plans are established, they also have strategic implications. This chapter formulates tools, techniques, and processes for developing and managing customer expectations for the operation of systems and applications. These techniques apply principally to the short term. They include processes for achieving service-level agreements between the IT organization and all its clients. These agreements establish acceptable service levels for IT's clients and include mechanisms for demonstrating the degree to which IT attains service levels.

One very important result of the service-level agreement process is that the entire firm clearly understands what is expected from and delivered by computer center operations. To achieve this result, the process must include an obvious means for recording and publicizing service levels delivered. Techniques that focus on how well the organization achieves its objectives are crucial, even though some objectives need modification and some are not completely fulfilled. In addition, clearly and unambiguously understanding how organizations intend to accomplish objectives helps eliminate debate and confusion within the firm about what the measurements really are. This chapter discusses tools, techniques, and methods to meet these goals by establishing agreements on service levels to be delivered and how to report service levels actually achieved.[4]

THE DISCIPLINED APPROACH

Service-level agreements (SLAs) are the foundation of a group of management processes collectively called *disciplines*.[5] Disciplines are management processes consisting of procedures, tools, and people organized to govern important facets of computer system operations, whether located in the IT organization or client organizations. Figure 12.1 depicts relationships among the processes comprising system operational disciplines.

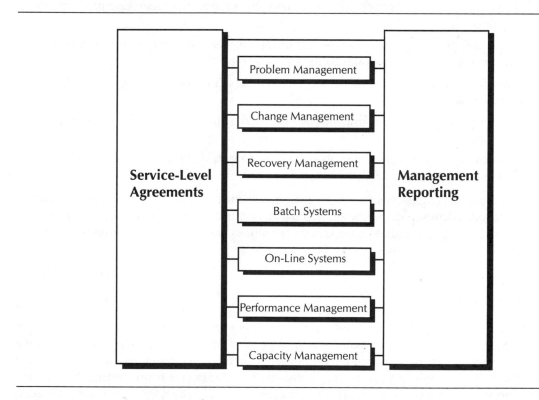

FIGURE 12.1   System Operational Processes

Computer center performance can best be judged against specific, quantifiable service criteria. Negotiated service agreements establish client-oriented performance criteria—the most realistic criteria for assessing performance. Service control and service management techniques depend critically on these criteria, so they must be clearly and carefully established.

Specific management reports reveal computer center performance to the organization. Although service-level agreements must involve all client organizations, reports focus on specific organizations. Management reports are an essential tool for computer operations managers because they organize and display results of specific intermediate disciplines. Managers of client organizations also depend on them. Reports are important to clients because they display information documenting the status of service they receive and actions taken to correct problems they experience and to improve system performance or capacity.

To provide superior customer service, computer operations managers must control all computer center activities. They must consider the following items or disciplines.

1. Operational problems, defects, or faults that cause missed service levels and additional resources expenditures.

2. System-wide change control, critical because mismanaged changes create problems and impair service.

3. Managers' plans to recover from service disruptions that inevitably result from system performance defects or uncontrollable outside events.

4. Plans for scheduling, processing, and delivering results of batch and online workload to client organizations.[6]

5. Maintenance of system performance or throughput at planned levels to meet service agreements.

6. Planned capacity needed to meet workload demands, which requires that disciplines 1 through 5 function correctly.

All these items are essential to successful computer system operation on mainframes, departmental, client/server, or corporate intranet systems. Subsequent chapters treat each in more detail. Collectively, these processes or disciplines are the tools and techniques that successful computer operations managers need.

SERVICE-LEVEL AGREEMENTS

Although leading-edge organizations have used service-level agreements for two decades or more, SLAs continue to gain popularity as management tools for establishing and defining the customer service levels provided by IT, other computer center operators, and other service providers. "Increasingly, service-level agreements are becoming the major underpinning for the way services are delivered to internal IS customers," says Naomi Karten, President of Karten Associates.[7] SLAs' expanding popularity stems from the need to reduce departmental conflicts, level users' expectations, place IT services on a more business-like basis, and reduce threats from competitive suppliers or outsourcing organizations.

Threatened by outsourcing and other options available to user departments, IT, now more than ever, turns to SLAs to smooth relations over service levels and their associated costs and to establish measurable business reasons to avert outsourcing. "Companies really need a business case to keep things in-house, and one of the best ways to do that is through SLAs," said Allie Young, an analyst at Dataquest.[8] Although various threats drive some organizations to SLAs, alert, insightful managers adopt them because they are indispensable to other essential management techniques.

Establishing SLAs necessarily involves intense discussion and negotiation between service providers and their customers. Customers, their managers, service-provider managers, and, in some cases, the firm's senior managers must be involved.[9] The negotiations' objectives are to obtain agreement on service levels, their associated affordable costs, and means for tracking and reporting results and progress. Service-level negotiations are usually iterative because business is dynamic: requirements change, new technology becomes available, and other variables are introduced. In addition, negotiations evolve and become less contentious as various parties' develop and level expectations. The management team may need several planning cycles before service-level development becomes second nature.

Negotiations culminate in a document describing mutually acceptable agreements on service levels each IT client will receive.[10] Producing a document, rather than relying on verbal agreements is important to avoid misunderstandings and to clarify mutual agreement to the terms. The agreement must be complete, documenting all services that the client expects at mutually acceptable and achievable costs. For a mutually satisfactory agreement, parties must negotiate a balance between service levels and service costs. A properly constructed service-level agreement establishes an affordable, cost-effective means for client organizations to use IT services.

Client organizations must justify information system service costs by the mission they discharge for the firm. For example, costs of improved services to computer-aided design system users must be less than the value of productivity improvements derived from service improvements. IT services to all departments must be justified similarly. Planning, cost/benefits discussions, and resolution of financial issues precede SLA negotiations. Careful, disciplined planning simplifies SLA processes.

IT organizations initiate service-level agreements, but client and IT managers must negotiate them. In especially complicated and difficult cases, negotiations occur between the CIO and other executive managers. Disagreements that cannot be resolved at lower levels require higher-level attention, whether due to valid differences of opinion, resource shortfalls, or parochialism. Frequently, senior executives can better justify increased service levels and their additional expenses, especially when costs are known but benefits are deferred, intangible, or difficult to quantify.

Services that IT agrees to provide must be expressed in terms meaningful to client managers. For instance, response times measured at the user's terminal mean more to clients than CPU seconds measured at the service provider's CPU. The only meaningful turn-around time (the time from job submission to completion) is measured at the user's workstation, not at the central processor.

The SLA process must include all client organizations and consider nearly all computer services. Applications infrequently executed, those run on an as-required basis, or those that place low resource demands on the organization need not be specifically included. Applications used throughout the firm require special attention.

An earlier chapter stated that each application requires an owner-manager to manage it and discharge other ownership responsibilities. One ownership responsibility is to negotiate service levels with service providers on behalf of all application users located throughout the organization. The owner is also responsible for using cost-effective trade-offs to obtain a balanced agreement with service providers. Negotiating managers who exercise corporate statesmanship, keeping the firm's interests foremost during negotiations, achieve the most effective service-level agreements.

WHAT THE SLA INCLUDES

Typical service-level agreements begin with administrative information, including the date the agreement was established, its duration, and its expected renegotiation date. The agreement may specify unusual conditions such as significant workload changes that require renegotiation. It describes key service measures that the client's organization needs and service levels that the service provider will measure and deliver. Also included are resources that the provider needs to deliver the service and their associated costs.[11] Agreements must also describe mechanisms to report services actually delivered. Table 12.1 lists service-level agreement elements.

TABLE 12.1 Service-Level Agreement Contents

Effective date of agreement

      Agreement's duration

      Type of service provided

      Service measures

            Availability

            Service quantities

            Performance

            Reliability

      Resources needed and/or costs charged

      Reporting mechanism

      Signatures

Negotiation of service-level agreements occurs while the firm prepares its operational plan and shortly afterward. At this time, the firm allocates resources for the coming year, and near-term requirements for IT services become clear. The agreement's effective date usually coincides with this portion of the planning cycle. For relatively stable services, the agreement will probably last one year, coinciding with the plan period. Payroll or general ledger applications are examples of such systems. The agreements for applications of this type remain in place until they're renegotiated during the planning cycle one year later.

Applications with volatile demands or growing processing volumes may require more frequent SLA renegotiation. For example, a new strategic system experiencing great implementation success may have rapidly growing demand for IT services. Its service-level agreement may need review and renegotiation every six months or more often. Well-written agreements spell out the circumstances that trigger renegotiation. Table 12.2 illustrates these concepts in a partially completed SLA.

TABLE 12.2   SLA Administrative Information

SERVICE-LEVEL AGREEMENT

The purpose of this agreement is to document our understanding of service levels provided by *IT Computer Operations* to *The Personnel Department*. This agreement also indicates charges to be expected for service described herein and the source and amounts of funds reserved for these services. This is not a fixed price agreement. Charges will be made for services delivered at rates established by the controller.

The date of this agreement is *October 15, 1998.*

Unless renegotiated, this agreement is in effect for 12 months beginning *January 1, 1999.*

Table 12.2 illustrates the administrative information included in an SLA between the IT organization and the firm's personnel function. The agreement notes that personnel will be charged for actual services used according to rates established by the firm's controller. (This is not a fixed price contract.) The manner in which firms recover IT costs varies widely. Table 12.2 illustrates one possibility. Chapter 16 covers the subject of charging methods in detail.

Negotiated in mid-October, the agreement does not take effect until the next calendar year begins. In this case, the negotiation occurred during the planning cycle.

SLAs must specify the type of service required. The service may refer to regularly scheduled batch operations, continuously operating client/server systems, or complex online transaction processing applications. Service providers should negotiate agreements by service class because capacity, performance, and reliability depend on total demand levels in the overall organization. Table 12.3 is an example of a service specification.

TABLE 12.3   SLA Service Specification

Specific Type of Service	Application Service	Hours of Use
Client/server system	Distribution center	6:00 A.M. 7:00 P.M.
Data entry, report prep	Inventory application	daily*

System availability:

Monday–Friday:	The system will be available at the distribution center from 6:00 A.M. through 7:00 P.M., Monday through Friday.
Saturday:	Operation can be scheduled 24 hours in advance.
Sunday:	Emergency service only.

* Four operator workstations and 10 bar code readers will be available in the distribution center. The agreement to subsecond response time is part of our understanding.

Table 12.3 outlines the service to be delivered to the distribution function for inventory control and other purposes. It specifies data entry and report generation during the week, with extra hours available on Saturday. Distribution personnel use workstations and bar code devices located in the warehouse and supplied by IT. The agreement documents subsecond response times from IT to the workstations.

Schedule and Availability

The schedule describes the period when the system and its application programs are required to operate. The schedule for online applications should describe availability throughout the day. It must also outline the availability conditions for weekends and holidays. Operations managers must allow for scheduled maintenance or make provision for alternate service during maintenance. Client managers should focus on peak demand periods, such as weekends at the close of an accounting period and during significant events. Both parties should consider reduced availability or reduced performance during off hours, if these considerations yield cost savings. Likewise, during especially critical periods for the client, negotiating exceptionally high service levels may be attractive. Negotiators who clearly understand system capabilities, limitations, and special user needs produce the most effective agreements.

Timing

Job turn-around time is usually the key service measure of batch production runs. Turn-around time is the elapsed time between job initiation and delivery of output to the customer. Many batch production runs are scheduled through a batch management system. The batch management system accounts for the many relationships and dependencies among applications and between applications and databases. Because input data for one job may result from several other jobs, timing and sequencing jobs becomes crucial. Usually, clients understand these dependencies, and, in most cases, negotiating service agreements helps them better understand scheduling constraints. Frequently, large production runs occur overnight. In these cases, the critical measure is whether output from all overnight jobs is available when the next workday begins. The output may consist of reports delivered to client offices, or, more likely, online data sets available for users' review through individual workstations.

Response time is the most critical parameter for online activities. Response time is the elapsed time from the depression of the program function key or the Enter key to an on-screen display indicating that the function has been performed. Because it depends highly on the type of function being performed, response time should be specified by service type. For example, trivial transactions should have subsecond response; i.e., the system should not constrain operator performance when performing simple transactions. Trivial transactions require so little processing time, usually several milliseconds or less, that the response appears instantaneous. If subsecond response time cannot be achieved for trivial transactions, productivity drops substantially, and the system becomes a source of annoyance to the user.

Many transactions are distinctly nontrivial. For instance, an online system to solve differential equations may provide subsecond response time when users enter parameters, but the numerical methods used for the solution may require many million computer instructions to execute. The client usually understands these conditions and appreciates the response time needed to solve the problem. Many other situations, particularly applications that query several databases to search for specific data, are decidedly nontrivial, even though the customer query seems simple enough. Everyone must understand these situations thoroughly before entering into agreements. Through education, the information center can help.

During negotiations, the IT management team must provide information on service availability and reliability. An example of an availability statement is: The system will be available 98 percent of the time from 7:00 A.M. to 7:00 P.M., five days per week. An example of a reliability statement is: The mean time between failures will be no less than 10 days, and the mean time to repair will be no more than 15 minutes. Availability and reliability measures must be clear and unambiguous so users and IT personnel can use them easily. Table 12.4 shows the remaining items in the sample SLA.

TABLE 12.4   SLA Final Items

Reporting process: Personnel will receive quarterly reports documenting the service levels delivered on the payroll application.

Financial considerations: The distribution center owns workstations and bar code devices. All costs are included in the center's budget.

Additional items: none

Signatures

by:_____   by:_____

             IT organization   Application owner

Workload Forecasts

Reasonably accurate workload forecasts that can be translated into IT resources are essential to meeting service-level agreements. Based on anticipated workloads, operations departments can install production capacity sufficient to process loads generated by all applications and deliver services promised to clients. The planned workload should be defined for the service agreement period in terms that client managers can understand. IT managers must then translate these workload statements into meaningful and measurable units for capacity planning.

Workload forecasts should cover batch workload volumes, online transaction loads, printed output quantities, and other resources that clients require from service providers. In many cases, average workloads suffice. SLA negotiations, however, must include workload fluctuations if significant daily, weekly, or monthly departures from the average are expected.

Because many of today's applications are network centric, leading managers now routinely prepare SLAs defining network services. Formerly unique to mainframe managers, today, SLAs are common among client/server, LAN, intranet, and other network managers. According to a study by Infonetics Research, San Jose, CA, 61 percent of network managers expect to establish SLAs by 1999.[12] IT managers also find service levels important when discussing network operations with third-party providers, too, many of whom gladly provide service-level information.

Online and network transaction volumes are especially important when developing service agreements because they can vary widely and impact service severely. Following patterns of human behavior, they frequently experience a morning peak, a noontime lull, and late afternoon peak. Many client/server and Internet applications follow this pattern. Network operators factor these variations into performance and capacity plans and service agreements.

In some cases, one application generates peak loads in the valley of another application's load. Firms operating nationally experience workload waves throughout the day. Departments on the East Coast open for business three hours before departments on the West Coast. The air traffic control system in the United States illustrates this phenomenon: flights begin to depart from Boston and New York airports around 7:00 A.M. local time, several hours before airport activity in Seattle or Los Angeles. Workload peaks follow the sun in international operations.

Coinciding workload peaks from many applications is a more critical issue for IT managers. For example, financial closings at month's end usually generate coincident peak activity. In some cases, many activities coincide, forming a huge workload bubble for the organization. Year end is a prime example of this phenomenon. Not only are December closings in process, but other year-end activities are occurring as well. Year end is also a convenient time to implement new applications to supply the customer with new and different functions for the coming year. Many times, annual changes in tax laws or other government regulations mandate these new applications. If the fiscal year differs from the calendar year, the workload pattern takes yet another form. Satisfactory service level agreements anticipate, and clearly focus on, all these events.

Unanticipated or unusual increases in workload frequently cause missed service levels. Rapid growth in and access to Web pages, results of very favorable response to intranet communication services, exemplify load increases that may lead to unresponsive systems. Generally, however, workload increases do not occur without warning. Alert organizations re-open service-level discussions when they first anticipate such increases. Everyone benefits from well-managed workload; the entire firm suffers from one organization's poor planning.

Sometimes workload is difficult to predict because the organization's needs change or because the organization itself changes. Mergers, acquisitions, reorganizations, or consolidations are examples of structural changes that usually change workload. In these cases, IT should reanalyze workloads and prepare new forecast for client organizations.

Often tedious and time consuming, reasonably accurate workload forecasting is essential for providing satisfactory service. History is a good guide for most forecasts. IT should provide clients with current workload volumes and trend information; with a cost accounting or charging mechanism installed, this task is particularly easy. IT charging or cost recovery mechanisms are valuable in the SLA process because they also help focus on cost-effective service levels. (Chapter 16 discusses IT accounting processes.) Successful workload analysis and load forecasting are important to service providers and their clients.

Measurements of Satisfaction

Key service parameters found in service-level agreements require routine and continuous measurement and reporting. Specific, quantifiable parameters are preferable to more ambiguous measures. When service providers report these critical measurements ambiguously, inaccurately, or erroneously, client organizations develop their own unique tracking mechanisms. Uncoordinated, redundant measurements and reports usually lead to conflict, accusations, and finger-pointing, and consume energy needlessly. The most effective approach is to use unambiguous, credible reporting techniques at the outset.

One additional, perhaps obvious, consideration must be addressed: Service providers must measure and report service from the users' perspective. Reporting job completion time doesn't help if the output isn't immediately available to customers. Measuring online transaction response times at the server only confuses matters—what the user sees at the workstation is all that counts.

Response times at users' terminals or workstations can vary significantly from those measured at the CPU or server. Differences arise because of internal workstation processing delays, contention at terminal control units or network routers and bridges, telecommunication line loading, and delays within communication controllers or servers themselves. Still, overall system response times are mostly a function of application complexity, system loading, or contention for secondary storage devices.

The systems department can easily measure user response time by programming a personal computer to execute a representative sample of interactions with the CPU or server through the user's network. The PC's program can log and report statistics on response times by transaction type. By switching the PC to various parts of the user's network, the system development can obtain and analyze valuable data on user response times at dispersed locations. This objective technique is useful for establishing response-time benchmarks and monitoring critical online system responsiveness. Responding to demand for monitoring tools, especially in networked systems, some manufacturers now sell performance measuring and reporting systems.[13]

The service provider's measurement system must be highly credible with client organizations. Reliable, accurate measurement data inspires trust and confidence and fosters open and objective discussions when events don't go as planned. For many reasons, small departures from normal conditions may occur. Correcting these deviations is easiest in an atmosphere of open communication and mutual trust. Precise, consistent measurements of delivered service build trust and confidence between service providers and their clients. Objective, credible reports of service-level achievements pave the way for improvements in other activities too.

USER SATISFACTION SURVEYS

Periodically, service organizations should solicit informal customer opinions on service-level satisfaction. Such surveys furnish valuable insights into customers' perceptions of service performance.[14] Performance measured against service-level targets should be tempered with user perceptions of service. Simply meeting stated criteria is insufficient; the perception of service must also be satisfactory. Service providers must know how satisfied users are. They can best obtain this information by asking the users directly.

Service providers can choose from several methods for obtaining this information.[15] One method for online users is to conduct an opinion poll periodically via a brief questionnaire that appears on their workstation screens when they conclude their session. The questionnaire asks for their perceptions of service during the session. To be most effective, the questionnaire must be optional, anonymous, and concise.

Another survey should question a wider audience over a longer period and within a broader scope. A survey like this can produce results useful in detecting problem applications. It can reveal areas of substandard service or of mediocre or poor service perceptions. Surveys are also useful for detecting improperly established service agreements and for identifying clients who need education in service-level processes.

Conscientious client surveys can enhance internal service measures, relate internal measures to perceived service levels, and improve actual customer service. Properly used surveys can help service providers achieve high correlations between internal and external service measures. Attaining this goal optimizes resource use, improves efficiency, increases customer satisfaction, reduces costs, *and* improves service.

Client and IT managers should act together to correct unsatisfactory situations that surveys or measurements detect. Managers can also use survey and measurement data to establish more effective relationships between service users and providers. They can act to increase and improve communication, cultivate better service, or expand education. Such actions benefit everyone. Most organizations can profit greatly by exploiting the benefits of well-designed survey techniques.

ADDITIONAL CONSIDERATIONS

In most complex environments, service agreements must usually include some additional operational items. For example, during infrequent critical periods, some departments may need dedicated systems, large amounts of auxiliary storage, or other unusual services. Others may require special handling for data entry, analysis, or specific one-time processes. Prudent IT managers recognize the value of responding enthusiastically to these important client requirements, even though the request is beyond the SLA's scope. Satisfied customers must be a high-priority goal for every manager.

Unfortunately, not all managers endorse service-level agreements. In user-owned client/server operations, managers tend to regard these activities as unnecessary formalities, typical of bureaucratic central-system operations. They believe that installing larger workstations, faster LANs, or more powerful servers can solve problems. But, as the applications eventually become more complex and operations grow, some formalism is necessary to maintain order and sustain reliable service. Operating instinctively and reacting to difficulties proves unsatisfactory in large or complex situations.

In some firms, political considerations may frustrate the process because SLAs are contrary to the corporate culture. In some organizations, strong-willed managers prefer to demand service regardless of cost. These managers believe their departments require high service levels purely because they perform important functions for the firm. They may also believe that other departments exist to serve their functions, and that IT must do likewise, no questions asked. Some managers abhor the scrutiny that service-level processes bring to their activities. They prefer to conduct their business in an atmosphere of near secrecy. These attitudes seriously jeopardize successful computer, network, or other service operations.

Some firms' corporate culture is not receptive to the detailed planning and commitment processes needed to reach service-level agreements. Discipline is required; some firms simply don't have it. These firms tend to behave in an *ad hoc* manner, reacting to situations as they arise. Firms lacking discipline continuously subject IT and other managers to myriad conflicting forces that make smooth operations and harmonious relations nearly impossible. Establishing and using service-level agreements require the firm's managers in all departments to cooperate, use discipline, and exhibit maturity. Consequently, responsibility for successful computer system and other firm services transcends service providers.

CONGRUENCE OF EXPECTATIONS AND PERFORMANCE

The goals of SLA management processes are to develop mutually acceptable expectation levels, cultivate an atmosphere of joint commitment, and foster a spirit of trust and confidence between organizations. In speaking of SLAs, Robert Van Dyke, assistant vice president of IT/automation at First Union Corporation, said, "The idea is to better define and manage expectations so that each of us is able to perform our jobs better."[16] All employees perform better when essential business information flows freely and openly, and expectations and performance are approximately balanced.

Highly desirable and easily stated, these goals are not always readily achieved. For example, how can firms handle the dilemma that arises when clients demand better service than IT can deliver? Should the SLA describe service that cannot be delivered, or should it describe achievable service the client finds unsatisfactory? Considering affordable costs offers resolution to this problem. Through management processes described in previous chapters such as strategizing, planning, budgeting, plan reviews, and steering committee actions, affordable IT costs should have been reached. A firm that can afford additional capacity should procure it and develop an appropriate SLA. If costs limit capacity, clients must accept and SLAs must reflect this condition.

In some cases, expectations and performance converge only after one or two plan cycles. In firms just starting to systematize operational activities, the first iteration may not satisfy everyone completely. In many firms, such as those in banking or insurance, however, the service-planning process has been an integral part of their planning cycle for years. For these firms, leveling expectations and performance is a structured, ongoing process.

SLAs are the foundation for disciplined processes used by successful computer operations managers. They are critical to computer and network service providers managing centralized IT organizations, departmental client/server operations, or corporate-wide extranet systems. Managing service operations satisfactorily is imperative because it is a critical success factor for managers of systems.

SUMMARY

Management systems for production operations rely on a foundation of service-level agreements. Essential and valuable tools in IT's overall management system, they are important for other reasons as well. The negotiation process itself develops and enhances understanding and mutual respect between service providers and their clients. Through the negotiation process, IT organizations more clearly perceive the business and their clients' contributions to it. Client needs driven by their missions become obvious to service suppliers. On the other hand, client organizations more clearly understand the opportunities and constraints intrinsic in IT and technologies it supports. High levels of trust, confidence, and respect mutually benefit all the firm's units and the firm as a whole.

Review Questions

1. What are the elements of the disciplined approach to managing production operations?
2. How do SLAs deal with the issue of expectations in production operations?
3. What are the elements of the Management by Results system at Security Pacific?
4. What is the SLA's purpose? Who participates in establishing it?
5. What are the ingredients of a complete service-level agreement?
6. What is the distinction between turn-around time and response time?
7. Can the service-level agreement concept be considered standard business practice? Why?
8. What considerations can be used to break ties or resolve deadlocks during the SLA negotiation process?
9. Why are workload forecasts needed for satisfactory SLA processes?
10. What difficulties arise in selecting units of measure for workload forecasts?

11.   Why is the IT organization's measurement and publication of credible service-level performance data essential?

12.   In what ways do user-satisfaction surveys benefit service providers?

13.   What are some social or political benefits that accrue to firms using SLA processes?

14.   What modifications to SLA processes may be valuable for strategic information systems management?

Discussion Questions

1.   Describe elements of the management system Singleton used at Security Pacific.

2.   Identify the advantages of centralized IS organizations perceived by Singleton at Security Pacific, and compare them with advantages of decentralization.

3.   Many computer system activities are operational or tactical. Some are strategic or long-range. Discuss some examples of strategic activities.

4.   Discuss financial considerations that must be addressed in establishing the SLA.

5.   One step in gaining line management involvement in information technology is institutionalizing the processes of IT/line management cooperation. How do SLAs help accomplish this goal?

6.   Customers of information technology organizations sometimes state: "I can't predict my needs for IT services more than three months in advance." How can you convince them otherwise?

7.   Discuss the relationship of service-level process to tactical and operational planning discussed in Chapter 4.

8.   In addition to those mentioned in the text, what means can you propose for obtaining customer feedback on IT service? Discuss these methods' advantages and disadvantages.

Assignments

1.   Design a screen that will appear at log-off time to gather customer feedback about online service quality provided during the session.

2.   Contact a service business in your area and investigate its service-level process. Is it equivalent to the SLA? Does it measure and report service levels? Does it conduct customer satisfaction surveys? How?

3.   Interview your school's IT manager focusing on customer satisfaction and SLAs. To what extent are SLAs used? What reasons exist for the extent of their use? In your interview, did you find any special reasons to emphasize or neglect SLAs at colleges or universities?

SERVICE LEVEL AGREEMENT

The purpose of this agreement is to document our understanding of service levels provided by _____ to _____. This agreement also indicates the expected charges for the service described herein and the source and amounts of funds reserved for these services. This is not a fixed price agreement. Charges will be made for services delivered at rates established by the controller.

The date of this agreement is _____.

Unless renegotiated, this agreement is in effect for 12 months beginning
_____.

Specific type of service   Application service   Hours of use

Describe agreements reached on

System reliability:

Reporting process:

Financial considerations:

Additional items:

Signatures of parties to this agreement

by:_____   by:_____

      Service provider   Application owner

[1] Byron Belitsos, "A Measure of Success," *Computer Decisions*, January 1989, 48. See also John P. Singleton, Ephraim R. McLean, and Edward N. Altman, "Measuring Information Systems Performance: Experience with the Management by Results System at Security Pacific Bank," *MIS Quarterly*, June, 1988, 325.

[2] Security Pacific employed nearly 41,000 and earned $740 million in 1989. It was merged into Bank of America in March 1992.

[3] For current examples of firms using SLAs and their reasons for doing so, see Jaikumar Vijayan, "Service Pacts Ease Conflict," *Computerworld*, September 1997, 1.

[4] Additional details can be found in Edward A. Van Schaik, *A Management System for the Information Business: Organizational Analysis* (Englewood Cliffs, NJ: Prentice-Hall, Inc., 1985), 162.

[5] This nomenclature stems from the author's experiences with these management processes.

[6] Batch processing (or scheduled production processing) refers to batching or accumulating transactions for processing. Accumulating hours worked and rates of pay for processing payroll checks is an example of batch processing.

[7] Karten Associates, Inc., is a management consultancy in Randolph, MA. Jaikumar Vijayan, "Service Pacts Ease Conflict," *Computerworld*, September 22, 1997, 14.

[8] See note 7.

[9] David R. Vincent, "Service Level Management," *EDP Performance Review,* April 1988, 3. According to this article, well-executed service-level agreements improve technology integration and help users find more IT opportunities.

[10] See the appendix to this chapter for a sample service-level agreement.

[11] John P. Singleton, Ephraim R. McLean, and Edward N. Altman, "Measuring Information Systems Performance: Experience with the Management by Results System at Security Pacific Bank," *MIS Quarterly*, June 1988, 325.

[12] *Computerworld*, January 19, 1998, 53.

[13] Patrick Dryden, "Demand for Service-level Tracking Tools on the Rise," *Computerworld*, December 8, 1997, 4, and "Tools Become Active Service-level Monitors," *Computerworld*, January 26, 1998, 14. These articles describe tools and their manufacturers.

[14] Some long-distance operators conduct statistically significant customer surveys daily and compare the results to internal service measures. Using survey data, they refine and improve internal service measures, thus improving service and reducing costs through improved resource utilization.

[15] Robert E. Marsh, "Measuring and Reporting User Impact of Substandard Service," *EDP Performance Review*, March 1989, 1. The article describes the construction and use of a service-level indicator to measure responsiveness, availability, and reliability.

[16] Jaikumar Vijayan, "Service Pacts Ease Conflict," *Computerworld*, September 22, 1997, 14.

13 Problem, Change, and Recovery Management

On September 17, 1991, failure of AT&T's auxiliary power system at its switching center in lower Manhattan caused a seven-hour shutdown labeled *the* worst air traffic disruption in U.S. history. Air traffic controllers in the Northeast were unable to operate, leaving 85,000 passengers stranded. The Federal Aviation Administration (FAA) vowed never to let this happen again.

Trouble began early that morning when Consolidated Edison asked AT&T to reduce commercial power consumption in anticipation of high electrical loads later, due to exceptionally warm weather. At around 10:00 a.m., AT&T cut over to its own diesel generators at its Thomas Street switching center in lower Manhattan. What followed, according to Kenneth Garrett, senior vice president of network services at AT&T, "was largely caused by a combination of aberrant circumstances. The problem was not a result of any systemic or fundamental failure of AT&T's technological and human resources," he said.

The switching and other equipment at Thomas Street operates on direct current (DC) converted with rectifiers from alternating current (AC) supplied by Con Edison. Upon switching to the generator, an electromechanical overload relay set too low shut off power to the rectifiers, forcing the Digital Access and Cross-Connect system (DACS) driving three large phone switches to operate on battery power alone. With this load, battery life is six hours.

Changing to battery power normally triggers audible and visible alarms on several floors of the building housing the switches. Visible alarms located in unoccupied areas of the 15th and 20th floors operated normally but were not seen. Earlier, employees had disabled the audible alarms on these floors. On the switching center's 14th floor, the bulb in the visible alarm had burned out, and wires to the audible alarm, accidentally cut during remodeling, had not been replaced. In short, all alarms in the building signaling the shift to battery power were either inoperable, or unobserved.

Meanwhile, conditions appeared normal at AT&T's remote monitoring site in Georgia; battery operation was not monitored there. Eventually, operating personnel at the Thomas Street station noticed one of the visual alarms signaling battery power operation; by this time, only a few minutes of battery power remained. The system shut down at 4:50 p.m.

Efforts to recharge the system and to restore the phone network's operation again began immediately and, by 11:00 p.m., about 90 percent of domestic and international lines were in service. Some phone customers had no service for more than six hours.

This incident provoked numerous federal inquiries by the House Telecommunications and Finance Subcommittee, the Federal Communications Commission, and the House Government Operations Committee. In addition, the FAA began to take action to improve its unreliable network.

During the previous 12 months, the U.S. air traffic control system around the country experienced a total of 114 network outages, about 8.7 per month, averaging 6.1 hours each. In July 1992, 10 months after the Manhattan phone outage, the FAA announced a nearly fail-safe national network for the FAA's National Airspace System. MCI developed the network and began phasing it in over three years.

The network provides digital links to 20 regional centers, 80 major airports, and 5000 other FAA facilities. The net is designed to offer 99.999 percent availability, meaning no more than 5.3 minutes downtime annually. The mean time to repair the net is less than 30 seconds. Built on redundancy, the system has at least two paths leading to every site, and the network management center in Reston, VA (itself duplicated in Sacramento, CA) monitors all sites.

According to the FAA, the 10-year outsourcing deal with MCI, worth $856 million, reduced operating costs and also improved reliability.

INTRODUCTION

Chapter 12 laid the foundation for a disciplined, systematic approach to managing computer operations, a critical success factor for service providers. This chapter describes the tools, techniques, and processes for managing three more essential disciplines: problem management, change management, and recovery management. Figure 13.1 shows how these processes relate to the complete operations management system.

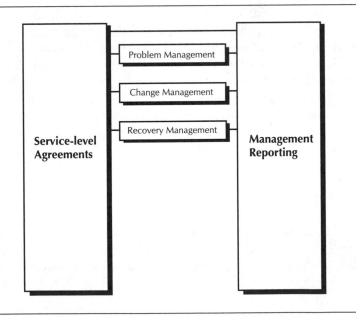

FIGURE 13.1 Disciplines of Problem, Change, and Recovery Management

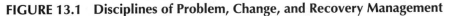

Problem, change, and recovery management processes are based on service-level agreements. In turn, they form the basis for management reporting processes. Service-level agreements are the foundation and management reporting is the capstone of the disciplined approach. The three processes—problem, change, and recovery management—form the management system to ensure service-level attainment. The next chapters describe the remaining processes.

Problem, change, and recovery management focus on departures and potential departures from acceptable operation. These management processes' purpose is to correct deviations from the norm and to prevent future deviations. Managers act to correct problems as they occur and to remove their sources. The problem-management discipline guides managers in taking corrective action when problems arise and ensures that the same or similar problems do not recur. Frequently, managers introduce system changes to increase system capability or take corrective action to overcome problems. Change, however, is a rich source of potential problems. Even remodeling changes can cause problems, as they did at AT&T's switching center. Change management concentrates on minimizing risks in system alteration and lets managers easily control change activities. Regardless of management diligence, uncontrollable events may nevertheless occur and lead to major operational problems. Recovery management deals with these eventualities.

These disciplines apply not only to IT and other service providers but must also be part of the user's management system. System users depend on system operations; they must be intimately involved in operational processes. In addition, user departments that acquire computing capability, such as servers, intranets, and application software, must assume responsibility for managing problems, controlling system changes, and developing recovery and contingency plans. Users who own and operate computing equipment and application programs must establish appropriate management disciplines or suffer degraded performance from poorly controlled systems. This chapter describes disciplined methods for system operations that apply equally to IT and user-operated systems.

PROBLEM DEFINITION

The word *problem* means different things to different people. In the business environment, problems are in the mind of the beholder because all observers do not necessarily perceive problems in the same way. For the purpose of this discussion and to prevent ambiguity, *problems* are incidents, events, or failures, however small, that negatively impact operating departments' ability to deliver service committed in service-level agreements. Problems include actual failures resulting in service degradation or potential failure mechanisms that could degrade service.

This definition necessarily excludes many other kinds of problems that computer operations managers may face. For example, next year's salary budget may be less than requested, requiring managers to revise their financial plans accordingly. Temporary construction activity may disrupt employee parking, creating problems for some, especially during inclement weather. Managers may need to resolve these issues, but, unless the issues lead to unmet service levels, they are excluded from the discussion of problem management. Although managers face many important issues, the disciplined management processes, including problem management, are designed to ensure satisfactory customer service.

Problem management is a method for detecting, reporting, and correcting problems affecting service-level attainment. Problem sources include hardware, software, network, human, procedural, and environmental failures that disrupt or potentially disrupt service delivery. The firm, client organizations, or the IT organization generate most problems, but external sources may originate problems as well. An application that fails because of insufficient online storage may originate from a hardware, application, or operator problem. The problem management system must handle problems from all sources that affect the delivery of satisfactory service.

Service providers hope to achieve several important objectives by managing problems, including reducing service defects to acceptable levels, achieving committed service levels, reducing defect cost, and reducing the total number of incidents.

Effective problem management systems reduce the number of incidents or events to acceptable levels and reduce their associated costs. Successful problem management enables service providers to deliver committed service levels to user groups. It minimizes the total number of problems needing attention. Fewer problems conserve organizational energy, reduce frustration, and lower costs. In addition, the problem management system gives managers tools to understand the root causes of the relatively infrequent failures that inevitably occur.

Successfully managing production operations requires systematic problem management, one of the first steps in operations management. Problem management, a necessary condition for success, is not necessarily sufficient for success.

Problem Management's Scope

Table 13.1 outlines many sources of potential and actual impacts to service levels.

TABLE 13.1   Problem Management's Scope

Hardware systems

Software or operating systems

Network components or systems

Human or procedural activities

Application programs

Environmental conditions

Other systems or activities

Unscheduled system restarts, unanticipated or intermittent operations, or other abnormal operating conditions usually indicate hardware or operating system failures. Difficulties may arise from bugs in the operating system or failures in peripheral devices such as tape drives, disk storage devices, or workstations. Failures can originate in CPU hardware or in system software. Operating system parts may malfunction or an important function may be unavailable to application programs. Operating system defects may also cause intermittent or unanticipated results.

Today, most major systems connect to networks that may cause problems by functioning intermittently or failing totally. Network failures can originate in hardware, software, or operating system communication programs. Incorrect human interactions with the network or improper procedures governing human interactions may also cause failures or problems.

Operator errors or incorrect or incomplete manual procedures can cause system failures. Procedures may create failures if they are incomplete, incorrect, or subject to misinterpretation, or if they do not cover all situations. Operator failures may also stem from not following established procedures or from executing them incorrectly.

The application portfolio is a frequent problem source, particularly during enhancement or maintenance activities. Chapter 8 noted that old applications with lengthy maintenance records are especially likely to cause problems. Applications may create unintended or unexpected results, terminate programs unusually or abnormally, or include incomplete or incorrect functions. These conditions usually cause unsatisfactory performance resulting in unmet service levels.

The environment is another source of difficulty. Power disruptions, air conditioning difficulties, or local network failures may disrupt service levels. On a larger scale, earthquakes, hurricanes, or other natural disasters infrequently cause serious difficulties. In addition to these causes, problem management's scope must also include any reasons for unmet service levels. Any situation that impacts or potentially impacts service falls within its scope.

Problem Management Processes

Problem management processes include techniques and tools specifically designed to detect, report, correct, and communicate important details of problems and their resolution. They involve a small informal committee with defined member responsibilities, which include using techniques for addressing and analyzing problems, and reporting their results. Organizations committed to minimizing incidents rely on defined processes for attaining and maintaining the highest possible service levels.

Problem Management Tools

Table 13.2 lists problem management's primary tasks as a series of steps.

TABLE 13.2   Tools and Processes of Problem Management

1.   Problem reports
2.   Problem logs
3.   Problem determination
4.   Resolution procedures
5.   Status review meetings
6.   Status reporting

A person who discovers a problem initiates a problem report by completing a form or an online questionnaire and sending it to the head of the problem management committee. The reports' purpose is to record the status of all incidents, from detection through solution. Beginning with problem detection, the report (frequently an online document) records all corrective actions taken to resolve the issue. Table 13.3 lists the items included in problem reports.

TABLE 13.3   Problem Report Contents

Problem control number

Name of problem reporter

Time and duration of the incident

Description of problem or symptom

Problem category (hardware, network, etc.)

Problem severity code

Additional supporting documentation

Individual responsible for solution

Estimated repair date

Action taken to recover

Actual repair date

Final resolution action

Problem reports are updated while the problem remains open or unresolved.[2] When an incident is closed (problem solved), the report is updated to include resolution data, posted, and filed (archived if online). The problem log contains a record of all problems.[3] The log is a document used to record incidents and actions assigned, and to track reported problems. As such, it contains essential information from reports on all resolved and unresolved problems during some fixed period; for example, the rolling last twelve months is a convenient and useful period. The problem log and its associated reports provide data for analysis by managers and other interested parties. Careful analysis of prior problems is very useful for anticipating and reducing future problems.

Problem Management Implementation

Problem status meetings are an essential part of the management process. Regularly scheduled status meetings are a forum for discussing and documenting unresolved problems, assigning priorities and responsibilities for resolution, and establishing target dates for corrective action. A daily status meeting is a reasonable approach.

Major problems require additional reviews. Some questions for these extraordinary reviews are: How could this problem have been prevented? and How could this problem's impact be minimized? Status meetings and reviews should focus on recurrent problems and careful analysis of unfavorable trends. Areas experiencing high problem levels must receive special attention. Trend analysis provides clues to weak or vulnerable areas. Rediscovered problems should receive special consideration because they reflect incomplete or ineffective problem resolution.

Representatives of each business function having open problems and key individuals representing IT or other service providers should participate in the status meetings. IT representatives are usually systems programmers, computer operators, and applications programmers. When the agenda includes hardware problems, a person representing the hardware service organization should also attend. The problem management committee leader should chair the meetings. When the process operates effectively, managers' attendance is not necessary. Managers are responsible for supporting the effort and for ensuring participation and results; however, meetings are generally more effective if managers are excluded. Problem meetings assess problems, but managers tend to assess people, an activity that should be reserved for another forum.

Classifying incidents by their source—hardware, software, network, etc.—and by their severity aids in problem determination and resolution. Committee members should perform the classification and assign problem resolution based on documented individual responsibilities. The documentation helps to clarify each individual's role and to ensure that proper skills are applied to problems. For example, if the problem resides in the operating system utility programs, system programmers responsible for these utilities should be assigned this problem.

Problem resolution procedures include action plans and estimated resolution dates proposed by the individual to whom the problem was assigned. The individual reports resolution status to the team leader and to users or members of the affected areas. After review with all concerned parties and with the team leader's concurrence, corrective actions are scheduled. When these actions involve system changes, the change management process should be invoked. In any event, when problem resolution satisfies the team, the problem report is closed and posted in the log. Figure 13.2 illustrates the flow of the problem management process.

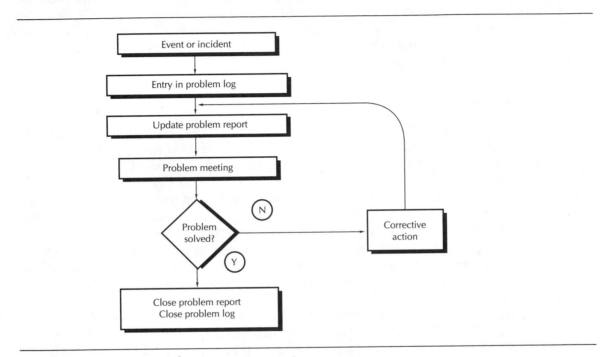

FIGURE 13.2 Problem Management Process Flow

Some firms develop interactive tools to help create problem reports and maintain the problem log, and others use vendor-supplied, automated, online tools to assist in problem management. Interactive, computer-based tools help team members analyze problems and let managers and others review problems, solutions, and their trends quickly and easily.

Problem management's success depends critically on managers' attitude toward the people involved in problem management. Because the process is designed to solve problems and help prevent incidents, managers must not use it as a means for individual performance assessment. Instead, managers must establish and foster an atmosphere of free and open communication among all participants. Finger-pointing and assigning blame should not occur in this environment. A productive environment encourages a spirit of mutual cooperation between service providers and system users. Jointly and separately, team members should feel responsible for resolving incidents promptly and for preventing incident recurrence. Accepting responsibility for resolving an incident is not equivalent to accepting responsibility for causing the incident. The most effective status meetings occur in a supportive environment. Managers' presence is unnecessary in a favorable environment.

PROBLEM MANAGEMENT REPORTS

Managers must receive summary reports documenting the effectiveness of the problem management process. The summary data should reveal trends and indicate problem areas that impact the delivery of satisfactory service levels. These summary reports are useful in establishing future service levels. IT plans should account for problem activity when establishing resource and expense levels.

Problem reports may be produced weekly, monthly, or on request. The reports highlight numbers and causes of incidents by category and severity. Duration of service outages, length of time for problem determination, and time needed for resolution must also be reported. An aged problem report is very useful in determining the organization's responsiveness to incidents. An analysis of duplicate or rediscovered problems is an important result derived from the report's data. The rediscovery rate is important because it measures problem resolution effectiveness. An analysis of problems caused by system changes is another product of the report's data. Problems caused by system changes measure the change management process' effectiveness.

Well-conceived reports enhance the process by providing concrete evidence of its effectiveness. Problem reports demonstrate service providers' responsiveness to customer needs. They lend credibility to service-level commitments and reinforce the alliance between service recipients and providers. Problem summary information is especially useful when renegotiating SLAs. Problem management processes are essential to operational success.

CHANGE MANAGEMENT

Change management is a management technique for planning, coordinating, handling, and reporting system changes that could negatively impact service delivery. Changes are the focus because they can greatly influence service levels. Changes can involve modifications to hardware, software, or networks; they can originate from human processes, manual procedures, and the environment. The objective of the change management process is ensuring that system changes are implemented with minimum or acceptable risk levels, not jeopardizing service levels. Service providers deal with many changes; however, the change management discipline includes only those that can potentially impact service.

Modifying complex human-machine systems raises significant, often unrecognized, risks of disrupting or degrading system processing and reducing service levels. Changes tend to generate problems. Frequently, well-intentioned individuals introduce small changes to improve areas that may appear remote from the operation. Nevertheless, even minor changes to sophisticated environments often lead to unanticipated difficulties. Change must be managed.

Rapid technological advances, coupled with information technology growth within the firm, make change a way of life for IT managers. Needing to manage evolving systems while maintaining service levels, IT and client organizations must follow a planned approach to change implementation that minimizes problems and achieves desired service levels during transitions.

Problem management is tightly linked to managing changes because resolving problems frequently requires system changes. Change management and problem management must cooperate to ensure that solutions to one problem do not cause another. In undisciplined environments, solutions to two problems commonly yield one additional problem requiring action. Some organizations report that 80 percent of corrective actions yield additional problems. An effective, disciplined approach to problem and change management can potentially reduce the ratio from 2:1 to 200:1 or less.[4] The organization benefits by improving effectiveness and reducing costs. In addition, other less obvious but important advantages exist. For example, minimizing problems and controlling changes enhances the IT organization's image and improves employee satisfaction.

Change Management's Scope

Change management addresses many activities that can negatively impact the delivery of committed service. Table 13.4 lists the major change sources affecting service levels.

TABLE 13.4   Change Management's Scope

Hardware changes

Operating system changes

Application program changes

Environmental changes

Procedural changes

Equipment relocation

Problem management-induced changes

Other changes affecting service

Most changes to computer-based information systems or to their environment risk service disruption. The change management plan must always include hardware changes, which frequently occur. Any alterations to operating systems, network software, utility programs, or other support programs must be considered. Application programs undergoing maintenance or enhancement are prime change-management candidates. These error-prone processes frequently lead to operational failures, so application program modifications must always be carefully monitored.

Change Management Processes

Change management includes tools, processes, and management procedures needed to analyze and implement system changes. Table 13.5 itemizes change management elements.

TABLE 13.5   Change Management Elements

Change request

Change analysis

Prioritization and risk assessment

Planning for change

Management authorization

Initiating alterations to computer systems or their environment requires a change request.[5] The change request document is an important record detailing change actions from initiation through implementation. It records the type and description of the requested change and identifies prerequisite actions. This document records risk assessment, test and recovery procedures, and implementation plans. It is also an approval form that authorizes individuals to make changes. Table 13.6 identifies the change request's contents.

TABLE 13.6   Change Request Document

Change description and assigned log number

Problem log number if problem results from an incident

Change type

Prerequisite changes

Change priority and risk assessment

Test and recovery procedures

Project plan for major changes

Requested implementation date

Individual responsible for managing change

Individual requesting change, if different from above

Management authorization

The change request document develops as the change is analyzed, planned, and implemented. When the work is complete, the change request is filed for future reference and analysis. Like problem management, aspects of change management can be automated with online, interactive forms and reports.

The change management team authorizes changes. Individuals representing computer operators, systems engineers, application programmers, and hardware and software vendor liaisons comprise this team. The facility manager must be represented because changes in the physical environment (heating, power, space, etc.) are critical to the operation. Usually a member of the operations group leads the team. Often, having some representatives serve on both the problem and change management teams offers advantages. The change team meets regularly for short periods to analyze, review, and approve plans for change.

The change management team focuses on risk analysis by categorizing all changes in terms of operational risk. Change analysis should focus on factors that quantify the magnitude and significance of alterations to ongoing operations. For example, changes that significantly impact system performance or capacity or that dramatically affect customers probably carry major risk. Changes that are interdependent on other pending changes or mandate future change must be reviewed very critically. Changes that affect several functional areas or require training are riskier than those that do not. Sound test and implementation plans greatly help to reduce risk.[6] Successful teams ensure low or acceptable risk in the stream of changes that are always a part of system operations.

Major or extraordinary changes require thorough project plans that identify people, responsibilities, target dates, and implementation reviews. The change team technically audits the process and, in special circumstances, may request that nonpartisan technical experts conduct additional reviews. The team approves changes only when it is confident they can be implemented with low or manageable risk. The team must also have confidence that recovery plans are satisfactory for unsuccessful implementations.

The goal of the change management process is to minimize risks associated with system alterations. This goal applies to all changes, whether major and substantial or relatively minor. Replacing a large mainframe may require a year or more of change planning and management. Installing the latest version of an operating system on network servers may take months, but another direct access storage device may be added in a week or so. Altering a batch program may require only a day or two. In any case, all changes, large and small, should be implemented only after considering and evaluating all the firm's risks.[7]

Changes must be capable of being tested, and the team must evaluate the test plan for thoroughness and completeness. Test cases and expected test results are documented in advance and compared with actual results upon implementation. Variances between expected and actual results require analysis and explanation. Variances may reveal incomplete preparation or ineffective implementation. Significant variances may indicate more problems or necessitate additional changes.

REPORTING CHANGE MANAGEMENT RESULTS

Successful change implementation must be communicated to all affected parties, and its details must be entered into the change log. The log should be reviewed periodically to ascertain trends and assess process effectiveness. Assessment focuses on problem areas, expectations vs. results, required emergency actions, and areas requiring numerous changes. One measure of process effectiveness is the percentage of changes implemented without problems. This number should exceed 99 percent in a well-disciplined environment.

Periodically, the service provider must analyze trend data and summary information from the change log review. (Change logs' formats are similar to the problem logs discussed earlier.) Managers who provide information services are responsible for effective operation of the change management process. To make assessments, they need information from the change log. Like problem management, change management does not need continuous, detailed, management involvement. The process does need management direction and support but not continuous intervention.

RECOVERY MANAGEMENT

Information technology managers recognize that the resources they manage are critical to the firm's success. They are keenly aware that unavailability of major online applications has immediate and serious consequences and that degraded operation of customer-support applications jeopardizes customer relations. Nevertheless, 25 percent of CIOs have no disaster-recovery plans, according to *Computerworld*.[8] The serious consequences of improper or nonexistent planning include lost or deferred sales and revenue, and possible damage to the firm's reputation.

The crash of the Sabre reservation system provides one very dramatic example of system failure. Downtime lasted approximately 13 hours. The failure of a utility program formatting large, direct-access storage devices caused the crash. The program went astray, destroying the serial numbers of 1080 volumes of data and making data access impossible. This system failure disabled American's load management process and permitted travel agencies and corporations to book unlimited deep-discount tickets and to re-book higher fare tickets at discounted rates.[9] American's losses from these activities have not been published.

In other, less dramatic situations, Merrill Lynch's CMA customers could not check account balances for several hours, AOL suffered from routing table errors, and other Internet difficulties severely impacted e-mail messages for most of a day. Change management and its ally, recovery management, gain increasing importance in networked environments. Because information systems are so essential to most businesses today, preventing problems, controlling changes, and recovering from unavoidable disasters is essential to business continuity.[10]

A variety of natural or human activities can damage or render vital information assets inoperable. To address this reality, IT managers must develop plans to cope with the potential loss or unavailability of information resources. Recovery management deals with the possibility of lost, damaged, unavailable, or unusable IT resources.

In recent years, major disasters caused loss of or damage to computer and communications facilities; more are surely in the offing. Florida experienced a major hurricane in 1991, the Chicago business district flooded in 1992, the World Trade Center and the London financial district were bombed and the Mississippi river flooded in 1993, earthquakes followed by floods and fires damaged parts of California in 1994, and an earthquake in Kobe, Japan, in 1995, destroyed more than 50 percent of NTT's service capacity in the region, causing some companies to go out of business.[11] Fires, floods, and natural disasters constantly threaten, underscoring the importance of recovery planning.

Many managers find dealing with recovery planning difficult. For example, they consider losing their computer center from tornado damage an extremely low risk. Or they consider solutions extremely complicated and unmanageable. For these reasons and others, some IT managers give recovery management planning low priority—so low that they devote little or no effort to recovery concerns.

On the other hand, emergency power outages or hard drive failures are more likely to occur. Because managing these contingencies is easier to visualize, managers usually appreciate the need to prepare contingency plans for them.

IT managers in financial institutions are required to have recovery plans. Federal regulations require banks and other financial institutions to maintain disaster recovery plans for inspection by federal auditors.[12] The Office of Management and Budget has directed federal agency attention to this subject since 1978 through Circular A-71, "Security of Federal Automated Information Systems." Updates to this circular have directed the Commerce Department to develop standards and guidelines that regulate recovery planning and assist other federal agencies in developing their plans.

IT managers must deal with a continuum of possibilities involving the likelihood of disaster vs. the seriousness of the event's consequences. The recovery management process deals with these uncertainties and articulates actions in cases where management believes the associated risk is worth the planning effort. Figure 13.3 shows the relationship between risk of loss, probability of occurrence, and required management action.

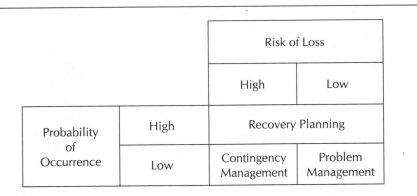

FIGURE 13.3 Recovery Planning and Contingency Management

Recovery plans deal with potential situations in which the risk of loss varies from low to high. Planning is appropriate because of the high probability that the event may occur. Virus invasion, server failure, and telecommunications link severance are examples of events which call for recovery plans. Contingency planning is the term reserved for high-risk events with low probability of occurrence. The problem management system can handle low risk events with low probability of occurrence as they arise.

Organizations must analyze their situation and place possible events in appropriate categories. For example, virus infection may not be related to geographic location, but floods or earthquakes probably are. Firms also differ in degree of loss from identical events. For example, firms with EDI or Internet links to suppliers and customers risk more from telecommunications failures than those using WANs exclusively for internal communications. Sound recovery management demands a thorough analysis of the firm's risks and probabilities.

Recovery plans provide backup resources to help restore service when disruptions occur. Most of these plans are like insurance policies—they're important to have, although one hopes not to need them. Prudent managers implement recovery plans knowing that complex and vital resources can suffer many types of loss. Recovering from minor emergencies is more than a once-in-a-lifetime experience: it's something no manager wants to experience, but an event for which every manager must plan.

Recovery management is a discipline of production operations. It involves people, processes, tools, and techniques. Recovery management originates from convictions that service providers have commitments to defined service levels. Recovery management is based on the realization that careful planning can help ensure service levels. Like the previous two processes, recovery management involves collaboration between managers and nonmanagers representing service providers and client organizations. Recovery management requires cooperation with user groups to achieve mutual goals.

Many examples validate the importance of sound recovery planning. For instance, in September, 1989, Hurricane Hugo blew apart Charleston, SC, destroying its electrical power supplies. A local credit union, which had a hot-site disaster recovery contract with Sun Data, used a mainframe computer for all its operations, including automatic teller machines. Without power, the computer could not operate just when victims of the hurricane needed cash in a hurry.

Under its contract's terms, the credit union transferred its data processing to Sun Data's hot site in Atlanta and, with the help of local communication links, was able to provide customer service. Credit union depositors needed money to get back on their feet in Hugo's wake, and federal officials wanted to know when funds would be available. By using the backup facility, installing some manual processes, and cooperating with other financial institutions in the area, the credit union provided critical service to its 52,000 account holders.

Because power problems usually result from and then cause other problems, firms are installing uninterruptible power sources to quiet systems and large data stores gracefully before switching to the hot site. Though expensive, these UPS devices also protect against damaging power surges.

Sound disaster recovery planning and effective plan implementation protect organizations and their reputations, maintain continuous customer service (sometimes when it's needed most), and indicate responsible IT management. No firm can afford to be without these important elements.

Organizations can survive almost any contingency if plans, equipment, backups, and money are available before disaster strikes. Protecting against disasters can be very costly; nevertheless, firms that elect not to prepare for disastrous events need to rethink their positions.

Contingency planning addresses high-risk events that have a low probability of occurrence. It ensures successful performance of critical jobs during periods when service or resources are lost or unavailable. Application owners are responsible for contingency planning to ensure their critical applications operate. For example, the production control manager who owns the materials requirements planning system is responsible for managing production if a major system hardware failure occurs. The service provider is responsible for correcting the failure. As in most other cases, the application's owner has more options and greater flexibility for handling the problem than the service provider does.

Split responsibility demands more, rather than less, interaction between application owners and service providers. Application owners should obtain critical planning information, such as probable outage frequency, likely outage duration, and anticipated service loss duration from the IT organization. Additionally, IT may recommend helpful action plans to the application owner. As in the disciplines discussed earlier, all affected parties must work together in a focused manner to serve the firm's best interests.

Crucial Applications

In consultation with application owners, IT managers should identify the organization's most critical applications. This important task is difficult because an application's criticality is subjective and can only be determined in context with other applications. That is, an application may or may not be critical, depending on whether or not other applications are available. For example, in a manufacturing plant order processing is an important function, but online order entry may be more critical. Order entry failure means business loss, whereas order processing failure may only delay the start of production. Frequently, interactions among applications are important in determining which applications are critical.

Timing is another important factor. Hardware and software supporting online applications such as air traffic control, oil refineries, and reservation systems, for example, are always critical. Payroll applications are generally critical weekly, but the ledger may be critical more often. Finally, the expected duration of the outage is another complicating factor. Most applications become critical during prolonged service outage.

Two reasons exist for performing contingency analysis in spite of obvious difficulties and ambiguities. The first is to prioritize applications that must be recovered early. In widespread failures, a predetermined priority sequence organizes recovery work and helps prevent chaos. The second reason is to understand the possible trade-offs in the prioritized recovery sequence. When hardware or telecommunications capacity is degraded, knowing what load can be shed in favor of higher priority applications is very useful. Defining and analyzing these trade-offs must involve system owners and users. A well-organized approach directs users' attention to the most critical tasks.

Environment

Each critical application requires a specific environment for successful operation. Table 13.7 identifies the important environmental factors.

TABLE 13.7   Application Environmental Factors

Hardware system components

Operating system and utility software

Communications resources

Databases

Sources of new input data

Knowledgeable users

Critical support personnel

Plans that include all critical environmental parameters must be developed for each important application. Interactions between critical applications and all other related applications or systems must be developed in the plan. Because this planning effort may be quite large, managers must understand the trade-offs between planning costs and the cost of loss or operational failure.

Emergency Planning

Emergency planning processes address situations stemming from natural disasters like floods or wind storms, and from events such as riots, fires, or explosions. Because these disasters usually affect considerable area, emergency plans must include the entire firm. At the corporate level, these are called business continuity planning or business resumption planning. Emergency plans must contain the steps needed to limit damage, solve problems, and resume normal business activities. Typically, uncertainty clouds these events, which have a low probability of occurrence.

The most effective mechanisms for dealing with such emergencies involve early detection and containment procedures to limit damage. Detection mechanisms include fire alarms and smoke, heat, and motion detectors. (Failures in alarm systems resulted in the outage that the Business Vignette describes.) In some cases, liaison with civil defense authorities is appropriate. Additionally, plans for handling emergencies usually include procedures for evacuation, shelter, containment, and suppression.

Plans must identify evacuation conditions, establish means to communicate evacuation plans, and contain procedures to ensure that evacuation is successfully completed. Shelter plans include protection from rain, leaking pipes, and water from other sources, in addition to protection from wind damage and flying debris. Containment plans usually involve storage rooms for fuel and other hazardous materials and for vital information such as tape or disk volumes and critical documents. These storage facilities are constructed with fire- and water-retardant walls, floors, and ceilings. Suppression plans describe procedures to extinguish fires, stop waterflow, clear the facility of smoke, and maintain security for the event's duration.

Emergency plans must be clearly and succinctly documented. Everyone potentially involved with emergency situations must know about the plans. Responsibilities must be established and directed, and individuals must be trained to respond effectively. Emergency plans must be tested periodically to ensure that employees and managers are completely familiar with their responsibilities and know how to respond effectively. IT managers are responsible for developing effective emergency plans for their facilities.

Strategies

Contingency planners have many options available. Among others, these options include manual operations, back-up systems, or data servicers. Back-up systems may be located at the same firm or may exist at cooperating firms. The firm may also use the services of hot-site providers, as the credit union described earlier did. Today, many firms maintain operating hardware and software configurations for others' use in emergencies.

Disaster recovery service is a large and growing industry. With growth rates of 20 percent per year, total industry revenue now exceeds several billion dollars. Services consist of hot-site providers, contingency planning and consulting services, and firms that sell software systems for disaster recovery planning. Many vendors are willing to work with IT managers to orchestrate disaster planning and recovery.

In most cases, manual procedures are usually the least effective disaster responses. Manual processing may be effective for a relatively simple system experiencing a brief outage. But manual processes are impossible to implement and largely ineffective in many instances. For example, manual processing cannot sustain an airline when a reservation system fails nor help engineers whose computer-aided design system fails. For relatively uncomplicated systems and short outages, however, some form of manual fallback is usually appropriate. For example, clerks can manually complete order forms for a low-volume, online order entry system while the system recovers from a brief outage. When the system returns to full operation, clerks enter data written on forms into the online system. Short-term manual methods may suffice for simple, noncritical systems or applications.

Severe outages demand additional back-up forms for all systems. Distributed system architecture may be quite helpful if broadband communications lines link a firm's geographically separated processing centers. Distributed architectures are especially effective if hardware and the operating system are highly compatible. Compatibility is important for backup or recovery purposes because it allows load sharing among and between systems. However, distributed systems may also be difficult to recover: their widely dispersed nature demands dispersed recovery operations.

Telecommunication networks are critical for transmitting programs and data between sites for backup purposes and for communicating results during recovery. IT managers, sensitive to recovery management issues, strive for network architectures that can help significantly during emergencies.

Sometimes mutually beneficial arrangements can be negotiated with cooperating firms because each needs to provide for contingencies. If technical details receive sufficient attention, cooperation may be mutually advantageous. Documents outlining the terms and conditions of agreements must be prepared and signed by representatives of each firm. Effective agreements outline procedures for testing back-up and recovery procedures and implementing the recovery plan.

Some firms provide data processing centers for other firms' use when severe outages occur. These "insurance installations" write a policy that provides fee-based system capacity for emergency use. Technical issues of hardware and software compatibility, and program and data logistics must be addressed. Network availability and capacity must also be factored into decisions to use these services. In addition, some major firms provide disaster planning tools or total disaster planning services for organizations unprepared to undertake this activity independently.

Service bureau organizations are another alternative in recovery planning. Frequently, these data-processing organizations are well equipped to handle additional processing loads. Your firm may already employ data servicers for peak-load processing or for special applications. Developing an emergency backup arrangement with them would be natural. In these cases, logistics of backup and recovery may be partially developed, thus simplifying emergency contingency plans.

Usually, telecommunication systems require special contingency planning because they are highly critical to the firm and offer unique planning opportunities for critical applications.[13] Firms with multiple locations linked via telecommunication networks must consider two issues: 1) how the firm maintains intersite communications in the event of network disruption, and 2) how the firm will use the intersite network to assist in backup and recovery for its individual locations.

Several considerations are important in maintaining intersite network systems. Usually, network redundancy, alternative routing, and alternative termination facilities are considered first. Modern networks usually provide ample opportunity to exploit these alternatives. In an emergency, it may be possible to exchange voice and data facilities, or to use the traditional dial network for data. Some firms employ value-added network firms for backup purposes. As an absolute last resort, manual processes may be considered. However, manual processes are very unattractive for network operations, even in emergencies.

Using the network for major application hardware and software systems backup and recovery must be addressed as part of the firm's system architecture. If recovery management is an architectural consideration in network design, considerable advantage can be obtained in recovery planning, and significant side benefits can be realized as well. For example, networks that exhibit recovery advantages usually permit efficient load sharing with corresponding cost reductions. Networks pose special problems and offer special opportunities for recovery management planners. Because of their critical role, networks require special consideration in the recovery planning process.

The goal of recovery management is to develop, document, and test action plans covering contingencies facing the IT organization and the firm. Actions must include all parts of the organization that engage in information technology activities. The firm must be prepared to cope with adversity in a logical, reasonable, and rational manner. It must protect its major information technology investments throughout the organization.

IT workers are a critical and indispensable resource. Their unique knowledge, skills, and abilities make IT personnel especially vital in times of emergency. Most recovery plans assume that employees will survive the disaster and carry out recovery activities. Because this assumption may not always be true, outside help sources should be considered. The personnel strategy must address the availability of skilled, in-house staff and adjunct personnel from external sources. Recovery plans must include provisions for notifying key people and for managing work assignments during the recovery process.

Disaster planning must consider the equipment and space necessary to conduct essential operations. Because some data processing equipment requires power, air-conditioning, and a raised floor, alternative space within the firm's facility should be earmarked for emergency use. Additional space, not currently owned or leased by the firm, should also be considered for planning purposes. Sources and availability of this space will change constantly, so the options should be reviewed periodically to ensure appropriateness.

Recovery plans must include written emergency processes outlining actions that recovery teams will take. Departures from standard operating procedures are expected during emergencies. For example, in an emergency, systems programmers may be assigned to all local and remote data centers to implement network recovery procedures. Because networks are critical, the first priority of key technical people will be to ensure network operation.

Emergency personnel rosters and recovery team assignments and responsibilities are important parts of recovery plans. All employees and managers involved in recovery operations must receive the written plan. To avoid destruction, key individuals must file the plans off-site rather than storing them at the firm.

Recovery plan testing is essential. Testing recovery plans periodically is as important as performing occasional fire drills. Testing plans and rehearsing recovery actions maintains employee awareness of their roles and responsibilities and focuses attention on this easy-to-defer activity. To test the completeness and validity of the recovery process, telecommunication links must be exercised and data retrieved and restored. Generally not all processes can be completely tested, but testing should be sufficient to ensure readiness. System owners and managers must be confident that recovery plans are thorough, effective, and ready to be implemented when needed.

"Companies should test their disaster recovery plan with a comprehensive drill at least once per year," according to Frank DeLuca, senior director of client services at Sungard Recovery Services. Although testing a plan may be difficult and costly, recovering from a disaster without a plan may be far more expensive. "There are economic ramifications if you don't have a good plan," states Mark Mizuhara, general manager at Iron Mountain, Inc., a disaster recovery consultancy in Boston.[14]

The following business vignette illustrates recovery management's importance and the extent to which it applies to the organization's daily operations.

A Business Vignette

Risk Management Succeeds

The 1989 San Francisco earthquake brought death and destruction to the Bay area. It damaged or destroyed roads, bridges, buildings, and utilities, but many computer installations survived virtually intact—in many cases less damaged than the buildings housing them. Computer system survival in an area known for its computer design and development activities resulted directly from disaster planning and advance preparation.

Most Silicon Valley computer-technology firms are critically dependent on computer systems for their operation—indeed, their survival—and they employ extensive risk management techniques.

Losing electrical power is one of the predictable risks of operating in the Valley. To protect against a momentary or prolonged power outage, companies use uninterruptible power supplies (UPS) that take over when routine power fails. Although quite expensive to install, UPS systems offer great protection against a common threat and are considered a good investment.

One Valley firm's mainframe computers continued to operate during the quake although less well-protected personal computer equipment sustained damage. A two-tiered back-up power supply supported mainframe operations—a huge battery system kicked in immediately, and diesel powered electrical generators activated later. The company closed one facility to check for damage and to make minor repairs but reopened the following day.

Another large company in Santa Clara County kept most of its facilities open immediately after the quake. Anticipating earthquakes, the company spent about $1 million to reinforce its buildings. This precautionary investment reduced the extent of the damage and saved money long term.

Another Silicon Valley firm considers these elaborate and expensive plans so essential that it appointed a director of corporate disaster recovery.

Problem, change, and recovery management are critical to managing computing facilities effectively. Built on service-level agreements, they support service commitments and meet some necessary conditions for success. Using these disciplined processes and others discussed later helps ensure successful computer operations. Computer operations is a critical success factor for IT managers.

These disciplined methods mesh well with daily computing system operations and with tactical and operational planning, so essential to system operations. The disciplines discussed thus far are based on the premise that thoughtful, conscientious individuals can work together to minimize the effects of inevitable difficulties that arise in complex environments. Service providers and service users benefit from this structured approach because it maximizes results to everyone's advantage.

As distributed computing evolves from departmental computers and client/servers to intranet and extranet applications, operational disciplines become more widely appropriate. IT organizations as well as service providers in operational departments must rely on systematic operational disciplines and procedures. Problem, change, and recovery management are vital disciplines for owners of data handling facilities regardless of organizational affiliation.

Review Questions

1. Define "problem" in the context of problem management. What is problem management?

2. Why are human and procedural issues within the scope of problem management? In what ways should these issues be treated differently from the others?

3. What sociological considerations surround problem management implementation?

4. What actions can management take to sponsor and promote effective problem management meetings?

5. Discuss the relationship between the disciplines of problem management, change management, and service-level agreements.

6. What are the tools and processes of problem management?

7. What items does the problem management report contain?

8. What are the ingredients of an effective problem resolution procedure?

9. Change management deals with understanding and controlling risk. What change management actions accomplish this?

10. How can managers ensure that they are sufficiently involved in the change management process?

11. What is recovery management? How is it related to contingency planning?

12. What interactions take place between service providers and service users during the development of contingency plans?

13. Why is developing a list of critical applications difficult?

14. What special problems and opportunities does a telecommunications network present in developing recovery plans?

15. Discuss the reasons why recovery strategies should be developed as part of the usual IT strategy and planning process.

16. Why is testing recovery plans difficult? What are some ways to overcome these difficulties?

Discussion Questions

1. Discuss the recovery management issues that the Manhattan phone outage raised for IT managers.

2. What special considerations enter into the recovery management process when a firm uses third-party service providers?

3. Because of the Manhattan incident and others, the FAA took action to correct its network defects. What similar options are available to firms? What actions can they take to reduce their vulnerability?

4. Relate the items in this chapter to the notion of critical success factors introduced in Chapter 1.

5. For the topics presented in this chapter, discuss the balance between bureaucracy, effectiveness, and efficiency.

6. How might one attempt to quantify the economic benefits derived from effective operation of problem and change management?

7. Discuss some approaches a firm might use to evaluate contingency and emergency plans and test recovery plans.

8. If you were chief information officer of a firm that relies extensively on client/server computing, how would you organize people to implement the disciplines discussed in this chapter?

9. What are the advantages and disadvantages of relying on informal organizations to accomplish the goals of these disciplines? If you were the firm's CIO, would you prefer using formal or informal organizations for these tasks? Explain your rationale.

10. Discuss some ways in which you could quantify the effectiveness of the processes discussed in this chapter.

11. Assume you are manager of a department and each of your employees has a workstation interconnected to all others in the department through a LAN. The LAN itself connects to the firm's central processing facility. Discuss important factors involved in the recovery management process for your department.

12. During which stage of growth is the subject of recovery management likely to arise? Why?

Assignments

1. Itemize the information elements from the problem management process that you think are important to report. Design the format for this management report. How frequently should you publish the report? To whom should you send it? As IT manager, discuss the follow-up action you would expect to take after issuing this report.

2. Visit your university computer center or the computer center of a local firm, and review its problem management system. Prepare a critique of your findings, and report them to the class.

3. Design a change log patterned after the problem log in Appendix B. Be sure the log clarifies the relationships between changes (e.g., the log must identify prerequisite changes, corequisite changes, and mandatory future changes).

Problem control number _____

Individual reporting the incident _____

Time and duration of incident _____

Description of the problem _____

_____

_____

_____

Problem category _____

Problem severity code (*circle one*)   1   2   3   4   5

Individual assigned to correct the problem _____

Estimated repair date _____

Actions taken to recover (*Append additional pages if required.*) _____

_____

_____

_____

_____

_____

_____

_____

_____

_____

_____

_____

_____

_____

Actual repair date (*problem closed*) _____

Final resolution actions, if any _____

_____

_____

_____

APPENDIX B   Sample Problem Log

Problem sequence number _____

Date of incident _____

Problem category _____

Brief problem description _____

_____

_____

_____

Severity code (*circle one*) 1 2 3 4 5

Duration of repair action _____

APPENDIX C   Change Request Document

Change log number _____

Problem log number (*if applicable*) _____

Description of requested change _____

_____

_____

_____

Prerequisite changes (*indicate change numbers*) _____

_____

Corequisite changes (*indicate change numbers*) _____

_____

Category of change _____

Priority (*circle one*) low   medium   high

Risk assessment (*circle one*) low   medium   high

Requested implementation date _____

Individual requesting change _____

Change manager _____

Management authorization _____

Attach test and recover plan.

Attach project plan if applicable.

[1] Anita Taff, "AT&T Relates Events Leading To Massive N.Y. Net Outage," *Network World*, October 14, 1991, 15; Ellen Messmer, "AT&T Outage Spurs Federal Action on Reliability Issues," *Network World*, September 30, 1991, 9; and Gary H. Anthes, "Phone Outage Drives FAA to Backup Strategy," *Computerworld*, August 3, 1992, 63.

[2] See Appendix A to this chapter for a sample problem report.

[3] See Appendix B to this chapter for a sample problem log.

[4] The author's experience with problem and change management is the basis for these figures.

[5] See Appendix C to this chapter for a sample change request document.

[6] John Kador, "Change Control and Configuration Management," *System Development*, May, 1989, 1. This article discusses how to test and implement changes with low risk.

[7] Application development frequently suffers from scope enlargement stemming from change requests. Many ideas in this section can be applied with minor modification to programming tasks.

[8] Although 98 percent of CIOs recognize the importance of disaster-recovery plans, 25 percent have none, according to *Computerworld*, May 26, 1997, 118.

[9] David Coursey, "Sabre Rattles and Hums, Crashes," *MIS Week*, May 22, 1989, 6.

[10] Julia King, "Business Continuity is Focus of Disaster Recovery," *Computerworld*, September 1, 1997, 17.

[11] "Rethinking Disaster Recovery After the Kobe Quake," *Data Communications*, July, 1995, 47.

[12] Ron Levine, "Disaster Recovery in Banking Environments," *DEC Professional*, January, 1988, 104.

[13] Common carriers must also have disaster recovery plans or risk customer loss to competitors. Some now advertise their ability to provide service in the face of natural disasters.

[14] Justin Hibbard, "Solid Plans Keep IS High and Dry," *Computerworld*, January 13, 1997, 59.

14

Managing Systems

Profile: Jerry Lenders
Position: Director of Technology, Infomart
Mission: "To learn enough from mistakes to prevent them from happening again."

Journalists are a demanding bunch, and nobody knows that better than Jerry Lenders, director of technology at Infomart in Toronto. He oversees what the company calls an "electronic printing press," an on-line database system that delivers news retrieval and library services to journalists at some of the top newspapers in North America.

"It has been said that we are the heart of the organization, and I think that is true," he notes. "If the system is not up, it causes aggravation for everyone, not just the people on-line at the time. Obviously, there is no revenue coming in, so it is crucial that we stay up because that revenue does not get replaced."

The database service, called Infomart On-line, also serves as a gateway to Southam News newswire, Dow Jones News/Retrieval Service and Datatimes, three on-line systems offering full-text retrieval of newspapers and other services. Lenders also manages an on-line system called Private File Service, a customized on-line database service used by 45 clients who need to store and manage text and numeric data.

The success of an electronic information business hinges on being able to offer subscribers ready, easy access to the information stored in its "electronic warehouses." "Newspapers don't want anything to get in their way," Lenders says. For Lenders, that means he must make certain that his system can provide more than 1,000 subscribers with instant access to information published in dozens of daily newspapers and from a variety of other sources round the clock: virtually 24 hours a day, seven days a week.

At the core of the Infomart system is a cluster of four DEC VAX minicomputers. Four other VAXs are available for overflow or as backup in the event that the primary cluster crashes. A triple redundancy in communications links helps ensure that lines stay open. Some 20G bytes of information, about half of it newspaper text, is stored on the system's disk farm of 15 drives. Managing the information that must be constantly available at a subscriber's fingertips can be daunting.

"This business has a lack of historical data as to what users really want: archival information or current information," Lenders says. "My gut feeling is that they only want to access current information, but we're monitoring it to find out." Lenders estimates that the database is growing at the rate of about 11M bytes per day. "We recently bought four DEC RA90 hard disk drives capable of storing 1.4G bytes each," Lenders says. "Those will last us until May or June of next year."

Keeping this growth under control is a challenge, he says. "We need to decide what information must be put on-line, what is put on high-speed storage drives and what will not be kept at all," he says. The newspaper files are backed up every night. "We operate the system for newspaper customers seven days on 22 hours; for others it is seven days on 20 hours," Lenders explains.

Lenders, who majored in computer science at the University of Waterloo in Toronto, began working at Infomart as a systems engineering representative (a programmer) in 1981, and two years later, moved up to become a systems engineering manager. In June of 1986, he became the director of technology with responsibility for the entire computer operation and its 16 employees.

He says that he constantly wrestles with providing adequate service to the company and its subscribers at the lowest possible cost. "The bigger challenge is that the technology is changing so rapidly," Lenders says. "That makes it difficult to stay on top of what is cost-effective. I am very conscious of the bottom line," he adds. "How do we maximize the bottom line and service availability? That is a fine, tricky line."

Since computer failure can be disastrous if its cause cannot be found and resolved quickly, guarding against that prospect is one of Lender's primary responsibilities. "We have an all-in-it-together attitude here, and we methodically resolve problems," he says.

"I don't mind when we make mistakes, but I always make sure that we learn enough from them to prevent them from happening again," Lenders reports. "Once is fine as long as we learn from our mistakes and plug the holes so that we can go on."

INTRODUCTION

Preceding chapters discussed service-level agreements and problem, change, and recovery management as the first elements in the disciplined approach to managing computer facilities. Managing computer systems skillfully is a critical success factor for managers; a well-organized process is essential for achieving success. Jerry Lenders understands this very well. His mission is to learn from problems and to prevent their recurrence while managing change in a rapidly growing environment.

This chapter builds on service-level agreements and the processes for managing problems, changes, and recovery actions. It explains tools, techniques, and procedures for implementing the remaining disciplines. Discussion of the disciplined approach to computer system management continues with the essentials of batch systems, online systems, performance management, and capacity planning.

Figure 14.1 displays the relationship among these processes and relates them to management reporting.

The underlying reason for developing and implementing batch, online, performance, and capacity management is to ensure attaining service-level objectives. Meeting these objectives is key to achieving customer satisfaction.

FIGURE 14.1   The Processes Continued

MANAGING SYSTEMS

Careful system performance management helps ensure the achievement of service-level objectives through effective and efficient use of systems resources. Users and providers of system services have expectations regarding service delivery. As discussed earlier, these expectations should be quantified through formal service-level agreements; nevertheless, whether they are specifically quantified or not, using system assets effectively and efficiently is a basic management responsibility.

And, as the Business Vignette suggests, system managers must provide service that meets customer expectations in spite of uncertainties. Jerry Lenders must provide continuous access to a large and growing database for customers who are uncertain about their specific needs. His success hinges on extraordinary system availability and adequate capacity in spite of growing but uncertain demand. Consistently high system performance is his primary objective.

Performance management defines processes for accomplishing measurably satisfactory output levels from hardware and software systems. Because both centralized and distributed programs deliver system performance to customers, definitions of batch and online system management processes must precede a detailed discussion of performance management.

Batch Systems Management

Managing batch systems includes processes for controlling the execution of regularly scheduled production application programs. Batch processing involves receiving incoming transactions, processing batch data and transactions, storing or distributing output data, and scheduling resources needed to accomplish these tasks. Typically, these regularly scheduled applications are defined and planned in service-level agreements. The quality of batch processing is measured against established service-level targets and reported to senior managers for review and analysis.

Large computing centers and most centralized computer facilities operate many important batch systems. Some examples of these vital systems include inventory reconciliation, work-center load analysis, and production requirements generation in manufacturing plants. Examples from the firm's finance department include daily accounts-payable processing or routine general ledger updating. Many product development laboratories use large simulation programs that are processed as batch systems. Usually, important work from personnel, sales, marketing, or service departments are regularly scheduled batch jobs, too.

Today, as central systems are downsized and client/server and other forms of distributed computing develop, some batch systems are being converted to online applications. Some provide online functions during the day and perform summary and other batch activities at night. Current batch systems and those planned for conversion to online operations require special attention.

Typically, an automated workload scheduling system—itself an application program—schedules and controls these production applications. The scheduling system maintains an orderly workflow through the computer center. Scheduling programs manage application dependencies by initiating batch jobs upon successful completion of prerequisite jobs. Automated job scheduling ensures the efficient use of operational systems, the orderly and organized progress of batch work, and the achievement of service-level agreements for batch operations. Only computerized scheduling systems can manage and control hundreds of batch operations in complex, large-system environments.

Routine computer center operating procedures include complete and unambiguous instructions for processing scheduled work in the absence of unusual occurrences. This means that job initiation information is complete, and that the operations staff knows and can meet conditions necessary to complete each job successfully. These instructions and information are part of the scheduling program's input data. In addition, instructions for handling potential difficulties must be part of these procedures, too. Examples of such difficulties include unsuccessful completion of a prerequisite job or job failure resulting from insufficient secondary storage space. Instructions for handling exceptional conditions are developed as part of recovery management.

Developing computer center processing schedules is part of operational planning. Operational planning consists of a long-range plan, a daily plan, and a current plan.[2] The long-range plan describes applications scheduled for the next several weeks or so. This plan includes daily, weekly, and month-end applications, and jobs scheduled on demand. A plan for each day's production is developed and entered into the daily scheduling system. In addition to managing work flow, scheduling programs report the current status of batch production work for system operators and managers.

To manage problems effectively, computer system managers must record deviations from established procedures that impact service levels or have the potential to do so. Unusual events such as unsuccessful job completion, unanticipated output, or atypical operator intervention must be logged and entered into the problem management system so that corrective action can occur. The change management team must review changes to production schedules, resource levels, or resource types required for production work. Managing production operations requires tight linkage to problem and change management.

In a multi-shift environment, system managers must ensure the smooth transfer of responsibility between shifts throughout the day. Usually a brief transition meeting helps communicate the system's current status to the incoming operations crew. Senior people from the oncoming shift should review an operations log maintained by the departing shift, so that orderly transition occurs.

Periodically system managers and their senior staff members must assess the effectiveness of the batch management process. They must evaluate process scope and methodology, job scheduling and resource management activities, and the relationship between batch, problem, and change management. Their review should develop recommendations and directives to improve current procedures for system management. In addition, their findings are very important to those negotiating and preparing service-level agreements.

Individuals responsible for hardware performance must also be tightly linked to those controlling batch management processes. Alterations to work schedules, variations in input transaction volumes, or changing requirements for online storage frequently affect system performance. These individuals may also request hardware or software system configuration changes. In addition, they may need configuration changes to optimize throughput, satisfy customer needs, or achieve service-level commitments. Solid teamwork and prompt, clear communication of changing conditions minimizes future problems and service-level disruptions.

IT managers and key staff members are vitally concerned with production operations because they are critical elements that strongly contribute to their success. To remain aware of operational status, successful managers must receive and scrutinize routine batch management reports and process improvement reviews highlighting operational trends. Variances from normal operations reported to the problem management system are high-interest items for IT managers, as are potential or developing work flow bottlenecks. Managers who obtain and evaluate critical status data skillfully achieve service levels and maintain satisfied customers.

In global businesses and electronic commerce operations, the batch window (the time when batch operations occur and online updates are prohibited) must decrease because online operations take more of the 24-hour clock. "While batch operations are very important and involve core business operations, there is tremendous pressure to have (online) systems available at all times," says David Floyer of International Data Corp.[3] Faster, clustered processors, improved data management systems, and new software tools to streamline batch processes relieve these pressures. Several important computer and database system developers build these software tools for their customers. Systems and techniques such as these become more important as online systems themselves gain importance.

Online Systems Management

Online systems management is a disciplined process consisting of information, tools, and procedures required to coordinate and manage online application activities. Online applications let internal and sometimes external users access important data through remotely connected devices such as personal computers, workstations, and various terminal types. System software and hardware needed for online applications are part of the application's configuration. Online management's objective is to ensure the achievement of online application service levels and the efficient and effective use of computer resources.

Today, many familiar applications operate online. Prominent examples of online systems include the order-entry and airline reservation programs discussed earlier. Other examples include inventory update programs, computer-aided design systems, and service dispatch and reporting programs. Many applications formerly operated as scheduled batch programs have been converted to client/server format and now operate as interactive, online programs. Typical examples include language compilers for application development, many database update and query applications, manufacturing floor control systems, and most decision-support systems.

As organizations install Internet technology, many new uses for interactive applications arise. These important new tools greatly increase internal communications (intranet applications) and also link critical business processes to outside organizations (extranet applications). These promising new uses for Internet technology are becoming mission-critical systems demanding skillful management.

To manage sophisticated online systems skillfully, people, tools, techniques, and facilities that monitor and control online application systems services must be well-coordinated. Common, well-defined control or interface points for users, vendors, and support groups accomplish this best. For example, control points must be defined for operations personnel, technical support individuals, and applications development and maintenance programmers, among others. Typically, an individual must be assigned to represent each important group. An individual in the operating-system support department, for example, might be the designated contact for operations personnel and for hardware vendors. Coordinating the operation of complex, online systems with several processors in various locations may require several individuals. A manufacturing plant in one country with subcontract operations in others linked by online systems is an example of this type of complex operation.

Specific persons at the control point(s) may be assigned several responsibilities. These responsibilities consist of user-interface functions and network-system functions. The user-interface functions respond to online application users and to programmers developing and maintaining online applications residing in the IT or the user organization. For example, an online application specialist from the system support department may be responsible for addressing technical issues with application programmers, or an experienced application programmer may give users technical guidance on application development. In each case, programmers have a place to obtain technical advice and express concerns.

The information center is a hub for coordinating these responsibilities. People seeking help should always have a fall-back contact point for resolving their concerns or contacting someone more qualified. This contact point, usually called the *help desk*, is generally in the information center. Table 14.1 shows some typical user interface functions.

TABLE 14.1   User Interface Functions

Receive user communication

Respond to user queries

Perform problem determination

Obtain technical assistance when required

Initiate problem management action

Provide guidance to users and support groups

The user-interface control point receives, records, and reports inquiries from online system users. At the help desk, callers describe current problems, seek assistance, or obtain advice regarding online applications. System users frequently inquire about the status of problems and changes, too. The person at the control point or help desk performs elementary problem determination, perhaps seeking assistance from appropriate support functions in complex situations. This person records and reports unresolved problems to the problem management system for resolution. In addition, the user interface person may guide users and technical support personnel for batch as well as online applications.

Table 14.2 lists activities of the online and network control point.

TABLE 14.2   Online and Network System Functions

Initiate, monitor, and terminate applications

Coordinate system terminal operator's activities

Monitor volume and performance indicators

Implement predefined network procedures

Develop problem determination procedures

Train personnel to use procedures

Coordinate application problem determination

The online and network control point can be an individual or group that initiates key online applications, monitors their operation, and terminates applications according to procedures. This individual or group coordinates activities of remote master terminal operators and system console operators. The control point also maintains volume and performance indicators for online applications. Examples of these indicators include the number of individuals using the application, current response times, and system resources devoted to the application. Both service-level management and management reporting require this data.

The online application control individual or group also implements predefined system and network back-up and recovery procedures. Technical support specialists in application development or network management departments usually develop these procedures. Control-point persons develop and test problem determination procedures and ensure that these procedures are customized for system users and their applications. They coordinate problem resolution associated with network devices such as servers, bridges, and communication controllers. Hence, they must cooperate closely with problem and change teams and with network managers.

To improve operations, online management processes must be reviewed periodically to assess their effectiveness. Findings and recommendations from reviews must be documented for managers' attention and action. Summary data documenting the effectiveness of online management processes should be directed to IT managers and, when appropriate, to online system owners/managers. Client managers are very interested in service levels and appreciate information on process effectiveness. Effectiveness reports help managers recognize trends in online services and service objective attainment. Openly sharing this information improves trust and confidence between organizations. Good relations are important to everyone.

PERFORMANCE MANAGEMENT

Performance management is a tool for defining, planning, measuring, analyzing, reporting, and improving the performance of hardware and operating systems, application programs, and system services. Performance management's objective is to use systems resources wisely to ensure the achievement of performance targets and service levels. Six elements form the performance management discipline: 1) defining performance, 2) planning performance, 3) measuring performance, 4) analyzing measurements, 5) reporting results, and 6) tuning the system.

Performance management applies to systems hardware, operating systems, systems utility programs, and application programs. The management process also includes application scheduling and other services that centralized or distributed computer operations provide. Because the overall performance of computer operations involves combinations of these resources, performance management must embrace them all.

Defining Performance

The volume of work accomplished per unit of time defines system performance. Performance measurements include CPU performance and associated hardware and software performance. Planning, measuring, analyzing, and reporting performance of large centralized processors focus on system throughput, the amount of work per unit of time. System tuning balances the CPU hardware, I/O channels and devices, networks, software, and application programs. The performance management objective is to keep all hardware components relatively busy while maintaining continuous, high-quality service to system users.

Service users measure performance by the number of jobs completed per unit of time, the number of transactions processed per unit of time, or job turnaround time for batch jobs.

Turnaround time is the elapsed time from job submission to job completion. Because the unit of work is not uniform for all these activities, planning and measuring processes are statistical. Reporting and tuning activity must account for the work's statistical nature, too.

Measurements can be very quantitative for large CPUs. Comparing actual instruction execution rates to its maximum rate determines the CPU's effectiveness. For example, if the CPU can execute 160 MIPs but remains in the wait state half the time (the hardware is not executing instructions but waits for work), the effective performance is 80 and the efficiency is 50 percent. The many other measures available for large CPUs are less important today because hardware prices have declined rapidly and online applications now dominate.

For the majority of today's interactive systems, such as client/server applications, user response time is the most critical performance element. Because hardware, software, and network components are specifically organized to improve users' productivity, the entire system (hardware, networks, and applications) must be optimized around users' productivity rather than hardware or software metrics. For critical online systems, users equate service levels to response time. For most of today's users, response time is *the* key performance measure. Therefore, all planning, measuring, analyzing, and tuning activities must focus on improving responsiveness to system users.

Performance Planning

Performance planning establishes objectives for human/computer system throughput and establishes procedures to ensure systems deliver the desired throughput. Performance planning is an integral part of the disciplines discussed previously; in particular, it is an essential step in service-level attainment. Performance plans must account for the amount and type of work to be performed as well as its distribution over time because workload characterization is the cornerstone of all performance and capacity planning programs.[4] The plan is based on known or assumed capabilities of combined hardware and software systems, including the performance characteristics of the applications themselves. For example, if online transactions require rapid response, system loading must be carefully controlled because responsiveness declines rapidly as system loading increases.

There are other examples. When program enhancement increases an application's efficiency, the entire system performs better. Reorganizing and tuning databases for specific applications improves response to data requests. Likewise, distributing the workload more evenly throughout the day also improves system throughput. In other words, system performance and system capacity are tightly linked. Improving system performance by tuning applications, for example, is equivalent to increasing system capacity. For this reason, performance management and all activities affecting system performance must be analyzed and clearly understood before capacity planning. Capacity planning before or without performance management is wasted effort.

Generally, no established parameters define system performance, but some measures are usually valuable: system response time, transaction processing rate, CPU time in the wait state vs. the system state, CPU time in the system state vs. the problem program state, and system component overlap measures. For instance, a CPU that spends an unusually high percentage of time in the problem program state usually indicates increased application throughput. Large system throughput also increases when input/output and processing operations overlap greatly. System programmers in the technical support department can identify many additional performance factors for large CPUs.

Regardless of the factors and methods used, detailed performance measures depend on the type of processing the system performs. For example, system performance measures for an airline reservation system differ from those used to evaluate a computer-aided design system. Rapid response to travel agents' requests is very important in reservation systems; however, instruction processing rates are more important for complex mathematical calculations that electronic design systems perform.

Technical support staff usually conduct performance planning for centralized and distributed systems using both general and installation-specific performance indicators. These indicators also form the basis of performance measurement.

Measuring Performance

Response time and throughput measurements under a variety of workload conditions are the basis for performance management. Key performance measurements include average transaction service time, transaction rates or number of transactions per unit of time, and average response time for transaction initiation. In addition, the amount of work or number of transactions the system performs for each user and in total is also an important performance measure.

The process for collecting measurements, establishing ranges of acceptable performance levels, and tuning applications and services must be well documented. Many performance measurements relate directly to, and are referenced in, service-level agreements. For example, the service-level agreement may specify subsecond response times for trivial transactions or may specify some number of transactions per minute for an online order-entry system. Usually, the organization's technical support manager has the responsibility and tools needed to collect data, analyze performance, and recommend corrective actions to enhance performance.

Large system performance management is concerned with key hardware components such as direct-access storage space, input/output units, main memory utilization, and paging/swapping subsystems. Performance management also includes controlling and managing system software components such as supervisor modules, program modules, and system input/output buffers. System programmers in the technical support department have the knowledge and tools to adjust system parameters and configurations to improve system performance.

Several methods can measure or monitor large computer system performance.[5] Widely available hardware devices or monitors can be attached to a computer system to accumulate statistics on system component utilization. For example, a hardware monitor attached to a computer input/output channel can measure the number of transactions processed and report the percentage of channel busy time. System programmers use this information to balance input/output activity among the system's channels.

Software monitors can also measure large computer system performance. Software monitors can either be part of the operating system or part of the applications themselves. Software monitors measure CPU activity such as CPU busy vs. wait state, CPU in the system state vs. the problem-program state, channel busy time, channel overlap time, and many other measurements.

Customized application program monitors can provide specific application data. This data may describe storage devices use or the frequency of module execution. From application monitors, programmers may determine transaction processing times in applications. They can generate frequency distributions of workstation usage by workstation type or location. The amount and type of performance data available is limited mostly by application programmers' imaginations.

Hardware monitoring devices—independent of the computer system operation—provide precise, detailed measurements for later analysis. Their use requires specific hardware system knowledge, and they are relatively expensive. On the other hand, software monitors degrade system performance somewhat, and they only measure information available to the computer instruction set. But software monitors are easy to install, easy to use, and relatively inexpensive. For these reasons, software monitors enjoy much more popularity than hardware monitors.

User response times typically measure performance of small, online systems or individual networked systems. Poor performance may result from overloaded workstations, inefficient or untuned applications, slow storage access, low modem speeds, or network constraints and bottlenecks. Most of these problems can be discovered and corrected; however, many organizations find that simply adding more capacity is quicker, although not always cheaper.

Finally, the user satisfaction surveys introduced in Chapter 12 are among the most useful tools for understanding system performance, especially for online and networked applications. Recall that user surveys should be developed and used to validate the credibility of performance measurements. Credible measurements reflect actual user experience. In addition to measuring user satisfaction, surveys' value lies in relating customer opinion to other current performance measures. Continual correlation of customer perceptions with internal measurements is an important technique for refining measurements and ensuring customer satisfaction.

Analyzing Measurements

Measurements yielding information to identify trends and bottlenecks in system performance are most useful. Data analysis, especially of peak workload periods, may reveal system bottlenecks. Reducing or removing these bottlenecks improves performance and increases capacity at no additional system cost.

When analyzing performance parameters, performance plans must be compared. Reasons for any departures from the plan must be understood to maintain service-level integrity and improve planning processes. Using performance measurements intelligently and conducting well-planned system tuning can greatly improve system performance and user satisfaction.

Performance measurements with comparisons to historical trends and the plan must be analyzed and published. System owners and/or IT managers need this information to assess service-level risk and the integrity of the performance management discipline.

System Tuning

When surveys or measurements detect that established performance levels are not being achieved, managers must take corrective action to improve system performance. These corrective actions include tuning hardware configurations (capacity increases, if warranted), input/output balancing, operating system performance improvements, system memory tuning, network optimization, and direct access storage space reorganization or reconfiguration. Other actions may be needed: tuning application programs that consume large amounts of critical resources and limiting the maximum number of concurrent users.

Periodically performance management processes themselves should be reviewed to determine their effectiveness. Findings, analysis, and recommendations should be presented to the management team for review and possible action. System managers, application owners, and users of IT service must regularly receive reports documenting performance levels achieved for all major application programs. Performance reports for applications and other services are prerequisites for capacity management and planning.

CAPACITY MANAGEMENT

Capacity management is the process of planning and controlling the quantity of system resources needed to satisfy users' current and future needs. Capacity management also includes forecasting physical facilities (electrical power, air conditioning, cable trays, raised floor, etc.) needed to install additional system resources. All disciplines discussed thus far are prerequisites to intelligent capacity management and planning. System capacity can be established only when managers deal systematically with problems and changes, and control the performance of batch and interactive systems. Built on previous disciplines, capacity management is the final step in achieving satisfactory service levels.

Capacity management's goal is to match available system resources with those needed to achieve service levels, while anticipating changes in system loads and service demands. Processing increased loads from current applications, additional applications, or service-level improvements may require additional capacity.[6] Meticulous capacity management also identifies surplus or excess resources, obsolete hardware, and unneeded software that can be removed from the installation. Capacity management relates directly to IT's financial performance and indirectly to the firm's financial performance. For many reasons, managing system capacity is a critical task for IT managers.

Capacity Analysis

The first step in managing system capacity is to conduct a detailed analysis of current system resource requirements. Based on present-day applications and services, this analysis establishes benchmarks for comparing proposed system configurations. Performance measurements calculate current system utilization. These measurements can be analyzed to determine average workloads and peak period workloads. Daily workload peaks and peak workload days should receive special attention during analysis. Additional analyses by service elements such as transactions processed, applications serviced, user-group workload, and department activity are needed as the basis for capacity planning.

Online storage is critical to computer operations and relatively easily measured. Therefore, it serves as a good example for discussing capacity analysis. If measurements reveal that dataset storage extensions during peak processing periods consume most of the available storage capacity, planners can predict job failure due to lack of storage as transaction volumes increase. These measurements tell planners to adjust the system configuration or workload parameters to prevent failure. Storage devices analysis can also detect datasets no longer used (a very common finding) or so infrequently used that they should be moved to offline storage.

Similar measurements and analyses can focus on CPU instruction rates, channel-busy percentages, input/output activity, user response times, and network loading, too. All system resources should be regularly subjected to detailed capacity analysis.

Capacity analysis tells system managers to what extent system components are used throughout the day. The analysis reveals processing bottlenecks, potential bottlenecks, and unused or underused system resources. Capacity analysis forms the foundation for capacity planning.

Capacity Planning

To satisfy future system needs, capacity planning must account for business volume growth, new application services, and improved service levels. Negotiating service levels with system users usually clarifies system load projections. The organization's business plan is the basis for establishing workload information—all workload data must be consistent with it and with current and projected service levels. Thus, service levels and the process for establishing them are essential to system capacity forecasting and planning.

System managers are responsible for collecting requirements and forecasting workloads. They can use many techniques, ranging from simple to complex, to forecast systems requirements. For example, if processing capacity is critical to the operation, then capacity management may hinge almost completely on CPU parameters. For most systems, however, capacity analysis is a composite of many measurements. Some cases may require system simulations for complete analysis. The key to capacity management is selecting the simplest technique that provides a satisfactory degree of accuracy.

Most successful computer installations maintain a database containing previous workload projections and capacity requirements. These historical databases are essential for studying trends in workload and capacity growth and for improving forecasting techniques. Combining service levels, system tuning, and capacity planning permits heavier loading of CPUs, less unused capacity, and lower costs, because new capacity purchases are deferred.[7] Capacity management is a cost-effective process.

In addition to volume changes in current applications, load increases from new applications or new functions, and planned changes in service levels, current capacity excesses or deficits from earlier analyses must be considered. For example, if present direct access storage is 90 percent utilized, and volume changes and new applications are expected to increase storage requirements by

20 percent over the planning period, then the system requires at least 20 percent more storage. Conservative planners may strive to increase storage 30 percent.

All other system resources must be similarly analyzed. For instance, if response times to online transaction processing is deteriorating, capacity planners have several choices. They can tune the online system to improve its performance or adjust the operating system configuration or system priorities to improve response. They can adjust or reschedule other work running in the multiprogramming environment to improve response, or they can increase system resources. They can increase system resources and improve response by installing higher speed storage devices for online programs and data, or perhaps they can reconfigure the network to improve performance. Finally, they can acquire faster CPUs or servers, or an additional server. System planners have many options for increasing system throughput.

The final result of capacity management is an optimized configuration of hardware, system software, and application programs carefully tuned to user needs. This optimum system configuration satisfies the organization's needs and achieves the performance promised in service-level agreements. Information derived from the capacity management process flows to the IT planning process where system alterations and their costs are planned for future years.

Capacity planning and management is the summation of all disciplines discussed earlier in this chapter because they all impact capacity requirements. For example, the method of handling problems directly relates to workload: good problem management reduces problems, lowers workload, and reduces capacity requirements. The same is true for change management. Efficiencies in batch and online systems' operations directly reduce needed capacity. In other words, poorly managed batch and online operations require capacity increases that well-managed operations do not. Finally, poorly handled or absent performance analysis and planning makes capacity planning ineffective. Proper capacity planning cannot occur when significant performance uncertainties exist.

Distributed System Capacity

Capacity analysis and planning in distributed systems, client/server systems, and Internet applications hinge on system performance measured by response time from the user's perspective. In many implementations, systems include powerful servers, massive hierarchical databases, and complex network components. Understanding these systems' operation is especially critical to avert potential bottlenecks. The system's operation is subtle: sound measurements are required.

Rapidly declining hardware prices tend to persuade managers that they can obtain additional capacity cheaply and easily, and that measurements are difficult to obtain and cost more than they're worth. For example, if work-station response declines, some managers elect to replace workstations, hoping that devices with faster microprocessors will improve response. They may or may not, depending on the bottleneck's initial cause. If the bottleneck is workstation speed, then a faster CPU improves performance. But if the bottleneck is in the server or network, a faster CPU leaves performance largely unchanged; the organization has expended resources for little or no gain.[8] Performance can truly be bought at the store, but knowing exactly where to spend money has no substitute.

As networks become more complex and firms become more dependent on them, soundly based capacity planning becomes more important. Bottlenecks are much more difficult to identify in complex networked environments, and up-front planning becomes more urgent.[9] Capacity planning for client/server, EDI, and extranet systems is more difficult than capacity planning for large, centralized systems, however, tools to help analysts and planners are becoming available.

"Seat-of-the-pants" methods cannot achieve satisfactory capacity planning in complex, user-oriented, distributed systems. Managers who substitute guesswork for solid analysis in complex systems pay the price of system failures. Network planners must rely on historical databases, network modeling, and simulation tools to assist them in the planning process.

Additional Planning Factors

Effective capacity analysis and planning requires antecedent processes. Some organizations try to determine capacity without completing the prerequisite steps. And some publications address capacity without recognizing the system's inherent dependency on changing business conditions and performance analysis and planning.

Senior systems managers must remain alert to conditions that potentially affect computer system demands. They must carefully scrutinize additional business information that bears on capacity forecasting. Managers must consider some important factors when planning system capacity:

1. Changes or alterations in strategic direction destined to improve or increase IT services

2. Business volume changes in either direction

3. Organizational changes (always a potential impact on IT resources)

4. Changes in the number of people who use IT services[10]

5. Changing financial conditions within the firm

6. Changes in service-level agreements or service-level objectives bearing on system performance requirements

7. Portfolio management actions such as new applications or changes to current applications that impact system throughput

8. System resources required for testing new applications or modifying current applications

9. Applications schedule changes initiated by operations or user managers

10. Schedule alterations for system backup and vital records processing

11. System outage data and job rerun times from the problem management system

The first five items result from changing business conditions within the firm; business plans usually reflect them. Changes in important factors such as volume, financial, or budget can occur between plans, usually as results of external conditions. IT managers must be particularly sensitive to these factors. For example, if the firm experiences unplanned growth in sales revenue, some client organizations may receive an unplanned budget increase. For example, client organizations may spend 10 percent of their budget on IT services, but their marginal spending for information technology may be 50 percent. This occurs because as revenues rise and functions receive additional discretionary resources, they spend more on systems to handle increased volume. The reverse is also true. Thus, fluctuations in demand for IT services are likely to exceed fluctuations in the firm's business activity. It is critical that IT managers and other service providers understand this phenomenon.

Information on the remaining six items is available from the production management disciplines or from individuals within the IT organization. Persons responsible for problem and change management, batch and online application specialists, information center personnel, and IT customer service representatives all gather information that helps to formulate workload projections. Clients also must be coached to alert service providers when they first anticipate changes in requirements. Sound capacity planning depends on a continuous stream of critical information. Validation procedures must accompany the continuous process of information gathering. As part of the validation process, the IT steering committee should examine and concur with unusual or unplanned workload projections.[11]

LINKING TO SERVICE LEVELS

Equipment plans developed from the capacity management process specify hardware components needed to meet promised service levels. Equipment plans translate into equipment, installation, and setup costs, and they also specify additional supporting facilities. After developing the optimal system configuration, the computer and network facilities must be evaluated to ensure that the proposed configuration can be housed properly. The configuration may require additional equipment and space. The capacity management discipline is highly important because it drives the budgeting and planning process and specifies resources needed to meet service levels.

Periodically, capacity management processes should be reviewed to assess their effectiveness. Close agreement between previous capacity forecasts and actual capacity requirements is one effectiveness criterion. Another is whether planned and installed capacity is sufficient to satisfy service-level agreements. Predicted vs. actual results of workload, capacity, and service levels should be compared for each client application. These comparisons should be made even if hardware or software has not been changed. Results of capacity reviews, including findings and recommendations, should be recorded for later scrutiny and analysis and retained for future reference. Subsequent service-level negotiations require reports detailing capacity requirements for each major application.

Service-level agreements are the foundation of the disciplines for managing computer operations, and management reporting is the capstone. The discussion of these processes has emphasized reporting's essential role. Widely communicating results of each process to managers enables them to make sound decisions and process improvements. Such communication greatly contributes to process improvements and increases trust and confidence among participants. It enables managers to succeed in a vital and critical area. Important, introspective, freely flowing information is essential to process improvement and operating excellence. Thus, for several reasons, management reporting is important.

IT managers and system managers are not exclusive users of management reports: IT customers find them essential as well. Mature, successful managers take every opportunity to share information, knowing that sharing important facts and insights improves performance. As Jerry Lenders observes, "We have an all-in-it-together attitude here, and we methodically resolve problems." Lenders' attitude is extremely important to all Infomart managers and to managers in other successful firms. Given the chance, managers in most firms prove themselves reasonably adept problem solvers. The system must give them that chance.

Management reporting and all informal communications are highly important to excellent operations even if reporting reveals flaws and invites criticism. Reporting problems, failures, and successes exposes individuals and organizations to criticism and praise. More importantly, open communication builds increased trust and confidence. These are important assets, even if attaining them makes managers vulnerable in the short run. Good managers understand the value of good communication. Effective managers insist on honest and complete reporting processes.

SUMMARY

Infomart's Jerry Lenders balances requirements for customer service, system availability, and recovery protection by providing double redundancy in CPUs and triple redundancy in communications links. The redundant capacity stems partly from uncertain customer demands and partly from the need for high availability. All capacity costs money, and Lenders' essential tradeoff is between system costs and system benefits to Infomart. In addition, Lenders must understand technology advances and relate them to business needs. Lenders exemplifies the role that many successful IT managers play in firms today.

This chapter presented tools and techniques for managing production systems—batch systems and online systems. In large centralized systems, departures from normal operations lead to problem-management action. Effective online operations require defined control points and responsibilities. Control points ensure answers to user questions, proper resolution or coordination of problems and changes, and coordination and control of online or distributed systems. Because distributed systems' users are often physically dispersed, control points function as communications hubs for system developers, operations personnel, and system users.

Centralized and distributed applications critically depend on computer capacity and on high performance of system resources. The link between system performance and system capacity is strong because improvements in performance, achieved through system tuning, effectively increase system capacity. In fact, this may be the cheapest capacity available. Capacity planning, however, can succeed only if all other disciplines operate well.

This chapter and preceding chapters describe the management processes that form a roadmap leading to successful computer operations—a critical success factor for service providers in IT or operational departments. Completely implemented, this management system ensures success. Implementation may seem arduous and time consuming, but shortcuts or circumventions lead to inefficiencies and, ultimately, to system failure. Failures in centralized or distributed operations lead to failures elsewhere. *Ad hoc* management of complex operations such as client/server or Internet implementations always causes managers extreme distress. Successful managers embrace the management system described here; they thrive on its benefits and enjoy its rewards.

Review Questions

1. Describe the relationship between service-level agreements, management reporting, and the disciplines this chapter discusses.

2. What are the essential reasons for developing and implementing the management processes this chapter describes?

3. Describe the relationship between system performance and management expectations at Infomart?

4. Define batch management.

5. What are some examples of batch systems? Do you think online systems are more difficult to manage? Why or why not?

6. Why are batch systems difficult to schedule? What computer program functions can streamline this task?

7. Describe the inputs that computer operators may provide to the problem management system.

8. Some computer systems operate around the clock. What special problems does this pose, and how are they solved?

9. Describe the relationship between operations personnel and the processes of performance management, capacity management, and problem management.

10. Define online management. What kinds of communications difficulties do online systems create? How can they be overcome?

11. What is the scope of online management?

12. Compare and contrast the user interface and the network system functions. How do their responsibilities differ?

13. The online and network system function initiates, monitors, and terminates applications. In your opinion, where in the organization should this function reside? Explain your rationale.

14. With whom do network control point personnel interface?

15. What is performance management? What are the processes of performance management?

16. How is CPU performance usually measured?

17. How do system users measure performance?

18. What does performance planning accomplish?

19. What tools are available to measure system performance?

20. What is the connection between system bottlenecks and system tuning?

21. Explain capacity management. What are its objectives?

22. How are capacity analysis and capacity planning related?

23. In what ways does capacity management depend on the previous disciplines?

24. What is the link between capacity planning and service levels?

25. Why is management reporting called the capstone of the management processes?

Discussion Questions

1. Discuss the essential difficulties that Jerry Lenders faced in his Infomart job.

2. Discuss the reasons why disciplined processes are essential in attaining committed service levels.

3. Discuss the connections between batch and online systems and the disciplines of problem management and change management.

4. Many batch programs are being converted to online systems. Describe the implications of this conversion to performance management and capacity planning.

5. You have read that the management processes must be examined periodically for effectiveness. In your opinion, why is this important? How might you do this?

6. Reporting results of the performance management process is important to several groups. Discuss the importance of reports to system operators, system programmers, application programmers, system users, and managers.

7. System performance is subjective. Its meaning to system programmers may differ from its meaning to application users. What accounts for these differences? Why must a common ground be found?

8. Discuss the advantages and disadvantages of hardware monitor devices and software monitoring tools. Why might combining devices and tools be most effective?

9. Analyzing performance measurements, reporting results, and tuning systems are iterative processes. Discuss the IT manager's role in these processes.

10. Discuss why the processes of problem management through performance management must precede capacity management. What difficulties might arise if the recovery management discipline does not precede capacity analysis?

11. Discuss the trade-offs among the processes of IT planning, service-level agreements, recovery management, and capacity planning. What risks are always present, regardless of the manner in which trade-offs are made?

12. Discuss the reasons why organizational changes should always stimulate a review of system capacity factors.

13. Information technology managers are change agents. How does this statement relate to your answer to Question 12?

Assignments

1. List the reports generated by the management system outlined in this chapter and Chapters 12 and 13. In your opinion, who should prepare these reports? Who should review them? Prepare a schedule for these reporting and reviewing processes, and describe the resulting management process.

2. The processes described in the disciplined approach to computer operations are more easily applied to mainframe installations. How can these principles be applied to minicomputer installations? What principles are relevant to client/server implementations? How would you implement them?

3. Assume your organization, which operates seven sites in the U.S., wants to install intranet systems to improve communication. How would you advise them regarding the disciplines discussed in this text? How would you suggest the organization handle recovery management?

ENDNOTES

[1] Michael Alexander, "Lenders Maintains Information Flow," *Computerworld*, December 12, 1988, 87. Copyright 1988 by Computerworld, Inc., Framingham, MA 01701; reprinted from *Computerworld*.

[2] Israel Borovits, *Management of Computer Operations* (Englewood Cliffs, NJ: Prentice-Hall, 1984), 191.

[3] Tim Ouellette, "Data Centers Close Window on Batch Time," *Computerworld*, November 24, 1997, 4.

[4] H. Pat Artis and Alan Sherkow, "Boosting Performance with Capacity Planning," *Business Software Review*, May, 1988, 67.

[5] Borovits (note 2) discusses various types of performance measurement techniques, 153-179.

[6] Borovits (note 2) describes a quantitative approach to hardware selection based on capacity measures, 55-64.

[7] Craig Stedman, "Capacity Planners Press CPU Limits," *Computerworld*, December 19, 1994, 57.

[8] Dennis Hamilton, "Stop Throwing Hardware at Performance," *Datamation*, October 1, 1994, 43.

[9] Craig Stedman, "Capacity Planning Takes Client/Server Steps," *Computerworld*, November 28, 1994, 75. This article lists some tools available for capacity planning in client/server environments.

[10] Douglas J. Howe, "A Basic Approach to a Capacity Planning Methodology: Second Thoughts," *EDP Performance Review*, October, 1988, 1. According to this article, end-user computing increases the complexity of capacity planning and extends the planning horizon to 10 or more years.

[11] The steering committee may not always agree with the projected additional load; nevertheless, the committee can help rationalize predicted workloads and additional capacity with business financial objectives.

15 Network Management

Web Warriors

Private firms now battle for a share of the package transport industry, a burgeoning, multibillion-dollar enterprise once largely monopolized by the U.S. Postal Service. Started in 1913 by the Postal Service, Parcel Post was the mainstay of U.S. package transport into the mid 1950s. Over the decades change at the USPS has been modest, but it still touts Parcel Post and Express Mail as reliable, low-cost alternatives to services that entrepreneurial companies offer—services that have overtaken Parcel Post's package delivery.

The Postal Service continued to see itself as a cash and carry business—you carry the parcel to the Post Office and pay cash, and they carry it to its destination city where they may or may not actually deliver it to the addressed location. Meanwhile its far more aggressive competitors served customers' needs from source to destination, from pick-up to delivery, guaranteeing schedules and nearly perfect reliability.[1] Since the 1980s, private firms have increasingly fought and won the battle for customers with high-tech weapons like 757s, satellites, and the World Wide Web. Today, sophisticated private firms struggle for larger shares of business, the USPS falls further and further behind.

Now that the Internet and Web technology virtually link the world, the big three of package delivery—FedEx, UPS, and DHL—tie their services ever more tightly to advanced technology, constantly striving to improve them for customers. Especially since 1994, the big three rely on the World Wide Web to offer advanced services unmatched by most other businesses. FedEx's Web site and core business systems are so entwined that considering them separate entities is essentially impossible. Averaging 26,000 package tracking requests daily, FedEx treats its Web site "just like any other mission-critical system," says FedEx's VP of network computing, Winn Stephenson.[2]

For these businesses, the adoption of Web-based technology has proceeded at a torrid pace. FedEx initiated Web competition in November 1994 by establishing a home page with a package tracking feature. UPS established its home page one month later and initiated package tracking in May 1995. Close behind, in July 1995, DHL launched a home page featuring drop box/office locator services, which are important to this international carrier. Package tracking by DHL debuted in April 1996. UPS initiated Web shipping features in March 1996; FedEx quickly adopted similar features in July 1996. Both services rely on the downloadable software provided to customers earlier in 1995. During this time these carriers also offered various other features.[3]

As in other mission-critical systems, maintaining Web sites that match or leapfrog competition is a full-time job that requires substantial resources from the big three. For example, Alan Boehme, DHL's director of customer access and logistics plans to spend nearly $800,000 to rebuild its Web site for increased function and future growth. "Anyone who thinks you can keep a Web site updated... for tens of thousands of dollars should recheck their figures," he claims.[4]

Although expensive, maintaining Web sites is only a small part of the big three's technology bill. FedEx has its Cosmos system (Customer, Operations, Management, and Services), and UPS began developing COMPASS (Computerized Monitoring, Planning, and Scheduling) in the 1980s for fleet and load management. To handle 12 million packages daily at more than 600 national and international airports, UPS operates UPSNet with 500,000 miles of communications lines and a UPS satellite linking 1300 centers in 46 countries. UPS spent $1.5 billion on technology between 1986 and 1991 and $3.2 billion between 1992 and 1997.

Meanwhile the U.S. Postal Service's Pete Stark says, "Our mission is to proceed in as deliberate and as careful a manner as possible so we don't jeopardize the security of credit card numbers, the information we're taking, or our own network in the process." In December 1996, USPS' introduction of a transactional Web service so customers can order free Express Mail supplies, suggested that it is still pondering security measures. Today, the idea of tracking parcels through the postal system is a fantasy that is unlikely to become a reality.

For most businesses, however, the Web is here and now. According to a 1997 study, about 80 percent of Fortune 500 companies have Web sites and five percent conduct Web-based business transactions.[5] These companies' median cost for developing a top-notch Web site is about $55,000, and annual operating costs average $50,000 according to this study. Nevertheless, as these Web sites develop mainstream operations like those of other Web warriors, both costs and benefits will rise appreciably.

INTRODUCTION

Previous chapters emphasized disciplined processes for managing system operations, discussed the importance of customer expectations, and described problem, change, and recovery management techniques. Performance analysis, capacity planning, and management reporting completed the management processes for system operations. In modern organizations, centralized and distributed computer operations are tightly linked, depending heavily on communications networks to transport most data arriving at, or sent from, their computer systems. Because networks are critical parts of most information systems, managing them deserves very special attention.

Networks are the unifying ingredient, the focal point of information infrastructures in modern firms. LANs connecting personal workstations are the most rapidly growing segment of the telecommunications market, indicating that small firms and departments in large firms are rapidly adopting networking technology. Ciba-Geigy, H. J. Heinz Company, G. D. Searle, and many other firms have undertaken major telecommunications projects to link operations electronically. Target Stores and Van Heusen are implementing T nets within and between their facilities. Merrill Lynch links its New York and New Jersey headquarters' facilities with 500 domestic branch offices and international branches in Tokyo, London, Singapore, Sydney, Bern, and Toronto. Networking's importance cannot be overemphasized in today's global economy. As we learned in the Business Vignette, Internet and Web technology are powerful weapons in an organization's strategic arsenal.

Networks greatly increase organizations' ability to process information. Networks add value as well as complexity to information infrastructures. Fortunately, many traditional information system management techniques and practices also apply to networks. For example, network managers find that developing strategies and plans, managing expectations, and exercising operational disciplines are all important.

This chapter addresses many network operational and management issues. It develops a management system foundation upon which network managers can rely for success.

THE INTERNET

No discussion of modern networks and network management would be complete if it didn't begin with the Internet. Fueled by powerful servers, advanced transport techniques, new programming languages, and intellectual and economic interests, the Internet has become the information superhighway's fast lane in less than a decade. From the beginning, the Internet embraced individuals and organizations globally; it now grows so rapidly that its participants are too numerous to count accurately.

One measure of the Internet's size and growth is the number of hosts advertised in the domain name system. According to data compiled by Network Wizards valid as of January 1998, the Internet has 29.7 million hosts, up from slightly more than 16 million one year earlier.[6] Continuation of such high growth rates is unlikely; nevertheless, the Internet in all its manifestations is still in its infancy—its importance will grow substantially in the years ahead.

The World Wide Web—the world's largest client/server network—and firms' traditional client/server applications converge as many organizations rapidly implement intranets. In fact, combining routine client/server operations and Web technology for internal business applications legitimizes the World Wide Web model for business communication applications. Today, the Internet heavily influences business computing models.

The Internet gives organizations important insights about using network-based technology productively. For example, the Internet and the World Wide Web teach us how to integrate many information types in multimedia form and develop online business-transaction systems that require little or no human intervention. From Internet technology, businesses learned to create cooperative efforts crossing organizational and national boundaries and to build systems encouraging cooperation among widely dispersed, diverse individuals and groups. For many organizational purposes, Web technology is unsurpassed as a multimedia communications vehicle.

Furthermore, the marriage of traditional client/server systems and Web technology fosters the type of organization that Drucker cited in the first chapter. Adopting Web technology further reduces information's value as a power source for traditional managers and accelerates trends toward participative management, self-managed groups, and empowered employees. These trends are irreversible. Powerful new communication technology changed forever traditional power structures and organizational relationships.

Introducing Web technology to internal applications adds new dimensions to network management, too. In addition to network managers' usual technical tasks, organizations' potential Web technology users need considerable support with Web authoring tools, HTML training, and other services. Network managers' traditional tasks can no longer be separated from developing and using Web technology. Who is more qualified to answer questions about firewalls, file transfer protocols, and audio streaming than network managers and their technical partners?

NETWORK MANAGEMENT'S IMPORTANCE

Today, many important information technology applications center around well-managed networks. They provide the basis for many well-known and emerging strategic systems such as Amazon.com and e-trade, and they stimulate growth and development of global information systems and international businesses. Network technology encourages and enables restructuring and business process improvement. Between firms, networks support alliances and joint ventures, link suppliers and customers, and facilitate many cooperative endeavors. Accordingly, networks, network applications, and their management deserve the utmost attention.

Networks Are Strategic Systems

Network management is increasingly important to businesses and their IT managers for several important reasons. First, intense interest in networks stems from their obvious and growing operational and strategic value. We learned earlier that most strategic systems rely heavily on telecommunications technology; for example, reservation systems, brokerage services, and many other information systems are telecommunication based. And new, rapidly emerging telecommunication technology such as groupware, enhanced video conferencing, mobile computing, Web commerce, and many others have great tactical and strategic importance to modern organizations.

Most insurance companies, for example, actively use telecommunications services. Security-General expects independent agents to sell its policies rather than competitors' because of superior agent service. Their system provides e-mail communications to headquarters, a policy forms library, and a facility that generates policies online. Security-General recovers its costs by charging agents a flat fee for using the online service. Agents recover their fees by writing more policies.

Thousand of firms, from trash recyclers to catalog merchandisers, from insurance companies to public utilities, extend real-time personal service using automatic number identification linked to workstations and customer databases. Boston Edison, for example, installed Computer Telephone Integration (CTI) to reduce workload, increase customers' convenience, and improve customer service in general.[7]

Well-planned networks, integrated into important business systems, generate advantage because they reduce costs, save time, establish service excellence, and upstage competition. Business processes enhanced with telecommunications technology, whether internal or external to the firm, deliver leverage to the firm through increased organizational effectiveness.

Networks Facilitate Restructuring

"Communications networking is an extremely important—some have argued the most important—part of a company's process information infrastructure. Once a network is established, a company often discovers many possible process innovations that might result from its use."[8] With these words, Tom Davenport underscores why decentralized business activities and IT operations are important current trends shaping corporate strategies.

Today, information technology, particularly telecommunication systems, lets business processes transcend the firm's boundaries or its country's borders. With electronic data interchange (EDI) or Web-based technology, technical and business information flows freely from one firm to another. Interorganizational systems link similar or supporting business processes with minimum human intervention. Today, sales and service data streams through LANs, volumes of business and management information travel commercial networks, and important company information cruises the Internet. Overestimating the value of this communication is nearly impossible.

The cooperative activities of five companies in five countries in building a new generation jet engine provides an excellent example of telecommunication-based enterprise. Located in the U.S., Great Britain, Japan, Germany, and Italy, the companies are jointly building jet engines, worth approximately $4.3 million each, for 2000 new aircraft over the next 20 years. This cooperative effort is possible only because a 24-hour communications network supplies data on design control, bills of material, parts catalogs, tool design, and many other items to the firms.[9] This example demonstrates that connectivity of systems, data, applications, and processes is the critical component of decentralized or allied businesses. Modern network technology accelerates restructuring and trends toward decentralization. Thus managing networks that make these important business transformations realizable is critically important to corporate strategies.

Other technologies, principally rapidly growing personal computers, make networks and network management more important, too. More than 40 PCs per 100 people operate in the U.S., about twice as many as in any other country, and many home PCs and business workstations connect to networks. However, PCs and workstations stimulate only part of network growth. The greatest stimulus comes from applications like electronic mail, facsimile, telex, telephone, voice mail, audio conferencing, computer conferencing, electronic document transfer, and video conferencing. To accommodate these services, an all-purpose workstation for personal audio, video, text, and image communication is in our future.

Network management is also important because networks are now more complex and difficult to manage. Network complexity grows rapidly because the number of nodes increases rapidly and communication paths grow longer and more convoluted as internetworking grows. Many more network products like concentrators, multiplexers, modems, and routers are now available, and customer equipment is also proliferating. Additionally, network and customer equipment suppliers frequently offer incompatible products. Although standardization progress is significant, incorporating multi-supplier products in networks poses severe technical and management challenges.

IT managers and their firms desire network integration, even though this objective strains network managers' present capabilities. Although most firms plan to spend more managing networks, availability of technical and management skills limits many. The large and complex installed base of networks and network devices rapidly grows larger and more complex. As a result, large pent-up demands for skilled network managers and effective network management tools and techniques exist.

Business and government executives are intensely interested in network capabilities and the opportunities networks offer their organizations. Additionally, the 1996 Telecommunications Act in the U.S., privatization in many countries, and dynamic worldwide industry repositioning have drawn executives' attention to telecommunications.

NETWORK MANAGEMENT'S SCOPE

Successful network managers employ processes and procedures to help achieve efficiency, effectiveness, and customer satisfaction in network operations. These management practices apply to hardware, system software, and application software that transports voice, data, or image information within and beyond the firm. Network management focuses on communication assets, including those in the firm's application program portfolio. However, most firms do not own all their communication assets; they usually lease some, like bandwith. Thus, network management's scope is broad and diffuse, rather than narrow and concentrated.

Physical network assets must be managed, including customer terminals, local cabling, concentrators, modems, multiplexers, and lines or links. Frequently, the physical system includes private branch exchanges and computer processors. Communication software, some application programs, and databases supporting servers, routers, and workstations must also be managed. Previous boundaries between telecommunication and information system management disappear rapidly as the technologies themselves become increasingly indistinguishable.

Network management and online management disciplines are also merging. Firms with extensive traditional data-processing operations use the management systems discussed earlier to manage these processes; those with extensive telecommunication systems use the management systems described in this chapter. Firms with both must blend network management.

IT managers must not make arbitrary distinctions that may annoy system users. For example, application users who experience problems want prompt solutions; they don't care whether the problem originates in RBOC equipment, local cabling, or their application. Their expectations of high quality service are entirely reasonable. IT managers can provide effective customer support only by integrating and coordinating portfolio, operations, and network management. IT managers must provide seamless, transparent telecommunication services to users. Successful IT managers meet this challenge.

Critical Success Factors

As in other segments of information technology, network managers must understand their critical success factors. CSFs developed in Chapter 1 for IT businesses apply to all its parts, including telecommunication systems. Sound business management, strategy development, and planning apply to network management, just as they do to other information technology areas. And, responding to customers' operational needs is always critically important.

However, some additional factors are critical for network managers, according to Kornel Terplan.[10] He considers the following items critical success factors for network managers.

Methodology:	How to proceed under certain circumstances.
Tools:	What are the most meaningful instruments for facilitating human work?
Human Resources:	The ultimate responsibility remains with people in any system managing communication networks.

Managing networks is easier when some general principles or broad guidelines apply. Critical success factors help in this regard. Methods, tools, and human resources are valuable considerations for network managers. This chapter expands on these concepts and explains their application.

MANAGERS' EXPECTATIONS OF NETWORKS

Technical capabilities and users' requirements for applications incorporating these capabilities drive the integration of voice, data, and image communications. Software developers, including those working on Internet applications, formulate new kinds of applications by linking computer screens to telephone systems and anticipating full motion video at workstations. PBXs play an expanded communication role in organizations, as customers expect to access databases through the telephone network for services ranging from health care to metal reclamation, and from loan applications to stock quotations. Integration occurs at the application and data transport level.

Network application growth demands increased network management capability. Appropriate tools, processes, and people must be available so that managers can maintain network functionality, security, and availability. Corporations willing to make sizable investments in large and sophisticated networks insist that networks be manageable when complete. Users expect vital network capability to be reliable, available, and cost effective. In short, investments in network hardware, software, and applications demand management tools, techniques, and processes that ensure successful network operations.

Executives expect to link employees and managers throughout their organizations with telecommunication systems expanding everyone's communication span. They want to tie customers and suppliers to the firm through sophisticated electronic data interchange (EDI) or Web-based applications. Corporate executives expect information technology to provide convenient business linkages to others, as needed. They have high expectations for achieving advantage from telecommunications technology.

As you learned earlier, unfulfilled expectations are an important source of difficulty for IT managers. Valuable, emerging telecommunications capability and its integration into the firm's mainstream operations further increase the attention that managers must devote to managing expectations. Management systems designed to cope with executives' and users' expectations throughout the firm serve successful IT managers well.

High expectations and an abundant supply of impressive technology fuel the demand for new products. Many network component vendors offer various options for managers' consideration. Even with growing attention to standards, many vendor-specific or proprietary system components complicate network management. Multi-vendor network systems increase network management difficulties and accelerate demand for open systems and improved network management capabilities.

THE DISCIPLINES REVISITED

Network managers require specific management tools, techniques, processes, and skilled people to maintain responsive, available network components that meet customer expectations. Communication network management is a system of people using management information and tools to maintain physical and logical control over network operations. Operational control involves solving physical and logical network problems and monitoring performance to attain contracted service levels. Like computer system operations, network management benefits substantially from disciplined techniques.

Network management disciplines apply to large and important networks, even global networks. For example, Merrill Lynch's worldwide telecom network, jointly designed and operated by MCI and IBM, gives network design, capacity planning, and disaster-recovery planning to MCI and performance, problem, change, and configuration management to IBM. Merrill will spend $200 million over 10 years on this global network.

MCI and IBM use disciplined techniques to manage Merrill Lynch's network, just as firms throughout the world use these disciplines for managing computer centers. Service-level agreements incorporating customer expectations or reliability, responsiveness, and availability form the basis for managing networks. Management reporting is the capstone. Network managers must concentrate on problem, change, and recovery management. Network managers also find performance planning and analysis, capacity planning, and configuration management crucial.

Some references define activities or disciplines of network management somewhat differently than this text. For example, the International Standards Organization (ISO) defines management-system functions as 1) fault management (problem management), 2) configuration management, 3) accounting management, 4) security management, 5) performance management, and 6) adjacent areas such as planning.[11]

This text and other sources consider change management and recovery management as essential disciplines.[12] Subsequent chapters broadly discuss accounting and security, which are also important network management concepts Terplan noted that computer-based network management systems incorporate most of these activities; however, ultimate responsibility lies with humans—they are a critical success factor. Figure 15.1 shows network management disciplines related to service-level agreements and management reporting.

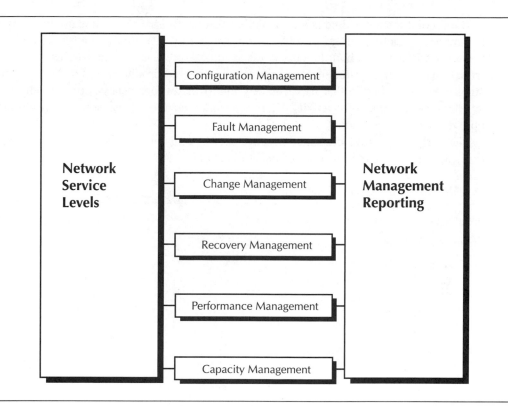

Figure 15.1   The Disciplines of Network Management

Network Service Levels

Network service levels, established much the same way as computer center service-level agreements, serve the same purpose. SLAs establish mutually defined service levels that network managers must deliver to client organizations. Usually, network service agreements are negotiated at fairly high levels in the firm; frequently, these negotiations do not involve individual user managers. For example, the network manager and the administrative manager responsible for office systems establish service agreements for the firm's office system network. Likewise, the order-processing manager and network manager resolve service levels for order-entry systems. Well-functioning network management is transparent to most individual managers and all network users.

Network service-level agreements describe the types of service provided and contain measures for monitoring network service levels. In general, this means that network and user managers must agree upon and document network availability, reliability, and responsiveness. The latter usually involves both response times and workload or volumes and is frequently referred to as network performance. Network managers must ensure that user managers understand the inverse relationship between workload volume and application response time.

Subsecond response time is a reasonable expectation for most online applications involving simple transactions. Because online transaction response time is highly visible to users, poor response degrades both productivity and morale. Research demonstrates direct, positive correlations between system responsiveness and improved productivity. For example, network tuning or reconfiguration that inexpensively improves responsiveness, substantially enhances unit effectiveness and productivity. Thus, network managers have considerable leverage to improve business operations.

Network SLAs are documented in much the same manner as SLAs for production operations. Table 15.1 lists the main ingredients of network SLAs.

TABLE 15.1　Network Service-level Agreement

Date of agreement

Duration of agreements

Network service to be provided

Service availability

Service reliability

Committed response times

Network cost/benefits statement

Service reporting

Signatures

Network SLAs describe the types of network services the client organization receives. Services could include a simple, local area net for the office area or a complex, broadband international net employing satellites for transoceanic communication. Services transcend physical implementation. They must include types of transmission (voice, data, or image) and transaction volumes. This part of the agreement intends to capture users' expectations regarding types and volumes of communication and geographic coverage.

The network SLA describes and documents users' expectations concerning availability, reliability, serviceability, and network service costs. Some networks limit availability to normal working hours. Others networks must be available continuously. For example, the office system LAN operates during normal working hours, but the service may extend these hours upon request. International operations may require almost continuous long-haul network capability. Response time requirements also vary widely. LAN users expect subsecond or near instantaneous response for simple transactions; users of international applications may be less demanding. The service agreement describes network service levels committed to users.

In many cases, firms purchase or lease all or part of their network from local phone companies or independent suppliers. AT&T, MCI, Sprint, and many others can meet most demands for long-haul services. Most suppliers are anxious to describe their high quality products. Network managers should obtain the information they need and factor it into their customer agreements.

It's important that network SLAs account for cost/benefits factors of network service. Just as SLAs for computing centers strive for cost-effective services, network managers and applications managers must also establish cost-effective service levels so that operations, frequently new or reengineered, attain their financial and operational objectives. As FedEx's satellite use and Merrill Lynch's client/servers demonstrate, telecommunication and computer technology advances valuably improve business overall, better customer service, and enhance response time.

Service levels attained vs. service objectives must be carefully reported. Reported results must be credible to, and verifiable by, all interested parties. Like computer system operations, satisfaction surveys and opinion polls are very effective in understanding and maintaining client satisfaction, so important to service providers.

Configuration Management

Configuration management includes the facilities and processes needed to plan, develop, operate, and maintain an inventory of system resources, attributes, and relationships.[13] It's an organized approach for controlling network topology, physical connectivity, and network equipment, and for maintaining supporting data. It also includes transmission bandwidth allocations to various applications and consolidation of low-speed traffic onto higher-speed circuits for more economical transmission. It is essential to sound network management. Table 15.2 list factors within configuration management's scope.

TABLE 15.2 Configuration Management's Scope

Physical connectivity

Logical network topology

Bandwidth allocation

Equipment inventory

Equipment specifications

User information

Vendor data

Configuration management is critical because it controls and allocates network assets. It is concerned with physical network elements, such as customer terminals, controllers, bridges, routers, modems, multiplexers, and links, and their physical and logical connections and topology. The process's purpose is to maintain accurate, timely topology configurations and equipment inventory, including physical location, technical capability, and intended customer application. It must include equipment and capability like bandwidth even though leased from suppliers rather than owned. Whether owned or leased, network managers must know all network capabilities.

Configuration managers control physical connections and interrelations of all telecommunications equipment. They manage the network's logical topology through routing tables and other devices and assign and control bandwidth allocation among applications. Databases describing the network's configuration at any given moment must be maintained for use by subsequent disciplines.

In addition to inventory data, databases of physical device addresses, application requirements, and other technical information must be readily available to reconfigure the network, if required. Defective component operation or changing bandwidth requirements may mandate network reconfiguration. Vendor information must also be available to track service calls, install additional equipment, and manage equipment enhancements and upgrades. Configuration management databases are essential for managing problems and changes, too.

Network Problems and Changes

Network problem management (sometimes called fault management) is similar in many respects to problem management for computer-based application systems. Network users, network operators, or automatic alarming devices can report faults or problems. (Recall the alarming devices at AT&T's switching center described in the Business Vignette in Chapter 13.) Faults are documented in problem reports and tracked in problem logs.

Problem resolution involves network technicians, vendor service personnel, network users, and application specialists. During problem resolution, information contained in configuration databases helps analysis and diagnosis. Problem management processes consist of identifying or locating faults, isolating faulty components through reconfiguration so the network can continue operation, replacing faulty components to restore the network to its original status, repairing the faulty component, and placing it in inventory.

Following satisfactory problem resolution, configuration databases are updated with pertinent new information. If a router fails, for example, a new unit replaces it, field inventory records are updated, and vendor maintenance activity initiated. When a link's performance degrades or fails, routing table information changes after proper analysis, traffic takes alternate paths, and link repair begins. After correcting the problem, the trouble report is closed and the problem log is updated with information on corrective action taken.

Not surprisingly, many sources contribute to network problems; however, physical problems are most common. Table 15.3 shows the interesting relationship between problem frequency and OSI protocol layers.[14]

TABLE 15.3 Network Problem Sources

Protocol Layer	Percentage of Problems
Application	3
Session	7
Presentation	8
Transport	10
Network	12
Data link	25
Physical	35

Network technicians use test equipment and many tools to isolate faults. Simple testers can discover faults in physical cabling, connectors, and switches. Table 15.3 shows that physical layer devices cause many network problems. Frequently, incorrect addressing or unplanned configuration changes cause network faults. Only network technicians should modify or service physical network components, because unskilled people making unplanned network changes, however well-intentioned, usually cause problems and network degradation.

Networks are usually designed to be highly flexible. Their physical and logical configurations change frequently. Changes result from fault correction, bandwidth allocation adjustments, application or user mobility, and network growth. If not managed carefully, changes to the network's physical or logical configuration are a prolific source of faults.

In some networks, bandwidth and configuration changes occur routinely and automatically. The network may include dynamic bandwidth allocation, for example, or it may change routing automatically. Long-distance telephone networks contain these features. In most networks, changes stem from problem corrections and performance improvements. Network expansion, increases in users and services, new equipment, or cost reductions also cause change. Chapter 13 describes change management processes that are good models for managing these nondynamic network changes. Because all network changes involve risk, disciplined processes to minimize and control them are essential.

Network Recovery Management

All risks in network operations cannot be eliminated: Some events are beyond network managers' control. In the Kobe earthquake of 1995, for example, many firms that took precautions such as dispersing critical facilities experienced severe difficulties because of the damage's broad geographic extent. The capabilities of Japan's national carrier, NTT, did not protect firms from the earthquake's enormous damage. Although many firms depend on strong, technically proficient third parties to supply vital network resources, major natural disasters or unpredictable events, such as the World Trade Center bombing, underscore the need for disaster recovery plans.

Network managers must plan to recover from local disasters that affect LANs and their connections to local mainframes, as well as the organization's phone system and PBX. Recovery plans for local phone systems and their connections to common carrier facilities or to centralized computer systems must be complete. Most, but not all, firms today have recovery plans for the phone network, according to some sources.

The Communications Managers Association reports that disaster plans are in place for networks in most firms. Disaster plans exist for data networks in 82 percent of firms, for voice networks in 68 percent, and for image networks in six percent of the firms. The report claims disaster plans covering data centers exist in 78 percent of firms; still about one-fifth of these firms lack disaster recovery plans for their network assets.[15]

Networks relying on common carrier services face some specific risks. The rapid deployment of fiber-optic links is concentrated along rights-of-way owned by communications companies, railroads, pipeline companies, and utility companies. Because rights-of-way are expensive and difficult to acquire, common carriers often install multiple fibers in one trench. To economize further, several carriers use the same path or share fibers in cable. These networks are dangerously vulnerable because an accidental breach of the trench may sever several carriers' links.

In an unusual series of difficulties, the Pacific Stock Exchange lost two t-1 lines to New York when construction workers inadvertently cut a major fiber backbone near Barstow, California. The backbone was quickly repaired, but the next day a third t-1 line near Baltimore was severed. Again the line was quickly repaired, but another failure near Barstow caused two lines to fail the following day. Then for three tense days, the exchange relied on a single line while the Barstow failure was under repair. Although different carriers provided each t-1 line, the exchange did not have the backup it bargained for. "A network outage of even 30 to 60 minutes could cost us millions in lost trades," claims Dave Eisenlohr of the Pacific Stock Exchange.[16] This example shows that network managers cannot be sure that leasing links from several companies offers backup: They must examine the routing as well.

Recovery management is an important and critical task for network managers. Many available, alternative network configurations, routings, and network solutions for computer system recovery planning make the task manageable.

Network Performance Assessment

Most network applications require network performance at or above certain minimum levels. For example, voice transmission requires a minimum bandwidth of 4000 Hertz to meet typical user expectations. Likewise, high-fidelity sound transmission requires significantly higher bandwidth, 20,000 Hertz or more. Network managers must maintain performance levels that satisfy the application and user expectations recorded in service agreements.

Network performance must be monitored and controlled. Monitoring lets managers understand performance; controlling lets managers adjust capacity and change performance. Traffic measuring tools let managers compare throughput to capacity. Managers must understand throughput changes, discover bottlenecks, and measure important factors like response time. In addition, network managers must measure and record mean time between failures and mean time to repair. These measures are needed to comply with availability, reliability, and serviceability commitments.

Network availability expressed in percentage is calculated by dividing the mean time between failure (MTBF) by the sum of the mean time between failure and the mean time to repair (MTTR). The following formula expresses this relationship:

$$\text{Availability (percent)} \quad = \quad \frac{\text{MTBF}}{\text{MTBF} + \text{MTTR}} \times 100.$$

For example, a network with a mean time between failures of 200 hours and a mean time to repair of one hour will be about 99.5 percent available. Note that availability approaches 100 percent as MTTR approaches zero.

As the formula indicates, availability increases as the time between failures increases, so reliable network components are important. However, because network components are never 100 percent reliable, most network managers strive to achieve availability by reducing repair time (MTTR) to the lowest possible number or by eliminating it entirely. Installing network redundancy (as the Infomart example in the Chapter 14 Business Vignette suggests), providing alternative capacity, or rerouting automatically can accomplish this. This strategy provides network reliability to the customer as well.[17] Reducing MTTR is the preferred way to achieve service levels; however, redundant capacity does not substitute for service agreements as some believe.

Transmission accuracy is usually not a concern with today's networks because of built-in error-detection and error-correction mechanisms. However, error rates must be monitored because retransmissions and error correction add to network loading. Error rates are also leading indicators of future failures or faults.

Capacity Assessment and Planning

Network capacity assessment and planning were relatively simple when terminals were directly connected to mainframes in star or hierarchical networks serving standard applications. Today's decentralized, client/server, or intranet operations operate far less predictably. New uses for creative applications can generate large, unplanned loads. Because client/server networks encourage user innovation and productivity, they diminish network planners' ability to estimate loads and plan capacity.

Today's dynamic networks and network applications demand that capacity be assessed regularly and capacity planning be nearly continuous. New technology introduction and adoption and rapidly growing applications mean that network managers must constantly monitor throughput, user-experienced response times, and network utilization. This data helps determine and quantify current loads. Analyzed over time, it also gives managers trend information needed to project future network bottlenecks and constraints so they can consider future design alternatives.

Load projections from current and planned future applications form the basis for accurately estimating future network loading. Growth in current application volumes may result from an increased number of users (end-user computing may be expanding its customer set) or from increasing transaction volumes with current systems (e-mail usage may be increasing). For example, new applications linking the distribution system and the firm's customers through EDI, or physically locating client/server workstations at suppliers' workplaces may place new, increased demands on networks. Business plans, functional plans, and service agreements may also create additional demands or new network uses.

Guesses about future loads and equipment requirements lead to excessive expenditures, poor performance, and inefficient network configurations. Large networks, such as Digital Equipment Corp.'s internal Easynet (more than 95,000 nodes), require computerized modeling and simulation tools to maintain satisfactory service levels.[18] Except for small, simple nets, fine-tuning and capacity planning cannot be accomplished manually.

New capacity plans must be developed to handle new incremental loads. They may specify additional bandwidth, more user-oriented devices, additional controllers or concentrators, different, more effective links, or a new, more optimum topology. Capacity planning's objective is to plan cost-effective network solutions that satisfy user needs specified in service-level agreements. Because telecommunication systems and applications are dynamic, network managers should refresh capacity plans quarterly and review them monthly.

Demand for network facilities always seems to rise, filling expanding supply. Additional bandwidth, more devices, and improved capacity usually lead quickly to increased load. Under changing conditions, network planners must focus on service-level agreements and cost-benefits analysis to optimize business.

Network planners should use strategic business plans as the basis for capacity plans and should implement capacity plans as part of tactical planning. Additional planning factors discussed in earlier chapters are valid for network planners, too. For example, network planners should not overlook changing financial conditions, organizational restructuring, and other items discussed in Chapter 14.

If possible, networks should remain simple and easy to understand. Even simple networks require skilled people and significant testing and reconfiguration time. Because network complexity grows exponentially with number of nodes, unnecessarily complex configurations of links and components increase installation and maintenance effort significantly. Nevertheless, when capacity enhancements are required, the change-management discipline helps greatly in implementing them.

Management Reporting

Many performance assessment tools provide data that managers can share with network users. Users must receive reports relating to their service-level agreements: they should be factually informed of service availability, reliability, cost, and system responsiveness. In turn, they should respond to satisfaction surveys. Like most situations, there is no substitute for high bandwidth communication between service providers and users.

Network managers and senior IT managers need additional information. Throughput, or the transaction-processing rate, network utilization, problem analysis, change-management results, and capacity plans are essential to their jobs. In addition, these managers need performance analysis results and capacity plans to formulate plans.

The techniques discussed above, from service-level agreements to management reporting, are network management's foundations. These disciplined processes are Terplan's first critical success factors for network managers.

NETWORK MANAGEMENT SYSTEMS

Many automated tools help managers and technicians operate networks. Computer-based tools to help manage networks consist of computer hardware; network-specific equipment such as performance monitors and software; and a variety of application programs. These network management systems help provide operational control, collect operational data, and monitor and report network usage and performance. Data collection, storage, and retrieval capabilities are valuable for managing problems, monitoring and managing performance, and planning network capability and growth. Terplan speaks about some of these tools; they comprise his second critical success factor.

Computer-based and other hardware tools to perform these tasks are increasingly more sophisticated and integrated. Network managers need automated tools that support multi-vendor hardware configurations and provide operational controls. Suppliers of network management systems work hard to support these objectives. But, despite an abundance of available products, the chances of finding a single product to manage the entire network from end to end are slim.

Until recently, network management systems were designed and operated to support specific network elements. For example, a LAN would receive management support from software installed in the server or the master client on the net. A large mainframe system would obtain support from system software in the network control program, part of the operating system. Support for each network element was separate from others and was vendor specific. The concepts of open systems and integrated network management systems aim to coordinate and integrate network management support across network elements and across vendors.

Automated tools to help manage networks range from simple display devices to relatively complex computer-based instruments to sophisticated network management systems. Elementary tools, called breakout boxes, contain indicator lights that display electrical signals like line status, data flows, and connection information. Breakout boxes give simple indications of line and interface operation for network technicians. More complicated instruments, called line monitors, display line status and actual signals traveling on communication lines. Line monitors help network technicians isolate failing or poorly performing communication lines.

Still more complex and valuable, LAN monitors observe all signals and messages flowing on a LAN. Frequently PCs or workstations on LANs contain vendor-supplied software that can record and interpret signal information in various formats and status information from several protocols. The workstations and software display message traffic and protocol information and can be programmed to summarize data on traffic volumes, error conditions, and network status. The most sophisticated LAN monitors perform diagnostic testing of network devices like terminals, modems, interface cards, and controllers. Many vendors supply software and hardware to do this.

The International Standards Organization's OSI model includes standards for measuring and monitoring networks. ISO's standards are called Common Management Information Service or CMIS. CMIS' standard protocol, called Common Management Information Protocol or CMIP, collects lower-layer data. CMIP reports data to CMIS databases for analysis. However, Simple Network Management Protocol or SNMP, a *de facto* standard originally designed for TCP/IP networks, is very popular today. Computers and other network devices report errors and status to SNMP, which collects and reports them to a monitor station. Although important, these tools and monitoring devices are relatively simple compared to the full-blown network management tools that prominent equipment suppliers offer.

IBM's system management software, SystemView, contains NetView, an element for managing networks. SystemView contains many system management functions like storage and data management, change management, performance monitoring, and client and remote network monitoring. Its remote monitoring feature checks device performance in multivendor networks. SystemView/NetView integrates system and network management for many IBM systems like S/390 MVS, OS/400, and OS/2, recognizing that managing networked systems includes managing the enterprise CPU as well as numerous critical devices attached to it through complex networks.[19]

Many important vendors offer products to help manage networks and networked systems. These include Hewlett-Packard's OpenView, Digital Equipment's POLYCENTER, Cabletron's SPECTRUM, Bay Network's OPTIVITY, and Sun Microsystems Solstice SunNet Manager. Integrated into large enterprise hardware and software systems, these systems help managers keep networks and their many devices performing satisfactorily as they grow and expand. Because they monitor and control complex networks and applications, network management systems themselves are necessarily sophisticated and complex. Without them, however, good network management would be impossible.

Tools for analyzing and monitoring internets grow in popularity and function as well. Remote Monitoring 2 (Rmon2) is a standard that defines performance statistics, connections, and application activity along networks. According to surveys, users of installations with many interconnected LANs find Rmon2 highly cost effective.[20]

Goals for integrated network-management facilities are yet to be attained, but progress is evident. Major vendors like Hewlett-Packard, DEC, IBM, and many others are assembling essential elements. Sales of network management products explode as managers search for tools to control and operate rapidly growing networks. One compendium listed and described more than 500 network management products.

Future network management systems will provide more than tools; they will contain expert systems to automate further some tasks previously reserved for network specialists. Expert systems are valuable for problem determination, recovery operations, and some operational control tasks. They may help design new network configurations and advise network managers during change management. Sophisticated expert systems may also analyze performance data in near real time and advise management about impending system overloads and capacity constraints. Because network monitoring systems gather enormous amounts of data with high short-term value, expert systems can potentially use information quickly to introduce system changes. They also can potentially optimize system resources and reduce system costs. Phone companies lead in sophisticated expert systems for network management.

INTERNATIONAL CONSIDERATIONS

Telecommunication technology greatly enriches international trade and commerce. "The globalization of commerce and communications is changing the way we work, the way we do business, the way we learn, and the way our nation competes in the global marketplace. Why global? We might as well ask: Why breathe?" stated William G. McGowan, former chairman of MCI Communications Corporation.[21]

Leading-edge, multinational firms recognize and capitalize on the opportunities. The Volvo Corporation uses international networks to link manufacturing and sales operations in Europe to those in the U.S.; without these linkages its business cannot operate. Kodak links tens of thousands of employees with its voice-messaging system and connects computers internationally through its communications facility. DEC uses international communications facilities to manage its worldwide operations; Cisco Systems, maker of computer network hardware, connects its 12,000 employees in 45 countries with intranets. Many firms use high-speed data pathways to develop and market hardware and software products internationally.

Recognizing the huge potential of international communications, suppliers are rapidly expanding their capacities. For example, optical-fiber capacity expansion underway in the Pacific basin seeks not only to carry additional traffic, but to provide backup in case of network failure. Additional cables linking the U.S. with Hong Kong, Korea, China, and the Philippines will augment the fiber-optic cables now connecting the mainland U.S. with Hawaii and Japan. Firms from several countries are constructing additional cables in the Atlantic, Caribbean, Mediterranean, and around Africa. Connections through the U.S. enable communication between the Far East and Europe.

As large quantities of fiber are being installed, wave-division multiplexing technology (signals on several light wavelengths transmitted simultaneously) increases the capacity of existing and new fiber by an order of magnitude.[22] In addition, satellite capacity planned for the next decade will add another order of magnitude to international capacity.

The global infrastructure for international communications is rapidly falling into place. AT&T, MCI, Sprint, and many regional Bells play a major role, usually in cooperation with important foreign partners like Deutsche Telekom, France Telecom, and others. Large segments of our economy already take advantage of the infrastructure: the travel industry and financial institutions are examples. More than $1 trillion are estimated to move through international telecommunications systems daily. Executives in many industries in all countries express intense interest in capitalizing on international networks' potential.

Although many of us take telephones for granted, worldwide fewer than one in four persons owns one. The highly uneven distribution of world telephone service results from affordability, government policy, population distribution, geography, individual needs, and worker skills. Some governments restrict communication to further their ends rather than their people's needs or desires. China, for example, intends to censor Internet traffic (it thinks it can!) to prevent "undesirable" outside information from reaching its citizens. In some countries geography is a barrier to phone communication. For example, in sparsely settled regions of Siberia only one access line is available for each 100 inhabitants, one-fiftieth the penetration of phones in Western Europe. In other countries poor economic conditions make telephone infrastructure development and phone ownership unaffordable.

First-rate telephone service is in great demand in Third World countries and in Eastern Europe because these countries know that not having it greatly impedes economic development. "A telephone system is highly capital intensive," says Drucker, "But technologies that replace the 'wiring' of traditional telephones with the 'beaming' of cellular phones are radically reducing the capital investment needed. And once a telephone service is installed it begins to pay for itself fairly soon, especially if well-maintained."[23] Leaders around the globe well understand the synergy between economic development and communication systems deployment. It's a factor in their thinking when developing national policies.

Rapidly declining prices for phone systems is coupled with global trends toward deregulation and privatization. As aggressive, well-financed firms install high capacity fiber lines and exploit digital wireless, telephones will become available to billions now without them. The trend toward international alliances, transnational investments, and global telecommunication systems accelerates, benefiting millions of new consumers around the globe.

The second important factor driving the growth of telecommunications systems is the steadily rising national income of many of the world's highly populated countries. With GDP growth rates exceeding five percent net of inflation, China, parts of India, some Latin American countries, and others now find telecommunication systems both affordable and indispensable. In parts of Eastern Europe, the number of people waiting for phone service exceeds the number who have it. Thus, several important forces conspire to accelerate the deployment of telecommunication systems.

Today, 120 years after the telephone's invention, about one billion people use it regularly—that number could double in the next decade. In advancing nations, telecommunication systems grow exponentially because telephone systems are themselves an important driver of economic growth. Indeed, telephones are a growth industry.

The dynamics of global telecommunications place grave responsibilities on international network managers to be knowledgeable about a broad range of topics. They must understand the characteristics of long-haul links including satellites, international services including value-added networks, tariff schedules in many countries, and many national and international rules and regulations governing transnational data flows. Telecommunication managers in international operations hold important, challenging positions.

Technological, economic, and policy considerations fuel the rapid growth of networks and network management businesses. Increasingly favorable economics drive rapid telecommunication technology advances evidenced by an explosion of products, services, and bandwidth. Policy and political changes cooperate to accelerate system adoption and implementation globally. Technology adoption both enables and mandates structural changes in organizations. Because structural changes usually require advanced systems, the cycle reinforces itself.

Management tools, techniques, and processes lag considerably behind technology adoption in many organizations. Thus, IT managers are not completely prepared to command the communication technology revolution. But command they must. Because network assets cross organizational, political, and geographic boundaries, network managers must be skilled generalists. Corporate networks place higher demands on IT managers' general management skills than do traditional information systems.

In fulfilling their functions and responsibilities, corporate network managers must develop network strategies to support the firm's business strategy. They must evaluate technical and regulatory developments worldwide and provide input to the firm's strategic plan so that it capitalizes on emerging trends. Using technical and business insights, they must shape the firm's strategic direction. In short, corporate network managers must be superior business strategists.

Corporate network managers must also be outstanding tacticians. They must clearly understand costs and benefits of current network systems and must evaluate emerging technologies and new products and services such as FDDI, Frame Relay, ATM, Web technology, and others. They must understand legal, regulatory, and business environment changes and must use this knowledge to develop and implement near-term plans. Their near-term plans must move the organization toward broader objectives set forth in strategies and strategic plans.

Corporate network managers must also focus on network operations by reviewing reports from the management disciplines and ensuring that the disciplines operate effectively. Operational control is one of their important responsibilities.

Corporate network managers have both line and staff responsibilities; they must consult with the firm's executives and must advise functional managers throughout the organization. Senior network managers have a unique responsibility to cooperate with the firm's executives in establishing organizational telecommunications policy. Governance rules for acquiring, applying, and operating network assets establish the basis for profitable network exploitation.

Network managers must establish a corporate policy that requires the owner or operator of any net for the business to follow the network management disciplines and tools discussed in this chapter. Effective policies unify network implementation, making it seamless and cost effective. Rogue departments that concoct *ad hoc* rules for departmental LANs brew trouble when the networks must be integrated later. Nevertheless, with networks, as Strassmann says of information systems, "It is good information policy to have a set of rules that are unusually permissive to innovation."[24]

Web technology offers splendid opportunities for network managers to offer their organization services that many find highly valuable. Firms that adopt and carefully manage intranets have reaped rich rewards. Network managers are well positioned to introduce servers and install Web authoring tools for many functional departments to use. Properly conceived and well-coordinated with senior managers, intranets offer network managers a once-in-a-lifetime opportunity to greatly influence the organization for the better. Although this requires initiative and extra effort, it rewards everyone substantially. It's an opportunity not to be missed.

And, as in all others endeavors, skilled and talented employees and managers are the sustaining force in network applications. Terplan's third critical success factor is Human Resources: ultimate responsibility remains with people in any system managing communication networks. Much more will be said about this later.

Most individuals trained and experienced in computer science, information systems, or telecommunications learned network management on the job. Many telecommunication managers were trained before deregulation. They worked with local telephone companies, arranging local and long-distance service; corporate executives considered these services utilities like water or electrical power. Today, their job is much more complex because many more products, services, and vendors interact. Telecommunications is much more than a utility. It generates revenue and creates strategic advantage while shaping and molding firms and perhaps industries themselves. Many firms today have merged telecommunications and information systems under the CIO as the technologies themselves merged.

Until recently, scant opportunities existed for students to learn network management during college. Although 35 colleges offer degrees in telecommunications, only a handful grant Ph.D. degrees. In contrast, about 360 colleges and universities in the U.S. offer Bachelors' degrees in information systems, and 60 offer the Ph.D degree. Today, academic interest in networks and network management is rising. Some universities let students combine telecommunications, information systems, and business courses while pursuing a Master's degree.

SUMMARY

Because networks are so important for most firms' information processing infrastructure, network management itself has gained importance. Network management includes methodology: processes and procedures to ensure efficiency, effectiveness, and customer satisfaction. Network managers need tools: instruments for facilitating the work of managing networks. And network management hinges critically on human resources. Ultimately, people are responsible for achieving telecommunications' potential.

Corporate network managers must strategize, plan strategically, manage tactically, and control operations skillfully. For those who manage, control, or use global networks, international considerations are critical as well. Today's network managers have an important, challenging, exciting, and potentially rewarding job.

Review Questions

1. What motivates package shippers to focus on Web-based applications so intently? What effect does adoption of Web technology have on current and potential competitors?

2. Why is network management important to most firms today?

3. What are the connections between network management and the disciplined approach to production management discussed earlier?

4. Describe some network capabilities and explain their strategic importance to businesses.

5. What technologies spur the growth of networks? How are some of these technologies cooperating to hasten network growth?

6. Review critical success factors listed in Chapter 1, and discuss those applicable to telecommunication systems and opportunities.

7. For what reasons are networks important to corporate organizations and to restructuring?

8. What are network managers' critical success factors according to Terplan?

9. Describe network management's scope.

10. What do users expect of networks? What do executives expect?

11. Name the network management disciplines.

12. What ingredients of configuration management are essential?

13. What databases are maintained in configuration management systems?

14. What processes does network problem management involve?

15. According to the text, most faults originate in the physical protocol layer. Why do you think this is so?

16. What network changes occur automatically in some networks? What types of changes occur through change management?

17. What risks are associated with using common-carrier networks? How can network managers minimize these risks?

18. How can network managers maintain high system availability?

19. What information is needed to develop a new capacity plan? From what sources does this information come?

20. Why is developing automated management tools for today's networks difficult?

21. Describe the specialized knowledge required of managers of international telecommunications systems.

22. What responsibilities do corporate network managers have? What responsibilities do they have regarding the disciplines?

23. How do network service-level agreements and applications service-level agreements differ?

Discussion Questions

1. Discuss the importance of telecommunication systems with respect to each strategic system discussed in Chapter 2.

2. From your reading or your experience with the Internet, discuss what you think are some profitable uses of intranet technology.

3. Discuss why international networks are expanding very rapidly. Consider the technical, economic, regulatory, sociological, and political factors that impact this expansion.

4. Discuss how a firm might use a telecommunication system as part of its implementation strategy for decentralization. What role might telecommunication systems play in the strategy of a firm that wants to grow by acquisition?

5. Discuss reasons why network management is difficult? Which reason is most important in your opinion? Why?

6. Define the boundaries of network management. Where is the boundary with respect to other internal firm operations? How far outside the firm does the boundary extend?

7. What sources of expectations may executives develop regarding telecommunication systems? What role should IT managers play in developing executive expectations?

8. Compare and contrast the service-level agreement for a traditional information system with a network system SLA.

9. Why is the configuration management discipline so important to network managers?

10. Why must configuration management be a real-time operation for managers of sophisticated networks? Mainframe computers may also have a modest configuration management system. Discuss similarities and differences between configuration management for mainframes and for networks.

11. Discuss the evolution of disciplined management systems as firms migrate from traditional, centralized data processing to client/server computing. What changes might occur as firms adopt EDI or extranets?

12. Referring to the formula for network availability in this chapter, discuss the relationship between availability, reliability, and serviceability. What actions can managers take to improve availability?

13. Discuss the capabilities of today's network management systems. What additional functions would you like such a system to have?

14. Discuss the potential applications of expert systems to network management. Why do expert systems have more potential in network management than in managing large centralized systems?

15. Discuss the network manager's responsibilities in an international firm by itemizing elements of his/her job description. What qualifications do you think an applicant for the top network management position at a firm like FedEx should have?

16. Consider a firm that wants to link its dispersed operations to its headquarters through networks. Discuss the implications of this strategic direction to application portfolio management.

Assignments

1. Using Chapter 13 as a guide, design a fault or problem report and a problem logging system appropriate for network managers. Develop a rationale for combining network and data processing problem meetings or for holding them as separate events. What factors are most important in this decision?

2. Study a major network management system like IBM's SystemView, Hewlett-Packard's OpenView, DEC's POLYCENTER, Bay Network's OPTIVITY, Sun Microsystem's Solstice SunNet Manager, or Cabletron's Spectrum. Develop a list of its functional characteristics, and discuss its strengths and weaknesses.

3. Review the Web sites for package shippers at www.fedex.com, www.ups.com, and www.dhl.com, and prepare a comparison of their contents. Discuss the features and limitations of each. For another comparison, review www.usps.com.

4. If your college or university has a Web site, review and compare it to www.purdue.edu, one of Computerworld's premier 100 sites. If your school has no Web site, browse Purdue's and itemize features that you would like your school to have.

[1] The USPS began to accept credit cards in 1996, lagging several decades behind common business practice.

[2] Natalie Engler, "Keeping Up with the Joneses," *Computerworld*, The Premier 100, February 24, 1997, 36.

[3] See note 2.

[4] See note 2. You can obtain considerable information about these companies from their Web sites—www.fedex.com, www.ups.com, and www.dhl.com.

[5] The Premier 100, February 24, 1997, 12, 14. See note 2.

[6] Network Wizards, at www.nw.com/, offers statistics on Internet hosts from 1981 to the present. A host is a domain name that has an IP address record associated with it, such as www.nw.com.

[7] Computer telephony integration (CTI), using phone information directly in computer applications, is growing rapidly. See "Anticipating Telephony," *LAN Times*, March 13, 1995, 53-63.

[8] Tom Davenport, *Process Innovation*: *Reengineering Work Through Information Technology* (Boston, MA: Harvard Business School Press, 1993), 254.

[9] Wayne Ryerson and John Pitts, "A Five-Nation Network for Aircraft Manufacturing," *Telecommunications*, October, 1989, 45.

[10] Kornel Terplan, *Communication Networks Management* (Englewood Cliffs, NJ: Prentice-Hall, Inc., 1987), 2.

[11] See, for example, William Stallings, *Business Data Communications* (New York: Macmillan Publishing Company, 1990), 722.

[12] See, for example, Michael J. Palmer and Robert Bruce Sinclair, *Advanced Networking Concepts* (Cambridge, MA: Course Technology, 1997) 173, 196.

[13] Stallings, 716. See note 11.

[14] Rolf Lang, "Diagnostic Tools Decipher LAN Ills, Reduce Downtime," *Federal Computer Week*, June 26, 1989, 48.

[15] *Communications News*, October, 1994, 6.

[16] Laura DeDio, "An Investment in Uptime," *Computerworld*, February 23, 1998, 45.

[17] David A. Stamper, *Business Data Communications* (Redwood City, CA: The Benjamin/Cummings Publishing Company, Inc., 2nd ed., 1989), 462-465. Stamper shows that redundancy in network components increases availability nearly, but not quite, 100 percent.

[18] Sheila Osmundsen, "Expanded Networks Need Expanded Planning," *Digital News*, August 3, 1992, 1.

[19] For more information about these devices and systems, see John Enck and Dan W. Blacharski, *Managing Multivendor Networks*, (Indianapolis, IN: Que Corporation, 1997), 227-241.

[20] Patrick Dryden, "New Monitor Broadens View of Network Service," *Computerworld*, February 23, 1998, 32.

[21] Willian G. McGowan, *Telecommunications*, April, 1989, S-6.

[22] Patrick R. Trischetta and Willian C. Marra, "Applying WDM Technology to Underseas Cable Networks," *IEEE Communications*, February 1998, 62.

[23] Peter F. Drucker, "Where the New Markets Are," *The Wall Street Journal*, April 9, 1992, A14.

[24] Paul A. Strassmann, *The Politics of Information Management* (New Canaan, CT: The Information Economics Press, 1995), 33.

Part
Five

Controlling Information Resources

Business controls are basic management responsibilities, required by law in many instances. Rapid growth in networked systems makes them increasingly important today. Managers must maintain financial and operational control as information technology penetrates organizations more deeply and increasingly links them to the external environment. This part covers Measuring IT Investments and Returns and IT Controls and Asset Protection. Because advanced technology is widely deployed, managers must be keenly aware of business controls. This part teaches managers how to develop and use refined tools, techniques, and processes to discharge their control responsibilities effectively.

16 Measuring IT Investments and Their Returns

A Business Vignette

The Real Language of Business[1]

Compared to the heavy accent now being placed on return on investment (ROI), competitive advantage was just small talk. Inquiring CEOs want to know: "How much bang is the corporate technology buck really buying?" If money talks, then ROI is one of its most common languages. Yet CIOs seem to have a hard time achieving fluency.

Despite all the money their companies have spent on information technology, CIOs still struggle to explain the benefits of an IT investment in hard dollars. They are more likely to justify IT expenditures by citing improvements in productivity, response time, customer complaints, error rates, and nonfiscal measures than by measuring the return on investment in information technology (ROIT).

Nonfinancial measures are useful and valid, but CEOs and CFOs who command the power of the purse rarely consider them sufficient. These ultimate arbiters frequently insist on proof that money spent on IT results in earnings and savings that surpass the investment. CIOs often feel caught between an irresistible force and an immovable object as they are compelled to provide financial justification for an investment that resists such hard measurements.

ROIT analysis can do more than keep CIOs in their boss's good graces. It can also direct them to the most lucrative opportunities for using technology. It can be a tool for evaluating existing and potential systems and a mental exercise to sharpen strategic thinking. And it can serve as a common language for explaining information systems to other executives.

For all these reasons, CIOs should learn to speak ROI more effectively. They need to find more techniques for measuring and demonstrating hard-dollar benefits and for persuasively articulating soft benefits—where possible—in hard-dollar terms. But CIOs should never become slaves to old, even obsolete, accounting formulas. After all, not every IS investment decision should be made strictly on the basis of ROI. An unbending insistence upon quantifiable returns can lead an organization to pass up investing in systems or applications that are ROI-resistant but still significant and worthwhile. Some of the most important advantages of IT, such as winning customer loyalty by improving quality and service, are hard to translate into dollars. An organization has to have the wisdom to know when to waive ROI for good reasons.

For the Perrier Group of America, a Greenwich, Connecticut-based company, ROIT combined restructuring and a supporting IT plan to yield a $20.4 million annual hard-dollar return on an initial investment of $9 million and subsequent annual costs of $5 million. Although difficult to evaluate financially, customer retention rates also are on the rise, yielding additional soft-dollar returns.

Perrier chose to overhaul its route operations, which distribute bottled water to customer businesses and homes, to bolster its standing in those markets. To accomplish the goals of reducing costs and increasing its customer-retention rate through improved service, the company consolidated redundant functions and departments at 40 standalone branch offices into regional centers.

Efforts to convert to a regional model of operation began in 1989 and ended in 1993. The benefits have been impressive, with a $5 million annual investment netting a $15.4 million annual return. Over the five-year life of the plan, Perrier pocketed a cool $77 million in savings, primarily resulting from reduced headcount and improved productivity.

Before 1989, each of the 40 branches operated by five Perrier companies (Arrowhead, Great Bear, Ozarka/Oasis, Poland Spring, and Zephyrhills) located throughout the country maintained its own distribution, data-entry, teleservice, and credit and collection personnel. In addition, each branch had its own decentralized minicomputer and phone system. "Essentially, it was like operating 40 individual businesses all operating their own departments," said James Waldeck, Perrier's senior vice president of the route division.

The technical agenda consisted of overhauling and enhancing the existing route-management software used by distribution personnel, creating common databases, providing handheld computers for drivers, implementing advanced telecommunications, and consolidating into regional data centers in Los Angeles and Greenwich. The consolidations standardized equipment, training, and maintenance.

The new technology has directly affected the bottom line. The route-management software overhaul alone has reduced headcount by 32 for an annual savings of $800,000. The handheld computers have eliminated 59 full-time positions, saving $2.3 million annually, while also reducing the cost of printing and storing invoices by $565,000. Consolidating the datacenters also reduced headcount; all told, the company was able to reduce personnel by 275, mostly in data-entry departments.

The improvement in customer service corresponds directly with an increased rate of customer retention and improved overall market share. Putting a dollar value on retention rate is somewhat tricky. "The key," says Rowan Snyder, corporate director of MIS, "is to recognize that it is very definitely costly to lose customers." Market share has grown 2 percent and revenues have increased significantly, according to company reports.

Waldeck also credits regional operations with helping Perrier stay ahead of the competition. "The water industry is in some tough times right now," he said. "While everyone is struggling to stay even, we're doing much better this year."

Perrier proved that regionalization and consolidation—business process improvement actions—supported by a comprehensive IT plan, could generate both hard- and soft-dollar returns on investments.

INTRODUCTION

Information technology resources comprise a large and, in some cases, growing portion of most firms' budgets. Because of cost pressures and increasing competition, many organizations today focus on ROI and intensely scrutinize IT budget items. Many firms spend two to five percent of revenue on IT activities; firms expect IT managers to explain these expenditures and, together with line managers, account for them to the firm's officers.[2] As corporations become more information intensive, firms spend more on networked systems; both the expenditures and the systems become increasingly visible. The consolidation of telecommunication departments with information processing departments, the growing importance of telecommunications to the firm, the widespread adoption of distributed computing and office and factory automation, and the increasing complexity of the application environment all heighten executives' interest in IT expenses.

Interest in IT expenses also intensifies because of the manifold order-of-magnitude declines in computers' and computing components' costs and steadily increasing labor costs. Thus the IT payoff stems from the benefits of using technology to reduce total labor costs, or to greatly increase the effectiveness of money spent on labor. Reducing computing cost by employing lower cost technology is much less important in the overall scheme of things.[3] To measure benefits, managers need to know the cost of using human and computer resources in business processes.

After decades of generous spending on computer and telecommunications activities, IT investments receive intense scrutiny today. Some experts consider these investments' payoff, once thought to be 20 to 100 percent or more, to be slight or even negative.[4] Planned investments in more powerful PCs, increased data storage, networked systems, scanners, color printers, and purchased software will increasingly depend on financial return on investment rather than on intangible notions such as keeping up with technology, expense-to-revenue ratios, and other ideas that do not address customer needs. In the future, accounting for IT investments and their returns will have renewed meaning.

Accounting for IT resources is complex and difficult. It is complex because resources are allocated in various ways. Some expense mechanisms tie directly to the internal workings of complex hardware and software systems, some accompany the operation of sophisticated networks, and others relate to increments of labor throughout the firm. For example, accounting for the operations of client/server systems or well-developed intranets requires detailed knowledge of operating and support labor, training costs, and hardware and network maintenance costs. Few firms' cost accounting systems can handle these details.

Virtually no organizations can account for the administrative time employees must spend dealing with personal computers, thought to be up to 10 percent of their time. Employees spend this time initiating programs, fixing hardware or software problems, backing up or restoring data, browsing the Internet, and communicating with others about these activities. The proliferation of personal workstations increases the organization's hidden administrative overhead.

Accounting is difficult because IT resources themselves take many forms and are generally widely scattered throughout the organization. In addition, knowledge workers and their equipment are inseparable in performance terms. Because the firm's accounting system treats people as expenses and machines as assets, it cannot track the performance of people/machine systems. When people, networks, and machines are entwined, benefits become more difficult to understand and to allocate properly. Although most people believe these systems increase organizational effectiveness, traditional accounting systems cannot capably measure the increase or its effects. Most firms possess no accurate ways to measure total costs and investments in technology and quantify returns.

Nevertheless, because sound decisions must be based on facts, measuring information system investments and returns is a critical task for corporate executives. They cannot ignore the challenge, nor can they expect it to disappear. Increasingly, CEOs and CFOs restrain IT expenditures, not so much because they consider returns unsatisfactory, but because IT executives and the line managers they support cannot quantify them. The questions surrounding IT returns and the absence of sound measurement systems warrant executives' reluctance to increase IT funding. Firms using information technology most effectively operate with the tightest budgets and are most likely to measure IT cost effectiveness.[5]

Measuring information technology investments and expenses and making these measurements available to these expenditures' beneficiaries is a critical first step in solving the ROIT problem. Accounting for IT resources and charging for their use alters incentives and impacts organizational behavior. In cost-sensitive organizations, high costs discourage use and low costs or free services encourage high, perhaps unwarranted, consumption. Consequently, accurate accounting and charging processes alter cost and expense flows, change individuals' and organizations' behavior, and favorably influence the economics of organizations within the firm. Although the administration of sound IT cost-accounting systems may seem arduous, it is critically important to the organization's financial health.

This chapter introduces IT resource accounting and discusses some common charge-back methods. It discusses several alternatives for handling application development, program maintenance, and centralized and distributed operations. This discussion of variations illustrates the subtleties of charging mechanisms and provides insight into evaluating the return on IT investments.

ACCOUNTING FOR IT RESOURCES

Skilled people, sophisticated networks, complex and strategically important systems, and modern hardware and software systems are very valuable. People with skills in operating system development, Java programming, or extranet implementation are in critically short supply. In some cases, acquiring and developing IT resources can take a long time; developed IT resources represent a large investment. IT investment decisions and allocating future IT resources are critical tasks for the firm's managers. To make these investment decisions intelligently, managers need a reasonably accurate knowledge of past expenditures and plausible projections of future expenditures. A relatively accurate cost-accounting system is essential for beginning to quantify the IT activities' costs so their net benefits can be ascertained.

There is no meaningful alternative to accounting for IT expenses. Some firms treat IT expenses as overhead and pool similar expenses together. For example, such firms may collect and summarize all labor expenses for IT, pool hardware expenses, and also summarize supply and other costs. Other firms may treat pooled expenses as corporate overhead that affects the firm's profit at a high level. Still other firms' expense categories for IT are granular: they pass expenses through to the ultimate beneficiaries of the service. Although many alternatives exist between these extremes, no alternative avoids accounting for expenditures. The choice of accounting methodology, however, has many important consequences.

Accounting is the process of collecting, analyzing, and reporting financial information about an organization. Accounting can have two forms: financial accounting and managerial accounting. Financial accounting provides information about the organization to outside individuals or institutions such as banks, stockholders, government agencies, or the public. Because the public uses the information, financial accounting is subject to carefully crafted accounting rules. Financial accounting's objective is to provide a balance sheet and an income statement for the firm.

On the other hand, managerial accounting focuses on information useful in the organization's internal management. Because the information is only used internally, organizations can establish their own rules and tailor them to meet their specific needs. Managerial accounting's goal is to provide the firm's managers with information that enables them to optimize the firm's performance. Accounting within the IT function gives IT managers important and valuable information useful in optimizing and controlling IT activities.

Traditional financial accounting methods do not reflect the importance of intangible assets in the service and information industries. Estimates indicate that more than $1 trillion is invested in software and databases that have been expensed and not capitalized. "Companies spend enormous amounts of money on internal computer systems for production, purchasing decisions, sales analysis," says Arthur Siegel, vice chairman for accounting and auditing at Price Waterhouse, "but general practice is to expense those costs as incurred. We need to rethink this."[6] At the least, firms can change their internal accounting systems to reflect their assets more accurately, even though current conventions require reporting differently to the public.

Innovative managerial accounting can overcome some deficiencies of traditional accounting systems. For example, application development costs can be capitalized and amortized over the application's anticipated life. Within the firm, purchased system software and applications can be treated like balance sheet assets, with depreciation taken to reflect technological and business obsolescence. From this perspective, managers can evaluate IT investments over their probable life, consider depreciation costs and obsolescence, and estimate their residual value. Although obtaining this kind of valuable information is not difficult, it usually remains undiscovered in most firms. Accounting for human or social capital is also a problem that needs a solution. These and many other aspects of accounting for IT resources are the main themes of this chapter.

As uses for information technology evolve, managers must be increasingly concerned about IT's financial aspects. Firms are increasingly dispersing IT assets and their associated costs and expenses throughout the firm as distributed data processing and intranet applications flourish. When important computer resources become widely dispersed, the budgetary responsibility for these expenses usually becomes dispersed too. In addition, the mix of costs and expenses changes due to the declining hardware costs, increased use of personal workstations, hidden costs of workstation overhead (mostly personnel costs), rapid growth of networks of all kinds, and trends toward purchased software. All these factors make understanding the total cost of information processing in the firm more difficult, even though the costs are increasing.

When information processing disperses, expense rebalancing occurs between centralized and distributed operations. Many central IS organizations tend to reduce spending on mainframe computers while purchasing more application programs, network hardware and software, and client/server systems. At the same time, personnel costs rise at a steady rate. As a result, most firms' total expenses for information processing rise even as costs to process instructions and store data decline dramatically.

As information technology becomes increasingly pervasive, identifying and quantifying costs and expenses become more difficult. Training costs associated with new information technology are a good example of this difficulty. When training in information systems activity was mostly confined to central IS organizations, accounting for it was relatively straightforward. As technology diffuses and engages more people, training expenses increase. As office systems, client/server computing, and Internet technology flourish, training and other startup expenses increase, but are less easy to identify. Dispersed startup efforts and labor associated with operation and informal maintenance make accurate accounting for this activity difficult. Introducing office automation may cost as much as $12,000 per workstation, for example, but about half of this money is spent on additional support staff, startup expenses, and individual training. In client/server operations, more than 75 percent of the long-term costs are people related. Traditional accounting systems, and the way in which they are applied, fail to account for many of these expenses.

Nevertheless, accounting for IT expenditures is not optional. What is optional, however, are the methods used in the accounting process. Current rules and regulations and commonly accepted accounting practices leave considerable room for customization. The degree of customization and the manner of allocating IT expenses to IT service users are important policy issues for the firm and the IT organization. IT resource accountability is a critical issue.

OBJECTIVES OF RESOURCE ACCOUNTABILITY

Accounting for IT resources is a logical continuation of the IT planning process described in Chapter 4. The management system for developing strategies and plans leads directly to the topic of resource accountability, because the planning process includes feedback mechanisms based on measurement and control. The IT planning model and the measurements and controls discussed earlier apply to applications, system operations, human resources, and technology.

The objectives of resource accountability are to help measure the progress of operational and tactical plans and to form a basis for management control. Control is a fundamental management responsibility and a critical success factor for managers. Specifically, IT managers must deliver services of all kinds, on schedule and within budget. IT resource accountability critically supports these measurement and control processes.

The IT planning process and the firm's planning process culminate in a budget for the firm and a budget for IT resources that support the firm's objectives. The firm's budget describes resource expenditures required to meet the plan's objectives throughout the plan period. Likewise, the IT organization's budget includes resources necessary for IT to meet its organizational goals and objectives. A successful planning process results in a consolidated budget that incorporates separate budgets for each unit within the firm. Budgets are tightly linked and collectively support the firm's goals and objectives.

Control is the process through which managers assure themselves that the firm's members are in accordance with the firm's policies and plans. In other words, policy setting and planning precede control. Generally, both planning and controlling the firm's operations involves the same people. Accounting systems are one way for the firm's personnel to communicate with each other regarding plans and actions. IT accounting information can also measure individual performance and appraise unit performance.

For example, consider a firm that embarks on a strategy to improve its product development and manufacturing effectiveness by introducing computer-aided design and computer-aided manufacturing systems (CAD/CAM systems). The firm's strategy expresses this long-term goal. The product development, product manufacturing, and IT organizations' functional strategies provide specific details. The firm's plans describe how each of these organizations will expend resources to accomplish this goal. Each organization's plans describe how it will apply resources to accomplish its part of the overall task. The IT organization develops strategies and plans to support the CAD/CAM installation, and the IT budget also allocates resources to accomplish the CAD/CAM installation. To complete the process and to control the installation activity, the IT organization needs a management system to measure and control the expenditure of budgeted resources.

Budgetary control is the process of relating actual expenses to planned or budgeted expenses and resolving any variances. All firms have a process for accomplishing this task, and this process includes the IT organization. IT managers periodically receive information from the controller describing IT's financial position relative to its budget. Using this information, managers can monitor and control spending. Usually, the controller's staff is available to assist.

The tasks of planning, controlling, communicating, and budgeting are all forms of management decision making. IT accounting information is very useful in making decisions regarding IT activities. These decision-making activities are also valuable because they cement relationships between IT and other parts of the firm. The budget plays an especially important role in this process.

IT accounting processes accomplish several important objectives. They 1) provide continuity between planning and implementation, 2) establish a mechanism to measure implementation progress, 3) form a basis for management control actions, 4) link IT actions to the firm's goals, 5) communicate plans and accomplishments, and 6) provide performance appraisal information. IT accounting is an essential activity.

Is controlling IT activities with budgetary processes alone sufficient? Are the firm's best interests served if client organizations are not financially involved in IT expenses? Because IT organizations provide valuable services to the firm, perhaps it is best to recover the costs of these services from the users in some way. Many firms believe so. Systems that charge clients for services delivered receive increased attention because executives focus more intently on alignment issues, IT managers want line managers to help defend IT spending, and CFOs need more granular data to support investment decisions and evaluate outsourcing proposals.

The tendency to link technology costs to operational departments accelerates as firms focus more closely on productivity, costs, and cost containment. IT managers want to demonstrate that operational requirements drive their expenses and want operational department heads to support these expenditures to senior executives. Thus, many firms find that charging users for IT services in one form or another helps provide important decision-making information.

RECOVERING COSTS FROM CLIENT ORGANIZATIONS

Most firms believe that the advantages of an IT cost-recovery process outweigh its disadvantages. The most important advantage of IT cost accounting and cost recovery is that they provide a basis for clarifying costs and benefits of IT services. A well-structured and smoothly running IT cost-accounting system attacks an important IT issue—the role and contribution of IT. An important secondary advantage is that the process strengthens communication between IT and user organizations and aids in organizational alignment. Table 16.1 lists some benefits of IT cost recovery.

TABLE 16.1 IT Cost Recovery Benefits

1. Helps clarify costs/benefits of IT services
2. Strengthens communication between IT and user organizations
3. Permits IT to operate as a business within a business
4. Increases employees' sensitivity to costs and benefits
5. Spotlights potentially unnecessary expenses
6. Encourages effective resource use
7. Improves IT's cost effectiveness
8. Enables IT benchmarking
9. Provides a financial basis for evaluating outsourcing

IT cost-accounting and cost-recovery methods focus managers' attention on services of low or marginal value. For instance, users paying for operating a regularly scheduled application of low or marginal value will analyze costs and benefits and may terminate the application. If production operations are free, marginal application programs continue to run, perhaps indefinitely. Cost accounting provides tools to detect unnecessary or cost-inefficient services. Inefficient services can be altered or eliminated, thus reducing expenses and improving overall effectiveness.

Users who pay for IT services have information they can use to maximize gains while minimizing costs. They may search for more effective processing schedules or increase their use of services of high marginal value. For example, an online application's operation may expand when users receive tangible evidence that additional services are financially attractive. Likewise, charging users for application program enhancement motivates them to request financially viable enhancements. Charging processes encourage effective use of scarce resources.

A well-designed charging mechanism also enhances the IT organization's effectiveness. Management more carefully scrutinizes well-known and sufficiently detailed IT costs. IT managers direct their attention to cost elements when they need to explain IT charges to clients. Attending to cost elements motivates IT to improve the cost effectiveness of its customer services. Insufficient attention to cost elements or insufficient detail in cost structures promotes ineffective resource utilization. For instance, grouping all telecommunication costs together masks the effectiveness of individual services; inefficiencies of some services will negatively affect the organization.

Pricing and billing users for IT services heightens the firm's employees' and managers' awareness of IT costs and benefits. IT personnel who are aware of the money the firm spends to support their activities become more cost conscious. Likewise, clients billed for IT services become more value conscious. Appreciating what the firm spends on their IT activities, they tend to use IT more wisely. Chargeback processes increase the cost and value consciousness of employees and managers alike.

Administrative overhead is the primary disadvantage of IT cost recovery. The cost-recovery process is not free and must be cost effective itself. Cost recovery may not be justified for organizations that are not information intensive. Also, cost recovery may be unacceptable to some organizations whose cultural norm is not to involve users in IT service costs.

Not all organizations favor user chargebacks. IT managers in some firms prefer to keep their finances clouded in secrecy rather than expose their operations to any criticism that may result from charging. In others firms, after budget approval, activities proceed with relative insensitivity to costs. Employees and managers in some agencies and organizations are relatively insensitive to expense: in these organizations chargebacks intended to foster cost awareness are ineffective. Thus, not all firms favor IT chargeback. Some are simply not intellectually prepared to operate an effective chargeback system.

One alternative to chargebacks is to distribute IT costs to users through a committee; another is to allocate them during the planning process. These simpler approaches suffer from many disadvantages: they are frequently political, inaccurate, and relatively inflexible to changing conditions, and they reduce line management accountability and responsibility. Usually this leads to the well-known condition in which all units argue for increased budgets at plan time, then spend all their allocations and maybe more in anticipation of the next planning cycle.

Another alternative is to bury IT costs and expenses in general corporate overhead, where they are relatively invisible. This approach, which turns information systems into a free utility, encourages expenditures and completely eliminates the advantages of chargeback systems. Cost-conscious firms do not favor this alternative.

Given the advantages of cost recovery processes, what goals and objectives should the firm strive for in the process? What can the process gain for the firm, and how must the process be established for maximum effectiveness?

Goals of Chargeback Systems

IT chargeback mechanisms should serve the firm and its constituent organizations by improving the firm's effectiveness and efficiency. The IT organization should benefit, but the whole firm should benefit as well. The goals of chargeback systems are 1) easily administered, 2) easily understood by customers, 3) equitably distributing costs, 4) promoting effective use of IT resources, and 5) providing incentives to change behavior.

Budgetary controls are important but subject to the same effectiveness criteria as other organizational activities. IT chargeback process must be easy to administer and easy for clients to understand. Administrative costs must be small in relation to the resources being administered. For example, if the cost to account for and bill an IT service is $5, then the total amount billed over the accounting period should considerably exceed $5. In other words, the chargeable services and the chargeback process itself must be cost effective.

For IT customers and managers to accept chargebacks, algorithms for generating or computing customer charges must be easy to explain and justify. A complex billing system based on obscure and hard-to-understand parameters alienates customers and creates planning and budgeting difficulties. Algorithms constructed from simple, easy-to-understand parameters make charging processes easier to administer.

Chargeback mechanisms must account for IT costs equitably. For example, the charge for programmer's time may be based on hours worked using two rates, depending on the programmer's skill level. Because programmers work at many different levels of productivity, this algorithm is not completely accurate, but most client organizations will consider it fair. It is also relatively routine to administer and, therefore, cost effective.

IT chargeback mechanisms alter financial incentives for organizations to use IT services. Choosing a chargeback method and the manner in which it is used influences usage patterns and relationships between service providers and clients. Therefore, another goal of the chargeback process is to promote cost-effective use of IT resources. For instance, when use of online direct access storage devices (DASD) is free, clients have no financial incentive to use storage effectively. Free DASD space doesn't encourage users to delete obsolete data sets, for example. On the other hand, pricing direct access storage at unreasonably high rates may prompt users to use inefficient date handling methods to reduce costs. They may use tape storage, for example, even if DASD is more effective for the organization.

When clients are sensitive to costs, charging mechanisms can steer the organization toward certain technologies and away from others.[7] New technologies can be attractively priced to encourage use, and older systems can be priced unattractively. Price may not reflect actual costs, because new equipment may be more expensive than older, fully depreciated hardware. In this instance, pricing policy provides incentives to operate more efficiently, not to recover actual costs.

Chargeback systems that are easy to understand and administer, distribute costs effectively, and promote effective IT resource use are valuable IT management systems. Successful IT managers rely on them extensively. Also, astute IT managers carefully consider advantages and disadvantages of alternative chargeback methods before selecting and implementing one.

Two major accounting alternatives exist for handling IT cost recovery: the cost-center method and the profit-center approach. Both methods distribute IT costs to users but vary considerably in other respects. The profit-center method is designed to generate revenue exceeding IT costs. IT organizations may use these profits for approved purposes. Of course, when costs and expenses exceed revenue, the profit center loses money, and IT must devise ways to recoup its losses.

On the other hand, the cost-center method seeks to break even financially. The revenue from services are expected to match costs very closely. Methods to make costs and revenues match exactly are often used. The following sections discuss details of these financial arrangements.

Profit Centers

IT organizations established as profit centers operate as a business within a business. By charging customers for services, the IT organization recovers its costs and expenses. Nevertheless, the usual relationship between revenue and expenses prevails, so the organization may operate at a profit or a loss.[8] Table 16.2 lists the chief advantages of the profit-center method.

TABLE 16.2   Advantages of IT Profit Centers

1. Easy to understand and explain

2. Promote business management

3. Provide benchmarks and comparisons

4. Establish financial rigor

5. Enable outside sale of services

Managers who operate profit centers develop important skills as business managers. They use many disciplines that independent business people use. They learn business management in a way not usually possible otherwise. IT profit center managers appreciate the sometimes invisible expenses of corporate overhead, employee benefits, equipment depreciation, and pricing strategies. Business management experience also extends to IT customers who may gain exposure to the financial consequences of previously hidden issues. Firms favor profit centers for this reason, among others.

Profit center prices reflect the real cost of doing business in the firm, and managers can easily compare them with prices of similar services from outside suppliers. These comparisons, or financial benchmarks, provide a mechanism for measuring IT effectiveness and provide incentives to improve IT. Firms whose policies let clients purchase competitive services usually have highly motivated IT service suppliers. Firms with noncompetitive IT organizations may elect to purchase many services externally. Benchmarks may also highlight inefficiencies within the firm because they direct attention to general overhead expenses and corporate accounting policies. Profit centers' benefits extend beyond IT and its clients.

Profit centers also provide a high degree of rigor and discipline to the financial relationship between IT and its client organizations. These centers establish costs more carefully, evaluate expenses more thoroughly, and determine prices and charges more insightfully. Such discipline assists other processes as well. For example, accurate costs needed to achieve service levels are available, thus negotiations on service level agreements proceed more objectively. Application portfolio management improves and make/buy decisions are made with heightened confidence. In general, financial management improves.

Establishing the IT organization as a profit center enables IT to sell services outside the firm, confident that the activity profits the firm. When firms intend to sell IT services to others, the profit-center methodology is a prerequisite to establishing an external revenue-generating service bureau. IT profit centers established as independent businesses with their own financial accounting systems, will link to and closely support the organization's financial accounting system. A profit center's accounting rigor lets the firm capitalize on its IT resources as sources of revenue and profit, if it chooses.

Profit centers also have some obvious disadvantages. Financial rigor generates administrative overhead—rigorous financial treatment may not be worth the price. Detailed knowledge of all expense ingredients and soundly based prices may not improve overall performance. In addition, the firm depends on successful management of the IT/client interface. Prices are not guaranteed to be properly established, and clients may not fully anticipate IT service costs during their budgeting process.

Profit itself may be a problem. If IT is highly profitable, its customers probably pay more than necessary, perhaps sacrificing alternative investments. On the other hand, not earning the profit anticipated may force some high-level adjustments. The degree of financial coupling between IT and its clients may be insufficient to avoid all these difficulties; nevertheless, skillful profit center executives have many tools to improve their operations and their firms.

Cost Centers

Cost centers are an another alternative for handling IT cost recovery. They are based on a process in which each client organization budgets for its anticipated IT services during its planning and budgeting cycle. The sum of the client-budgeted amounts for IT is the cost center support from its customers. Managers expect this amount to approximately equal anticipated IT expenses for the planned period. Because planned IT expenses and client-budgeted support may be mismatched, this approach encourages brisk interaction between organizations during the planning stage. Planning and budgeting are intense and iterative. Table 16.3 lists some other cost center characteristics.

TABLE 16.3   Cost Center Characteristics

1. Promotes intensely interactive planning and budgeting

2. Establishes prices in advance of known support

3. Forces mangers to handle variances

4. Exposes the planning process to manipulation

5. May lead to conflict (this may be beneficial)

6. Forces decision making

7. Reinforces the SLA and capacity planning processes

To complete the annual planning process, client organizations must know their IT requirements for the planning period. They must also know IT prices so they can submit their budgets. Their budgeted support for IT is their anticipated usage multiplied by the service price. During the same planning stage, IT also prepares a budget that anticipates all customer demands and includes plans to satisfy them. The total of all client budgets for IT must approximate the budget IT prepares.

If budgeted support is not close to planned IT expenses, the process begins again in the IT organization. This cycling means that the IT budget is the last to be completed. Also, because support levels change, prices may change. Price changes alter client budgets by small amounts. Usually this phenomena is handled by permitting some variance at plan time between the sum of all budgeted support for IT and the expense budget prepared by IT.

In the ideal situation, customer budgets for anticipated IT services approximately equals the IT budget for anticipated expenses. Ideally, actual customer demand for services equals planned customer demand, and prices from IT equal planned prices. In actual practice, many variances must be handled during plan execution. Demand changes generate price changes during the year, creating financial consequences for the clients. Small demand increases for some services with high fixed expenses, such as centralized production operations, bring IT a profit; small demand decreases result in an IT loss. By year-end, the variances between revenue and expense must be resolved.

There are several ways to resolve these variances. One approach is to carry the variance at a higher level in the firm, generating a small profit or loss at that level. Another approach is to distribute the profit or loss to customers as a retroactive rate adjustment. This may result in overages or underages in customer budgets. A third way is to adjust the rates or prices slightly during the year to keep the running variance close to zero throughout the plan period. Thus, managers have several means for dealing with cost center variances.

As one might expect, cost centers are subject to customer manipulation. For example, a client organization underbudgets for IT computer services and later increases its demand; if satisfied, the extra demand lowers rates for all customers. In this case, the gamble pays off for that particular client. If many clients underbudget, IT may be unable to satisfy true demand and customers—including those who budgeted accurately—receive poor service.

If corporate policy is to cover variances at a higher level at year-end, clients tend to over-budget for IT. IT then acquires greater capacity than required, clients obtain better service than planned or justified, and the higher level must cover excess expenses. Clients may spend the excess in their IT budget on other items, justified or not. There are other variations of this process as well. Given this scenario, cost centers work well when the firm's financial discipline is strong and the controller acts to protect IT and the firm from potential abuse.

Cost centers may be contentious and lead to conflict, some of which may be healthy. Cost-center methodologies tend to reveal latent demand when capacity planning processes are ineffective. When capacity planning is well orchestrated, cost-center planning proceeds smoothly with satisfactory results. In any case, cost-center planning forces decision making. This may be advantageous if other management processes within the firm are not completely effective.

In addition to profit-center and cost-center approaches, other alternatives combine aspects of each or use other methodologies for certain IT services. For example, some services may be priced separately, partial recovery of certain costs may be allowed, or costs may be based on long-term contracts with users. What are some alternatives and enhancements to cost recovery methodologies? The following section addresses these items.

ADDITIONAL COST RECOVERY CONSIDERATIONS

IT services are a varied lot; no single scheme for recovering costs satisfies everyone. For example, compare the labor-intensive process of application development to operating online mainframe applications or a data communication network. Application development has only slight economies of scale. With generally little excess or latent capacity, it struggles with a variety of people concerns and issues such as communication and motivation. The mainframe operation, however, may have economies of scale, may have large amounts of latent capacity, and is less susceptible to people issues and concerns. Telecommunication networks differ from each of these. These differences permit optimization of cost recovery methods to take advantage of service characteristics.

Funding Application Development and Maintenance

Portfolio management and managing application development lend themselves very well to customized or unique cost-accounting and cost-recovery approaches. For example, the profit or cost center can recover costs of application program maintenance and minor enhancements using rates established for application programmers. A programmer working 10 hours to perform minor enhancements bills the program owner for 10 hours of service at established rates.

Because minor enhancements are extremely common for important applications, it's appropriate to recognize this condition in advance and prepare a long-term contract describing ongoing support for the application.[9] The contract for "period" support may state, for example, that the programming department devotes 20 hours per week to program enhancements that the client manager requests. The contract period is probably a year or more and rates are sufficient to recover the programmer's expenses. This type of contract has inherent flexibility.

The programmer's hours are accounted for in a general way. The client manager has the opportunity to make minor enhancements without contending with other users for programming resources. Both the IT organization and the client department benefit from this arrangement.

Application development projects may also benefit from alternative cost-recovery methods. Consider, for example, a variation of the pay-as-you-go method for recovering costs on a major programming project. At the beginning of each phase, cost to complete the next phase is established. At the end of the phase, IT bills the customer for the contracted expense and recovers its costs. This approach closely relates the level of effort to the objectives established during the phase review. In effect, IT commits to producing the function on schedule *and* within budget. IT usually does not agree to functional changes during the phase, so client and IT managers must produce high quality plans. This approach reinforces the discipline of the phase review process with financial incentives. Charging for application development by phases is generally preferred to recovering costs by the traditional pay-as-you-go method.

Charging for application development by phases exposes poor planning, reveals the financial impact of change, and discloses costs of poor implementation. In addition, if terminating the project becomes a possibility, this approach makes the decision less difficult. Continuing the project, making up for past excesses by trimming future planned expenses, is less tempting.

Another method is to recover the cost for application development from its owner over some predetermined application life. This approach recognizes that clients realize application benefits after installation. It relates development expense to benefits realized over time. This method's advantage is that some higher level in the firm takes early risk, thus making application development more attractive for users. The firm's advantage is that it can develop applications of high potential value that may not be financially feasible otherwise. To the client, this method appears to capitalize development efforts, although it may be expensed at higher levels.

Some projects should be funded at the corporate level. The firm, not any single client manager, should financially manage applications that have great strategic value. Although one customer class may primarily use the application, strategic decisions regarding the application should be made at higher levels. For example, product managers in the development laboratory cannot effectively direct electronic design-automation programs that give advantage to firms engaged in designing and building electronic products. Product managers are likely to make short-range decisions; they focus on the product, not the firm, and their interests are diverse. The solution is decision making at a level where long-range considerations are preserved, the firm's interests are paramount, and divergent interests and motivations can be reconciled.

Widely used applications, such as office automation systems, must also be financially managed at some high level. Office managers should establish requirements and provide justification, but office system funding is best performed by a central administrative function. This approach recognizes the firm's office automation costs and benefits, but it avoids the difficulty of allocating costs to individual workers or departments. Some telecommunication systems and large database applications fall into this service category as well.

Cost Recovery in Production Operations

Many algorithms exist for recovering costs in production systems. A common theme in most of these algorithms is to charge for resources used to produce useful output. Common usage parameters include CPU cycles used, the amount of primary memory occupied and for how long, channel program utilization, pages of printed output, and other measures of production resources. Generally, these various parameters link together in some fashion to form a charging algorithm. IT establishes a price for each resource so that the charging algorithm reflects the cost of work performed. In some cases, this type of charging algorithm becomes quite complex—it may also be very accurate.

Although these measures can be quite precise, they are not very satisfactory to IT customers. They are also difficult to develop because they require a detailed knowledge of complex systems and the economics of their parts. Because of these complexities, precise charging algorithms are difficult to explain and hard for most clients to understand. Clients have trouble relating some measures to the useful work the system performs for them. Simple methods, such as charging for the application's elapsed time, are probably more effective.

For applications or processes that occur over extended periods, elapsed time is a common denominator of charging algorithms. For example, a continuously operating online application may be priced according to the fraction of the CPU resources it consumes. If the online order-entry system requires one-third of a major CPU, then the service price recovers one-third of the system's operating costs. Frequently, large online data stores' price is so much per track per day. The idea is to recover online storage costs from users in proportion to their storage volume. If large online data stores are associated with continuously operating applications, then the elapsed-time charge includes both CPU and DASD costs.

Many other variants are useful in charging for production operations. One is price differentials for classes of service. For instance, users of prime-shift capacity (8:00 A.M. to 6:00 P.M.) may pay more than off-prime-shift users. The price differential can be justified because prime-shift time is more valuable to most users than off-shift time. The price differential encourages system use when there is usually surplus capacity. Shifting workload from prime to off-prime time is equivalent to increasing capacity; it represents a real cost saving for the firm. Using the same rationale, weekend work may be processed at reduced prices, while priority or emergency work may cost more. These examples of price incentives encourage more effective use of IT resources.

Services dedicated to one user class should be charged directly to that class. For example, if one CPU and its associated equipment and support personnel are devoted to one client department, that department should pay the full cost of that service. This charging methodology is equitable and easily justified. Dedicated telecommunication links are another example.

In production operations, planning ahead and recognizing the effects of technological obsolescence are important. These can be accomplished by basing costs on a multiyear plan. The multiyear plan prevents wide fluctuations in prices when equipment is replaced or additional capacity comes on stream. Advanced planning recovers costs early and matches long-term revenue and expenses. Successful IT managers use this approach to everyone's advantage.

Finally, some services should be sold outright. End-user computing and office automation benefit considerably from this approach. Rather than billing customers for terminals or personal workstations, IT should sell the hardware to its clients and bill network time based on system use. This simplifies the accounting process and is easy to justify and understand. Most client managers are comfortable with this approach. Upgrades or additions to the customer's personal hardware or software are best handled on a purchase basis, too.

Recovering Costs in Distributed Systems

As firms rebalance resources by moving from centralized to customer-controlled systems, methods to understand and account for costs and expenses in distributed environments may also shift. Changes in financial treatment may vary widely depending on the firm's policies regarding distributed processing. For example, if the IT organization continues to own the hardware and develop or procure applications for the client department, little change occurs. On the other hand, financial accounting changes when client departments own the hardware and develop or procure their software. Sometimes the situation falls somewhere between these extremes.

As a general rule, when distributed systems serve one function, marketing for example, that function should justify and pay for the system. It should also fund software it develops or procures, informing the firm of these costs or expenses. This is easy for purchased items but more difficult for client developed programs. Policies requiring records of program development costs are essential; otherwise client program development becomes another hidden cost of distributed processing, which increases the difficulty of sound decision making.

In distributed processing environments, what matters most is identifying, controlling, and justifying costs; who does the work or owns the assets is far less important. As IT line activities decline and staff activities increase, financial rebalancing accompanies responsibility shifts. One staff responsibility that increases is ensuring that corporate executives have the information needed to make decisions. Establishing policy guidelines that account for IT expenses and benefits in reasonable and effective ways, wherever they are incurred, is an increasingly important staff responsibility.

Network Accounting and Cost Recovery

Networks significantly complicate IT cost-accounting and cost-recovery processes. Production computers connected to networks consider them information sources or sinks for application programs. IT accounting routines deal easily with application accounting, but, when applications in several computers communicate with each other and place computing load on various CPUs, accounting processes become very complex. To account for all costs accurately in ways that clients understand is impossible.

Accounting becomes even more complex when public carriers and value-added networks enter the picture. International operations add yet another level of complexity. To deal with these complexities and understand operational costs and expenses, corporate network managers must have a solid understanding of expense impacts on service levels and must know expenses to the component level. They should keep detailed records of cost, performance, and utilization trends. Knowledge of network costs permits managers to trade off various types of services and technology. They must be able to relate new technology costs to capacity increases and performance and service improvements.

IT managers must be very careful about charging clients for telecommunication services. Some services, such as long-distance dial-up voice or data communication, can be charged to users' accounts directly. Other network services, like those for EDI applications, are probably best recovered at the corporate level. Other applications must be analyzed individually. Network expenses that readily correlate with applications should be charged directly to users. Small or co-mingled expenses should be recovered indirectly.

Accounting for and recovering IT expenses from IT customers benefits from a businesslike approach. The alternatives discussed earlier can be used with either the profit-center or cost-center structure. Ultimately, methodology and variations depend on the firm's information technology maturity, IT's relative importance, corporate culture, and objectives the firm's executives hold for IT. As these factors change over time, the cost-accounting and chargeback systems also change, transforming as the organization matures.[10] In addition, relationships between IT and client organizations mature as the accounting process itself matures.

RELATIONSHIP TO CLIENT BEHAVIOR

One important goal of user chargeback processes is to encourage and promote cost-effective use of IT services. Many financial motivations result from cost-recovery actions; other motivations can also be developed.

Consider, for example, a firm that underutilizes information technology and wants to promote increased technology adoption. What strategy should this firm adopt? How long should the strategy be employed? One strategy for this firm is to increase IT funding and services but accumulate increased expenses at the corporate level. For example, a firm that wants to encourage intersite communication can install a network linking several of its sites and cover these expenses at headquarters. Thus, the network is a free utility, encouraging user managers to develop intersite communication applications.

This funding approach is also appropriate for firms lagging behind the industry in technology adoption or those in the initiation stage with a new technology. This technique hastens the firm's migration into the contagion stage where financial incentives are perhaps no longer needed. When the firm reaches the contagion stage, it can reduce special incentives and use another approach.

Various cost-accounting and cost-recovery approaches contain many incentives to modify behavior. As firms mature in their use of information technology and managers become proficient in using IT management systems, firms can employ variations to maximize the chances of attaining corporate goals. For example, the organization may abandon the IT cost center for the profit center when it recognizes opportunities to sell IT services outside the firm.

Mature, sophisticated firms with attractive IT assets may want to sell services to generate revenue. Many firms make their production facility available as a service bureau. Some firms sell their applications, and some sell the use of their applications—airline reservation systems are an example. Other firms engage in contract programming. In cases like these, corporate goals for the IT function may require the adoption of alternative accounting methods.

IT accounting systems having carefully constructed chargeback methods are fundamental to effective IT operation. They provide the basis for using IT services cost effectively. Just as importantly, they greatly enhance communication between IT and client organizations. The firm's performance improves through motivations present in accounting and chargeback mechanisms. Successful IT managers fully exploit these concepts.

SOME COMPROMISES TO CONSIDER

Cost-recovery approaches discussed thus far in this chapter are subject to many variations that can be expanded further for many purposes. For example, the firm may not require cost centers to recover its costs fully or to return overages to using organizations. The managerial accounting system's purpose is to serve managers, not to devise highly precise accounting processes. If the system controls, communicates, motivates, and helps measure and plan as management desires, then additional accuracy and precision in accounting processes may not be justified. Sacrificing accuracy can reduce administrative overhead and improve efficiency. Likewise, the accounting methodology can change as the firm's information technology use matures. Precise comparisons over time can be traded for improved motivation or control as business conditions change. The system's purpose is to serve management, not to be a pinnacle of accounting purity.

EXPECTATIONS

IT clients expect cost-recovery systems to accomplish certain goals. They expect that systems be easy to use and understand. Cost-recovery systems must be designed for clients and must help them advance their business relationship with IT or other service providers. Clients expect fair distribution of IT costs, and they demand consistency. Frequent or unusually large price changes upset their plans and diminish their confidence in IT's ability to manage its affairs properly. Cost-recovery processes are valuable to organizations, but their use and administration require sound management skills.

Still, IT managers and other executives have larger expectations for IT's accounting system. They expect to understand how IT spends resources to advance the firm's goals: well-designed accounting techniques help executives understand the IT organization as a business. Executives also expect to measure IT's value to the firm. IT accounting and chargeback systems are invaluable in this regard.

However, as the Business Vignette noted, improved organizational effectiveness may be the only measure of many IT expenditures. Perrier increased its customer retention rate, for

---

example, and used information technology to restructure. Generally, traditional accounting systems inadequately measure organizational effectiveness. Innovation in IT accounting systems can overcome some of these difficulties; nevertheless, many challenges remain for IT managers or CIOs and for the firm's senior executives.

CIOs can challenge the firm's accounting system. They can raise questions about the firm's ability to measure organizational effectiveness. And, they can act to correct deficiencies in traditional accounting systems by installing innovative managerial accounting systems for the IT organization. In the process, they forge stronger ties and linkages between IT and the rest of the firm. In addition, they strengthen their business relationships to operating executives and to the firm's corporate staff.

MEASURING IT INVESTMENT RETURNS

Thoughtful CIOs must consider improved ways to value investments in machines, software, and people. Improved valuations of these resources leads to improved productivity, performance, and organizational effectiveness measures. IT's role and contribution has long been a leading management issue. Innovative accounting for and measurement of costs and benefits have the potential to resolve this issue, in part.

Traditional ROI calculations, discussed in the section on application program investments, are a good starting point for evaluating IT investments' worth.[11] But for many IT projects, such as investments in technology infrastructure or those leading to improved quality or customer service, strict ROI calculations frequently show a negative or low return because of uncertain results or the inability to fathom the business dynamics resulting from the investment. Typical ROI calculations are based on a static business environment, but frequently IT investments are designed to alter the environment. Under these circumstances, traditional evaluation techniques are weak.

To help overcome these difficulties, John Henderson built an evaluation model adapting techniques from stock-trading analyses that permit multiple outcome scenarios from the initial investment.[12] Henderson's Option Model augments ROI calculations by permitting risk assessments of various investment decisions as business strategies and systems requirements change. The model recognizes risk explicitly and encourages making investments in stages to determine consequences before committing to major expenditures. As investments are put in place, future decisions are based on information gleaned from recent results and altered views of the future.

The Options Model does not remove all uncertainty about IT investments, but it does provide additional insights. The model is complex and difficult to use; some managers may avoid it for these reasons. In addition, people, systems, and business strategies are not subject to the same financial rigor as stocks and options; this introduces additional uncertainty. The model's use should supplement traditional evaluations and management judgment, not substitute for them.

During the past decade, information systems' value to business in general and their worth to our economy's service sector in particular has been extensively discussed. During the 1980s, service industries invested more than $1 trillion in information technology, yet productivity rose less than one percent annually, according to U.S. government statistics. Some questioned

IT's value, reasoning that IT investments offered intrinsically low return; others claimed the problem was one of measurement or lack of it. The dilemma, called the *productivity paradox*, is a serious matter and the subject of many studies and analyses.

Efforts to resolve the issue for individual companies led Paul Strassmann and others to search for new measures of business performance that go beyond earnings per share and revenue growth.[13] Strassmann believes that "There is no demonstrable correlation between the financial performance of a firm and the amount it spends on information technologies. What matters is not how much you spend but how well you spend it in supporting business missions and contributing to productivity." His thesis is that management quality is more important in technology and service businesses than assets. Therefore, according to Strassmann, measures that evaluate management's added value must replace return on investment measurements.

To make this evaluation, Strassmann developed an Information Productivity Index based on publicly available data that measures corporate management's effectiveness. The index is a ratio of economic value-added (calculated by subtracting the value of shareholder equity from operating profit after taxes and multiplying the difference by the cost of capital) divided by the cost of management (cost of sales, general overhead, and administrative expenses is a reasonable approximation of management costs). Measures such as Strassmann's Information Productivity Index gain popularity in determining managers' effectiveness and their use of information to improve firms' productivity.[14] As competition increases and productivity becomes more important, measurements of returns on information technology investments increase in importance.

SUMMARY

Information technology resources are a large and growing portion of many firms' budgets. Firm expect their managers to use these resources effectively and account for them properly. In many cases, widely dispersed IT resources make accounting processes difficult. In addition, the complex and varied character of the resources themselves makes accounting tasks more difficult. For all these reasons, accounting for IT resources is important to IT managers and their firms.

Accounting for IT resources assists in planning, controlling, communicating, and assessing performance. Doing so is essential to effectively operating IT or other service organizations. When IT organizations elect to recover costs from customers, additional elements of motivation and effectiveness come into play.

Chargeback methods are very important to IT managers. They must be handled skillfully, with a blend of psychology and accounting, tempered by practical implementation considerations. Accounting systems for IT organizations that intend to sell services outside the firm must be more rigorous than others. Although highly desirable, IT accounting systems are not an end in themselves. The purpose of managerial accounting systems is to permit business managers to operate more effectively. Rules for their construction and implementation must be designed to meet this goal.

Accounting for IT resources and measuring the return on IT investments raises some difficult but very important questions. IT investments affect business operations at the organizational level but the effects may not be directly financial. Dollars saved or costs avoided are relatively easy to understand, but most IT investments have a broader impact. Increasingly, for example, application developers design and install Internet-type systems to improve market position, provide superior customer service, or improve internal communication. However, improved organizational effectiveness is often the only measure of their return on investment—traditional accounting systems cannot provide this measurement. Clearly this area requires a creative approach from CIOs and other senior executives. Innovative managerial accounting systems are the first step in the process of measuring organizational effectiveness.

Review Questions

1. Why is IT resource accountability important to the IT organization and to the firm?

2. Define financial accounting.

3. How is managerial accounting different from financial accounting?

4. What trends in information handling make IT accounting more difficult and more important?

5. What purposes do the IT accounting systems serve?

6. What are the benefits of IT cost-recovery systems?

7. How does a well-designed charging mechanism enhance IT's effectiveness?

8. What is the primary disadvantage of IT cost-recovery processes?

9. What are the goals of IT chargeback systems?

10. What are the advantages and disadvantages of profit centers for recovering IT costs?

11. What are the characteristics of cost-center methods of IT cost recovery?

12. Cost centers can be contentious. Under what circumstances is this advantageous?

13. Why might using several methods to recover various IT costs be best?

14. What does "period support for application programs" mean?

15. What alternatives can IT managers use to charge for application development? What are the advantages of each?

16. The accuracy of cost-recovery systems is not necessarily very important to IT managers. Explain why. When is it very important?

17. What is the productivity paradox?

18. What do clients expect from IT cost-recovery methodologies?

19. What is the CIO's role in IT accounting and chargeback systems?

Discussion Questions

1. Discuss the differences and similarities between financial accounting and managerial accounting for IT organizations.

2. Discuss the relationships between planning, budgeting, measuring, controlling, and accounting for IT activities.

3. How can the activities listed in Question 2 work together to help implement a new market analysis program?

4. How does a well-designed cost-recovery system improve IT's effectiveness?

5. Discuss the reasons why IT cost recovery need not be accurate, but must appear to be equitable to clients.

6. The text discusses some instances in which charging methods alter user behavior toward the consumption of IT resources. Can you give additional examples of this phenomenon?

7. Compare and contrast profit-center and cost-center methods for recovering IT costs.

8. Application development and maintenance costs might be recovered differently than production operation costs. Explain why.

9. Identify goals in managing application development and maintenance that the alternative cost recovery approaches satisfy.

10. Discuss advantages and disadvantages of price differentials in recovering costs of production operations.

11. Describe how firms might price computer services based on multi-year plans that include replacing a major CPU.

12. What type of cost recovery system would you recommend for a firm that believes that IT's role and contribution is a major issue?

13. Describe possible changes in IT cost-recovery methods that may occur as firms grow from modest technology users to data servicing operations.

14. Discuss client expectations of IT accounting systems. What expectations might the firm's executives have of these systems?

15. Accounting for networks is difficult, and charging users for them is even more difficult. How can configuration management databases help account for some network fixed costs? Discuss advantages and disadvantages of accounting for data networks as corporate overhead.

Assignments

1. Draw a flowchart of the budgeting process that takes place in firms using cost-center methods to recover IT costs. What prevents the iteration from continuing endlessly?

2. Visit a local firm to determine what methods it uses to recover IT costs. Why did it adopt the approach it uses? Why is it effective for the firm? What improvements would you suggest?

3. Analyze the accounting package for one network management system, and prepare a report for the class. Indicate in your report what goals this accounting system accomplishes.

ENDNOTES

[1] Allan E. Alter, "The Real Language of Business," *CIO*, January 1993, 37, and Megan Santosus, "A Watertight Case," *CIO*, January 1993, 40. Reprinted through the courtesy of CIO. (c) 1993 CIO Communications Inc.

[2] The 100 most effective users of information systems as measured by *Computerworld* spend 2.4 percent of revenue on IS. As networking's importance grows, it accounts for 12 to 60 percent of the IT budget, according to *Computerworld*, The Network 25, September 29, 1997, 5.

[3] Paul A. Strassmann, *Information Payoff* (New York: The Free Press, 1985), 80. "Payoffs are realized by managing the benefits. Costs are important, but secondary," according to Strassmann.

[4] For a thorough analysis of this phenomenon known as the productivity paradox, see Paul A. Strassmann, *The Squandered Computer* (New Canaan, CT: The Information Economics Press, 1997). In particular see Chapter 3, "Spending and Profitability," 23-40.

[5] Michael L. Sullivan-Trainor, "Best of Breed," *Computerworld Premier 100*, September 19, 1994, 8.

[6] Richard Greene, "Inequitable Equity," *Forbes*, July 11, 1988, 83. Alfred Rappaport, an accounting professor at Northwestern University, commented, "As we become a more information-intensive society, shareholders' equity is getting further away from the way the market will value a company."

[7] David A. Flower, "Chargeback Methodology for Systems," *Journal of Information Management*, Spring, 1988, 17. This article discusses the chargeback system at Prudential Insurance.

[8] "Many, many years ago I coined the term profit center. I am thoroughly ashamed of it. Because inside a business there are no profit centers. There are only cost centers. Profit comes only from the outside. When a customer returns with a repeat order and his check doesn't bounce, then you have a profit center. Until then you have only cost centers." Notes from an informal talk by Peter Drucker as quoted in Charles Wang, *Techno Vision* (New York: McGraw-Hill, Inc., 1994), xvi.

[9] Flower (note 7). Prudential Insurance uses this approach.

[10] Jeanne Buse, "Chargeback Systems Come of Age," *Datamation*, November 1, 1988, 47.

[11] Various examples showing how to calculate IT investment returns can be found in Bill Bysinger and Ken Knight, *Investing in Information Technology* (New York: Van Nostrand Reinhold, 1996), Chapters 9 and 10.

[12] Jeff Moad, "Time for A Fresh Approach to ROI," *Datamation*, February 15, 1995, 57.

[13] Paul A. Strassmann, *The Business Value of Computers* (New Canaan, CT: The Information Economics Press, 1990). For a thorough discussion of the productivity paradox, see Paul Strassmann, *The Squandered Computer* (New Canaan, CT: The Information Economics Press, 1997) Part III, Chapters 9-14.

[14] In a departure from methods used in previous years, *Computerworld's* Premier 100 in 1994 were determined using Strassmann's methods. See Paul Strassmann, "How We Evaluated Productivity," *Computerworld Premier 100*, September 19, 1994, 45.

17

IT Controls and Asset Protection

Robbery on the Information Superhighway[1]

At 2 a.m. February 15, 1995, FBI agents arrested Kevin D. Mitnick at his home in Raleigh, North Carolina, after weeks of sleuthing on the information superhighway. Mitnick, 31, described by prosecutors as the nation's most-wanted computer hacker, was accused of infiltrating numerous computer systems from New York to California and stealing information worth more than $1 million, including thousands of credit card numbers and software controlling cellular telephone operations.

Tsutomu Shimomura, a computational physicist and outstanding cybersleuth, picked up the trail leading to Mitnick on Christmas Day 1994. He discovered that someone had broken into his computers near San Diego from an unknown remote location. The hacker, who used sophisticated techniques to steal several thousand files, angered Shimomura, who promptly terminated his vacation and dedicated himself to apprehending the cyberthief. Aided by California lawmen and the FBI, Shimomura established monitoring posts to catch the thief at work.

At his beach cottage north of San Diego, Shimomura discovered someone had systematically looted his powerful workstation of hundreds of files of information particularly useful in breaching computer networks and cellular phone systems. This act was not random but a deliberate attempt to obtain important information, including advanced security software that could be exploited for illicit purposes. The intruder posed as a familiar node on the Internet; he commandeered a computer located at Loyola University in Chicago for his attack. He also heckled Shimomura, leaving him a computer-altered message on his voice-message system. Shimomura's interest in this case intensified.

The attacker clearly infuriated the wrong person. Shimomura, 30, is a valued consultant on computer security to the Air Force, the National Security Agency, and the FBI. Over the years, he designed security tools for networked systems and developed a reputation as a computer security expert.

On January 28, Bruce Koball, a computer programmer and an organizer for the public policy group Computers, Freedom and Privacy, linked a newspaper account of the break-in with a puzzling message he received from an online service called the Well in Sausilito, California. The group's file on Well's system had grown by millions of bytes, Well's officials told Koball. Upon investigation, Koball discovered his files now contained Shimomura's stolen information. Well notified Shimomura, who recruited two colleagues and established an around-the-clock monitoring system on Well's property. With the help of another security expert, the team discovered another break-in and theft of 20,000 credit card numbers from Netcom Communications in San Jose, California.

On February 9, the team moved from Well to Netcom where they set up equipment to capture the hacker's every keystroke. Calls came from many locations, leading the FBI to conclude that the intruder was in Colorado, Minneapolis, or Raleigh. Meanwhile, the U.S. attorney in San Francisco, aided by subpoenaed phone records from GTE, Sprint, and others, determined the calls were being placed to Netcom's phone bank in Raleigh from a cellular phone modem. The intruder cleverly disguised his call path by altering software in the phone company switches, but after hours of tracking records, they determined the calls originated near the Raleigh-Durham airport. The action quickly shifted to North Carolina.

Shortly after midnight on Monday, February 13, Sprint technicians and Shimomura cruised the area with a radio direction finder, attempting to locate signals from the attacker's cellular phone. Quickly they identified an apartment complex near the airport as the location of the intruder's cellular calls. On Monday, the FBI sent a surveillance crew to assist in pinpointing the calls. They needed a precise address in order to obtain a search warrant.

By Tuesday evening, February 14, agents had determined the specific address. At 8:30 p.m., a federal judge issued a warrant from his home in Raleigh. FBI agents knocked on the door to apartment 202 at about 2:00 a.m. After five minutes, Mitnick, who had been living under the assumed name Glenn Case, opened the door, claiming he was on the phone to his lawyer.

Mitnick was jailed without bond and with carefully controlled phone privileges. He was charged with computer fraud and illegal use of a telephone access device. In 1989, Kevin Mitnick had been convicted of stealing software from Digital Equipment Company. He had also been convicted of hacking MCI phone computers.

Six years earlier in November 1988, the first celebrated case of superhighway mayhem occurred: Robert T. Morris released a virus infecting more than 6000 computer systems from Massachusetts to California, causing millions of dollars in damages. Speaking of the perpetrator, a computer scientist at Argonne National Laboratory said, "He's somebody we would hire. The right to hack is held higher than the right of someone to tell you not to. It's an inalienable right."[2] Society clearly believes differently. Morris was caught, prosecuted, convicted of felony acts, and sentenced to a $10,000 fine, three years' probation, and community service.[3]

INTRODUCTION

Managers who own or use information resources are fundamentally responsible to protect these assets from theft, damage, and misuse. Nearly all firms find controls become more important as information technology penetrates the organization. In most instances, the law requires business controls. Information systems in today's fast-paced business environment must be grounded on a solid base of operational and accounting controls. Applications, databases, networks, and hardware must be carefully protected against loss or damage. Successful IT managers understand these issues and are prepared to demonstrate effective controls to auditors and the firm's senior executives.

As part of their staff responsibility, IT managers are expected to lead in establishing control policies for the entire firm. Their knowledge of technology and its risks and exposures is critical in developing vital security and control policies. They must guide their firms in this difficult and important activity.

This chapter discusses in detail business control and asset protection issues associated with information systems. It identifies participants in this activity, defines their responsibilities, and explores their activities in depth. Controlling and protecting application programs begins in development and continues throughout their life. This chapter develops control disciplines for applications and presents mechanisms for auditing and reporting control status. The firm's auditing activities rely partially on control and protection policies developed in this chapter.

THE MEANING AND IMPORTANCE OF CONTROL

To operate under control, IT managers must first understand their mission—what they are supposed to do. They must know what is expected of them—what is acceptable and what is not. To operate in control when discharging their mission means that managers know the details of all significant activities taking place within their organization. The details consist of knowing the particulars of what, when, where, why, how, and who as they apply to all important organizational activities.

Control also means that managers must have routine methods for comparing actual and planned performance. Deviations from planned performance are obvious in a well-controlled organization. Not only must managers routinely acquire information about plan deviations or out-of-control conditions, they must respond quickly. Managers must also be able to detect performance improvements resulting from variance corrections. Managers operating under control have this knowledge and these capabilities for all activities within their jurisdiction.

Control is a primary management responsibility. All successful managers establish and maintain effective business controls as part of their routine activities.

Today, business controls become more important in automated organizations because lack of control or out-of-control conditions are more subtle, less obvious, and potentially more damaging to the operation. For instance, insufficient controls in a manual accounts payable activity may lead to unauthorized payments; however, an uncontrolled accounts payable program can write thousands of unauthorized checks per hour. (This actually happened at one organization.) Failure to control computer system migration at Oxford Health Plans caused the firm to overestimate revenue by $111 million during one quarter and underestimate medical costs. This resulted in a $78 million loss and the CFO's resignation.[4] Many real-life examples demonstrate the serious consequences of out-of-control computer activities.

Control is especially important to IT managers because many organizations that IT supports rely on computer-generated reports and other automated tools to maintain their controls. For example, an inventory audit at a distribution center quickly leads to computer-produced reports showing transaction activity and inventory status. IT control weaknesses can directly affect inventory control and other controls throughout the firm. Table 17.1 summarizes reasons why business controls are important to the IT department.

TABLE 17.1 Why Business Controls Are Important to IT Managers

1. Control is a primary management responsibility.
2. Uncontrolled events can be subtle and very damaging.
3. The firm relies on IT for many control processes.
4. Law requires control in public corporations.
5. Controls assist organizations in protecting assets.
6. Environmental and executive pressures require controls.
7. Technology introduction requires controlled processes.

The law mandates control requirements and accurate recordkeeping for publicly held corporations. Specifically, organizations must provide proper transaction authorization and perform recordkeeping in conformity with established accounting principles. Public corporations must provide and maintain asset protection and physically verify and reconcile assets with inventory records. Managers must document the extent to which they have followed corporate accounting principles.

Managers must evaluate the sufficiency of controls and appraise actions taken to correct control weaknesses. The firm's officers must certify that they have taken these actions. Management performance and judgment in controlling assets also require assessment. The following executive statement is typical of those found in annual reports of publicly-held companies:

> Management maintains a system of internal control over the preparation of our published financial statements that provides reasonable assurance as to the integrity and reliability of the consolidated financial statements, the protection of assets from unauthorized use or disposition, and the prevention and detection of fraudulent financial reporting. The internal control system provided appropriate division of responsibility, and written policies and procedures are communicated to employees and updated as necessary. Management is responsible for proactively fostering a strong ethical climate so that the company's affairs are conducted according to the highest standards of personal and corporate conduct.[5]

Controls are also important because legislatures and executives are concerned about control activities. With the growing complexity of business today, executives expect both manual and automated control mechanisms to be maintained at peak efficiency. Business executives demand that complex hardware and software systems and applications crossing internal boundaries, interrelating business activities throughout the firm, and sometimes linking it to suppliers and customers be well controlled. Because most organizations critically depend on information systems, they must carefully control them if the organization itself is to be in control.

New technology frequently creates the potential for greatly improving employee productivity and the effectiveness and efficiency of the operation overall. Computerized automation also increases the firm's ability to control errors and omissions and to prevent fraud. New technology, however, also brings new and increased control risks. To mitigate these risks, managers must install controls preceding or at least accompanying technology introduction.

Because information technology and business controls are so entwined, introducing new information technology usually coincides with the need for new control and protection measures. Deploying personal computers, for instance, and using new networking technology greatly increases data security and physical inventory control requirements. Still, many organizations introduce new technologies much more rapidly than they institute control mechanisms. For these firms, weak control or out-of-control situations frequently accompany new technology. Anticipating future conditions, alert and skilled managers plan control systems before introducing new technology.

BUSINESS CONTROL PRINCIPLES

Managers' primary job is to take charge of assets entrusted to them, protect those assets, capitalize on them to advance their part of the business, and grow, develop, or add value to them. Therefore, managers who have custody of tangible or intangible information assets must control and protect them. To do so, managers must know what assets they control and understand their value.

Asset Identification and Classification

Tangible information assets include physical property, like CPUs, servers, routers, cabling systems, and personal workstations. Many information assets are intangible, intellectual assets such as operating systems, application programs, and databases; often they are more valuable than the physical assets. In almost all instances, program and data assets are worth much more than their storage devices. In most firms, the value of information assets generally greatly exceeds the annual IT budget. Sometimes, the enterprise's value is considerably understated on the balance sheet due to the tremendous value of intangible or expensed information assets.[6] Consequently, IT managers have considerable responsibility to protect information assets from loss, damage, or improper use. Information asset protection is an important task for them.

Conducting an asset inventory is the first step in controlling and protecting assets. Managers must develop an organized list of assets for which they are responsible. Confining the list to information technology items, some examples are computer hardware, system software, application programs, databases, documentation, and passwords. The list of possible assets is lengthy because documentation includes many additional items like strategies, plans, designs, and algorithms. An accurate inventory identifies those assets that managers must control and protect.

Secondly, managers must establish each inventoried item's value. This generally reveals three types of assets: assets with intrinsic value such as money, stock certificates, or checks; assets with possible proprietary value such as new product designs; and assets valued because they control other important assets. (The payroll program is an example of a controlling asset.) When assets are organized by value, managers can establish a rational basis for controlling and protecting them. This step is called *asset classification*. After completing this step, managers can develop and implement sound controls.

Most organizations have an asset classification scheme for proprietary information. One familiar classification structure has four information categories: top secret, secret, confidential, and unclassified. Unclassified information is public and available to anyone. The remaining information is available *only* on a need-to-know basis. (Need to know is a critical, often neglected, asset protection principle.) This means that individuals cleared for secret information can access such information only if their job requires it. Having secret clearance does not mean that the individual can access all information classified secret.[7]

In addition to these classification categories, most firms have a fifth category for personal information. This information includes salary, performance, and medical data. Some firms classify this information as personnel, others identify it as personal and confidential. Access to this information is also restricted to those with a need to know.

For protection to be meaningful, the classification category must be indicated in the document itself or contained within the dataset; i.e., the material's security classification must be obvious to anyone viewing it. Labeling material with its security classification is an important, often neglected, information control principle.

Separation of Duties

One of the most effective control measures in business operations is the separation of duties concept. Separation of duties means that several individuals are involved in transaction processing, and that no single individual processes transactions from beginning to end. To understand how this works, consider payroll processing: one person prepares time cards, another validates totals and transmits the information for processing, another controls blank checks and supervises processing, another validates processing through check register data, and yet another distributes the checks. In order for payroll fraud to occur, several people must act together. Separation of duties greatly reduces opportunity for fraudulent acts.

Separation of duties is relatively easy to administer and control. Managers can validate control mechanisms at each interface periodically. Additionally, managers can make control even more effective by periodically changing the individuals responsible for each task.

Validating the output with the input is also important. Payroll, for example, can accomplish this periodically by hand delivering payroll checks. The person delivering the checks must work outside the immediate organization, positively identify the recipient, and verify hours worked with the employee. This theme has many variations for accounts payable, customer shipments, incoming inventory, and other activities.

Efficiency and Effectiveness of Controls

Controls that operate simply and are easily understood are most satisfactory. They are most effective when routinized and operating in a timely manner. To be totally effective, controls must cover all possible exposures. When difficulties are sensed, control mechanisms must respond and generate action in a timely manner. For instance, when input data errors are detected in an application's operation, managers must be notified promptly. They must quickly verify the errors' cause and initiate corrective action. Managers must invoke the problem management system to take final corrective and preventative action.

In all cases, however, managers must relate the cost of control and protection processes to the expected frequency of unfavorable events and to the anticipated loss resulting from these events. Excessive control and excessive expenditures on controls are both possible. Controls must be cost justified. To properly balance these conflicting forces, managers must analyze the application and use good judgment.

CONTROL RESPONSIBILITIES

Business controls operate most effectively when clear responsibilities are assigned to specific individuals. For application systems, several individuals or groups are important in this process:

1. The application program owner (almost always a manager)

2. Application users (some applications have many)

3. The application's programming manager

4. The individual providing the computing environment

5. The IT manager (either with line or staff responsibility)

Each individual has definite responsibilities that must be discharged correctly for application controls to be effective.

Owner and User Responsibilities

The application owner manages the department that uses the application to conduct its business activity. For instance, the perpetual inventory program owner in a manufacturing plant is the materials manager; the accounts payable owner is the organization's accounting manager.

Application owners are responsible for providing business direction for their applications. The owner/manager is responsible for establishing the application program's functional capability and for justifying or analyzing benefits of any expenditures related to the application. The owner authorizes the program's use, classifies the data associated with it, and stipulates proper program and data access controls.

The payroll manager, for example, owns the payroll program and controls access to it and to its data. The owner authorizes payroll program processing. When the program requires modification for any reason, or if the firm decides to obtain payroll processing from a vendor, the payroll manager provides the business case and establishes the application's business direction. The payroll manager is totally responsible for the payroll program. Application owner/managers throughout the organization are responsible for applications needed to operate their parts of the business.

Application users are individuals or groups authorized by owners to use applications and related data according to owners' specifications. They are required to protect the data according to the owner's classification. Users are responsible for advising owners of operational difficulties and functional deficiencies. Payroll department individuals who update payroll records and initiate payroll processes are application users.

Some applications have many users. One example is the personal workstation network comprising the firm's office system. With networked PCs, one administrative manager is named the application's owner and all secretaries and administrative personnel are considered users. The inventory control system for a manufacturing plant is another example. The production control manager owns the inventory system, but employees throughout the plant, including production planners and shipping clerks, are users. In these examples, owners and users all have responsibilities for the applications and their data.

IT Managers' Responsibilities

Responsibility for organizing and managing application development, maintenance, or enhancement resides with programming managers. They are accountable to the owners for meeting programming objectives. These objectives include functional capability, schedule attainment, cost control, and quality performance. Programming managers are also custodians of applications and associated data during development, maintenance, or enhancement. The programming manager who modifies the payroll program as directed by the payroll manager is an example of an individual performing a custodial role.

The supplier of computing services is responsible for providing the computing environment within which the application is processed. The services supplier must negotiate service-level agreements with the owner and achieve the service levels. Service suppliers must maintain secure environments for applications and data. For example, for payroll and other applications processed in the central computer center, the operations manager supplies service and is responsible for the computing environment. When operational department managers own client/server or distributed computing systems, they are responsible for the computing environment.

In their staff role, IT managers are responsible for ensuring that these individuals receive proper guidance regarding these responsibilities. IT managers must establish procedures for developing and using applications. They must routinely evaluate business controls of these applications. In some cases, they may need to conduct audits validating business controls.

These assignments and responsibilities are mandatory for operating application programs under control. Not establishing these positions or operating them ineffectively weakens control and makes failure possible. IT managers must ensure that these positions are established and managed properly.

APPLICATION CONTROLS

Application controls are necessary to ensure that applications routinely function properly. They are most effective when applications themselves generate documentation validating proper operation. Application owners should use this information in conjunction with other control information to certify their processing. Controls themselves should be well documented to ensure prompt, accurate, application audits. Automated and manual control mechanisms should be confidential information and handled accordingly.

Application systems benefit from most business controls principles. Separation of duties, for example, applies to applications as it does to other activities. In addition to general controls, however, some control principles stem from the intangible nature of application programs. Thus, managers can best control application programs using a combination of programmed and manual controls. With application program assets, the assets' characteristics themselves are useful in their control, as discussed later.

Application Processing Controls

Application control and protection consists of two parts: 1) ensuring that application programs perform according to management-established specifications, and 2) maintaining program and data integrity. The first part, dealing directly with program operation, focuses on the application's correct operation and proper handling of input and output data. The second part, dealing with security and protection of program and data assets, focuses on controlling access to programs and data files and on the integrity of information in the files.

Correct functioning of applications and proper handling of input/output data requires that applications have auditability features and control points. These control devices are most effective when designed into the system. (Chapter 9 pointed out that application control and auditability is a design issue.) Operating within the phase review process, managers must establish requirements and specifications for control elements that later become part of the application design. They must audit development processes during phase reviews to ensure that design specifications contain their requirements.

System Control Points

Figure 17.1 illustrates system control points. Controls can be applied when a transaction originates, that is when entering data in the system. Once in the system, control can be exerted when data enters or leaves storage, when it moves through teleprocessing systems, or when the CPU processes it. Lastly, control should be maintained over the results exiting the system.

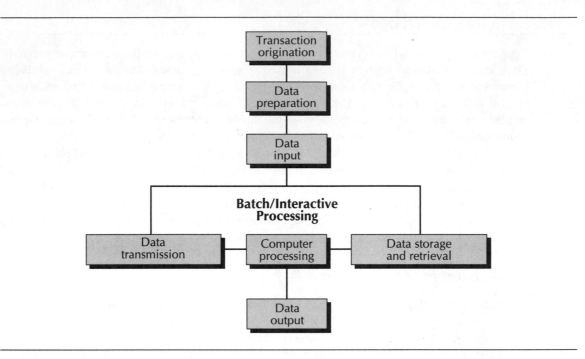

FIGURE 17.1   System Control Points

Figure 17.1 identifies points in the data processing system where control exposures exist and controls must be exercised. Exposures occur when transactions originate and when data is electronically entered. Input data can be stored for later retrieval and processing or it can be sent to the computer system for immediate processing. In either case, the system processes the information and generates output for the application department's use. Computation results may be stored for later retrieval or additional computation.

Table 17.2 displays some important control actions that must be considered when a transaction originates. Transaction origination usually involves one or more input documents that must be controlled and monitored. Specifically designed source documents usually have separate, printed squares for each character or number corresponding to the application's data requirements and are prenumbered for transaction control. Preprinted information and electronic scanning greatly reduces clerical errors.

TABLE 17.2   Considerations at Transaction Origination

Document design

Manual review of source documents

Authorization

Separation of duties

Transaction numbering

User identification

Transmittal logs between organizations

Error detection and correction

Document retention and storage

Only authorized users should handle source documents. Transaction documents contain identification numbers so logs recording document transmittal between units can be maintained. Because user departments develop input documents and transmit them to computer operation, transmission logs validate that the proper documents have been processed.

During input processing, error detection and correction require special attention. Input data resubmission must be recorded and verified. Error handling must be performed according to carefully constructed procedures to prevent additional errors or fraudulent transactions. When input processing is complete, source documents must be stored and retained according to established retention policies.

Batch-data input and interactive-data input controls are very important and must be clearly established and carefully monitored. Table 17.3 indicates some common means for establishing input process controls.

TABLE 17.3   Input Data Controls

Batch Data Input	Interactive Data Input
Input processing schedules	Terminal access security
Source document cancellation	Terminal usage logs
Editing and validation	Data editing and validation
Control totals	Display and prompting formats
Batch control processes	Interactive control totals
Error handling procedures	Error detection and correction

Batch processing includes scheduling batch runs and ensuring that run input matches source-document data. Usually batch applications edit the input data and validate it through control totals. For instance, financial data input contains a count of the number of transactions and totals of important data in an accounts summary. This prevents adding extra input or losing valid input data. After input validation, batch processing begins and source documents are marked as processed.

The production operation department controls batch processing. It is responsible for advising user departments when improper batch execution occurs. Error handling is especially crucial because correcting errors through rerun procedures is itself an error-prone process. Errors must be carefully examined because they sometimes indicate fraudulent processing. Operations departments must maintain error logs, and IT and user managers must review them regularly.

Although more complicated, interactive processing involves many of the same control considerations as batch processing. The complications stem from using remote devices for interactive operations. Proper physical controls over these terminal devices is a mandatory first step. The devices must be inaccessible to all but authorized users. The system identifies authorized users through passwords or personal identification numbers. Specific details of password construction and use are discussed later.

Input data must be edited for reasonableness and tested for validity. Constructing algorithms for editing and testing requires sound application knowledge and considerable ingenuity. Because many complex rules for editing and auditing usually exist, designing interactive displays and prompting formats cleverly is challenging; nevertheless, well-designed displays are very effective control mechanisms. As a simple example, if the field to be entered is numeric and can never exceed four digits, the screen design should allow for no more than four characters, and the editor should test for non-numeric characters. If the dollar figure entered cannot exceed three figures to the left of the decimal point, screen design and prompting must enforce this limit.

For some important applications, expert systems help perform these front-end processes. Expert systems are especially valuable for interactive data entry because individuals can receive expert assistance and make corrections immediately. As a part of system design, these elements must be documented in the user's functional requirements.

Tables 17.4 and 17.5 list the control and audit tasks that must be incorporated in processing, data storage and retrieval, and data output.

Processing, Storage, and Output Controls

Operating systems and applications themselves enhance program processing validity. System software performs specialized functions to help control the use of data stores like tapes and disk files. (See table 17.4.) The system checks tape-label information to ensure that correct tape files are being accessed and also verifies control totals to ensure that the correct amount of information has been processed. For all types of secondary storage, the system validates that the correct version of the data set has been processed. This is especially valuable for applications that create today's results from yesterday's processing and today's transactions, and when previous data sets must be preserved.

In carefully crafted applications, programmed subroutines that validate complete processing and correct results accompany program execution. Processes or algorithms for accomplishing this are highly application dependent and must be specified before the application begins. Their critical nature demands that some algorithms be confidential. Application owners must be intimately involved in developing error detection and correction processes and must review error correction activity carefully for improper or fraudulent conduct.

TABLE 17.4   Computer Processing Controls

Validate the input dataset.

Validate the dataset version.

Verify IT processing correctness.

Verify processing completeness.

Detect and correct errors.

Data storage and retrieval must provide full protection for classified data and secure update procedures for application program source code. Classified data must be protected so that only those with a need to know can access it. Application program source code and object code must always be treated as classified information. Program changes are similar to data changes and are handled in much the same manner. Special procedures for reviewing and altering classified data are discussed later.

Finally, data output and distribution processes require control mechanisms. (See Table 17.5.) To verify that processing occurred as planned, totals should balance, input and output volumes should reconcile, and manual output processes must be under document control. Output documents should contain control totals and transaction counts that can be reconciled with input records. Mechanisms within applications to validate accurate processing should produce reports for the application owners. Different individuals in the organization should receive routine output documents and these special processing reports. Some records must be retained for future audits or other purposes.

Output records must describe how error-handling activity was conducted. All error-recovery actions must be documented and recorded for future review by the problem management team. Controlling data output activity is vital because it offers managers the opportunity to reconcile and validate the entire process.

TABLE 17.5   Data Output Handling

Reconciliation of output to input

Maintenance of transaction records

Balancing of transaction volumes

Control of error handling

Records retention requirements

Output distribution control

Operations personnel must exert control over output distribution to ensure privacy, confidentiality, and security.

Some information is personal and must be divulged only to authorized individuals. For example, only an employee's manager and certain other individuals should see private employee salary or performance records. Information like marketing data, new product designs, and proposed pricing actions is confidential because improper disclosure may harm the firm's competitive position. Other information like checks and stock certificates must be secured because of their intrinsic value. Inventory records should also be protected because they describe the location and value of tangible property.

Application Program Audits

Auditability features in business application programs are essential to the organization's control posture. What makes an application system auditable?

An application system is auditable if the application owner can establish easily and confidently that the system continually performs specified functions. In addition, the owner must be able to verify that it processes exceptional conditions, discrepancy handling, and error conditions according to prescribed specifications and procedures. Auditability, therefore, includes manual and automated procedures in processing and user organizations.

Auditable systems contain functions and features that let owners easily determine whether they are processing data correctly. Features providing input editing capabilities and journaling or logging functions throughout the processing stages are important ingredients of auditable systems. Auditable systems must verify that processing occurred according to specifications and that results can be compared to known or expected operational standards. For example, payroll programs produce a check register detailing each check produced by identification number. Payroll program owners, who maintain physical possession of blank checks, can account for each check processed, including any used to align the printer. Additionally, control totals ensure owners that the program calculated the payroll as expected.

Application owners must have reliable mechanisms to determine that authorized users accessed the system for legitimate business purposes. They must have assurance that their data, used in conjunction with application systems, is protected. Routinely generated documentation or reports must make these mechanisms, assurances, features, and functions available to application owners. Systems including these features are auditable and considered under control.

Auditability begins with sound system development techniques and practices. Well-documented programs written in high-level languages are easier to audit than others. Sound application development techniques, such as structured design, modular programming, and complete documentation, are particularly important when maintaining or enhancing programs. Program modification, part of nearly every program's life, complicates the task of maintaining system integrity.

However, programming disciplines and standards ease the task of maintaining program integrity. Program testing processes are vital; preserving test data and results makes subsequent maintenance and enhancement easier, too. Ideally, inserting auditing features into applications or adding reports to improve audit trails occurs during program development. Maintenance and enhancement activities must preserve and, if possible, improve original auditing features.

Controls in Production Operations

Application production processes must also operate in controlled environments. Internal application controls are only effective for disciplined and controlled production processes. Controlled and highly disciplined production environments must complement application controls and audits.

Disciplined operations during application processing greatly enhance application integrity. Well-disciplined production operations display sound control over performance objectives, maintain sufficient system capacity for application operations, and let batch and online systems process as anticipated. Detailed scheduling and rigorous online management provide controlled environments for application processing. The disciplines form the basis for complete confidence regarding application control and auditability. Required for other reasons too, they are vital ingredients of business controls.

Controls in Client/Server Operations

Some firms' distributed processing is partially motivated by operating departments' desires to escape central IT bureaucracy, including control and security procedures. Thus, some distributed processing environments that lack satisfactory processing controls or audit trails are vulnerable to unauthorized penetration. Many small operations do not implement separation of duties well. Nevertheless, departmental CPUs, client/server systems, or other distributed processing implementations require the same controls as centralized operations, for all the same reasons.

Because client/server systems include more points of vulnerability, control and asset protection is necessarily more difficult. But because well-designed client/server systems reduce manual operations, paper documents, data tapes, and removable storage media, audits and controls can be implemented in programs. For example, workstation number, operator code, and automatic date and time stamp identify transactions originating at client workstations. Recordkeeping at the server can be fully automated as well. Like centralized systems, client/server applications can be well-controlled and inform management of successful operation, too.

Organizations that move applications from secured centralized systems to distributed systems must be aware of new, different exposures and vulnerabilities. Client/server operations need not be less secure than centralized operations. They must be secured and protected according to their asset's value and worth to the organization. When installing distributed operations, the chief risk is granting increased access before implementing controls and security. Business controls are a basic management responsibility—they must not be an afterthought.

NETWORK CONTROLS AND SECURITY

Network managers face many business control and security challenges. Their network operations grow rapidly through departmental LANs and interdepartmental WANs, intranet technology adoption, connections to common carriers and value-added networks, and long-haul national and international network use. These interconnections provide faster, more direct coupling between organizations for transporting valuable information assets within the firm and between it and others. This phenomenon so pervades the financial industry, for example, that daily electronic funds transfer exceeds the dollar reserves of central banks.[8] Because these rapidly developing networks are so critical to private and public institutions, network control and security in today's environment gain importance rapidly.[9]

Networks face passive and active threats. Passive threats are attempts to monitor network data transmission to read messages or obtain information. The intruder hopes to profit from information or to identify information sources. Active threats are attempts to alter, destroy, or divert message data, or to pose as network nodes. As we learned in the Business Vignette, Kevin Mitnick posed as a network node and actively diverted valuable information. In this extreme case, Mitnick became an active network participant, exchanging information with legitimate sites and obtaining free network services.

To minimize these threats, network managers must control system and data access and protect data in transit. The first step in controlling system access is securing the physical system. Facilities housing user devices like workstations, facsimile machines, and phones must be secured, admitting only authorized individuals. Rooms containing controllers, routers, bridges, or servers must also be tightly secured. Cables connecting user devices to network systems should pass through restricted access passageways. Transmission media between bridges, routers, servers, network gateways, and external communication links like the phone system must also be protected.

The second step in securing system access is establishing user identification and verification processes. For most systems, this means that users sign on to the system with their name followed by a password. Sometimes maligned, properly implemented and used user identification-password procedures can be quite effective. For example, sound procedures require users to change passwords at least monthly and establish six-character passwords that include two numbers separated with alphabetic characters. Unfortunately, even with these restrictions, people tend to choose passwords that are easy to remember and therefore easy to duplicate. In well-protected operations, users receive system-generated, frequently changed passwords from their managers. Password systems alone, however, are insufficient protection.

Network systems should also have automatic disconnect capability. When password validation fails after five attempts, the system disconnects and logs the event. This restricts hackers from using automatic password-creation programs. Many other security measures can accurately identify legitimate system users. System managers should review all recorded attempts to gain unauthorized system access or use the system in an unauthorized manner.

System users must receive training in security procedures and understand the need for system protection and policies on unauthorized use; in many firms this is not the norm. In a recent survey, 78 percent responded that employees have little or no knowledge of computer misuse and 61 percent said written policies are loosely enforced.[10] Network and user managers should not hesitate to withdraw access from employees who abuse system controls or violate policies on unauthorized use. In some firms, personal or unauthorized use of information systems is cause for dismissal.

The third step is data security. Properly secured data means that each system dataset has an owner and the owner's identification and dataset classification are part of the data. Owners must specify who can access datasets and what kind of access is permitted. The types of dataset access are read, write, alter, delete, and execute. When data security works properly, users access datasets with owner authorization and the security system identifies their actions, i.e., when a user tries to open a dataset, the security system validates the user and permits only authorized access levels. Managers must investigate recorded unauthorized attempts to access data. Because application programs are also datasets, execute authorization permits users to operate them.

Data Encryption

As the Business Vignette demonstrated, for several reasons, access security can never be perfect. Because common carrier links can be tapped and microwave or satellite transmissions can be intercepted, it is necessary to protect critical data in transit. Message encryption is the most satisfactory means for protecting data and maintaining network security. Before transmission, encryption programs use an algorithm and a key to change message characters into a different character stream. When received, the algorithm and key decode or decipher the message. The encryption process may operate at the character level, but it usually operates on the message bit stream.

Properly used, data encryption can make telecommunication very secure. Methods exist to authenticate transmissions and validate signatures. A third party can validate especially sensitive traffic and guard against lost or stolen keys. In addition, current research attempts to improve encryption algorithms by increasing their operational speed and security. Network managers must use encryption and authentication techniques to secure critical network traffic against passive and active threats. This is especially important for organizations using insecure facilities such as the Internet.

Data encryption is the ultimate protection for stored information too, not only because it protects against sophisticated hackers like Kevin Mitnick. It also thwarts unethical employees bent on theft or forms of sabotage. According to J. Russell Gates, the partner in charge of Arthur Andersen's Computer Risk Management group, computer crime and other security issues are four times more likely to be attributed to insiders.[11] For this reason and others, many private corporations and public agencies encrypt critical, locally stored information.

Controlling business operations and securing firms' intangible assets depends heavily on control and security of telecommunication systems. Today, intranets link the firm's employees to important internal information sources, and client/server systems link employees to business applications. EDI and extranets integrate physically dispersed operations and link applications to customers and suppliers. Networks are the pipelines for business information. They must operate under control, and they must be secure.

Firewalls

Networks have become pervasive within firms, and between the firm's employees and the outside world. As a result, network intensive organizations have adopted special precautions, in addition to those noted above. The most common of these is called the *firewall*.

A firewall is a computer system inserted between internal and external networks through which all incoming and outgoing traffic must pass. Because all traffic in either direction must pass through the firewall computer, it is uniquely positioned to validate and authorize information entering or leaving according to the organization's rules. A firewall can divert traffic not satisfying the ground rules for further analysis.

Firewalls perform other functions, too. They log messages and analyze traffic flow patterns so engineers can manage internal and external networks more effectively. Message log analysis helps detect changing workload patterns and identify system usage trends. Firewall computers can also encrypt and decrypt sensitive material if not done elsewhere. Firewalls can handle other forms of security surveillance and management, too. Still, firewalls only protect against external threats—much more probable internal threats require other measures.[12]

Because networks proliferate so rapidly and are so important today, firms increasingly use firewalls to manage the electronic interface between organizations. They are an important adjunct to many other valuable security techniques. Table 17.6 lists the network security considerations we have discussed.

TABLE 17.6 Network Security Considerations

Physically secured workstation devices

Physically secured network components

User identification and verification

Processes to deal with unauthorized use

Dataset protection mechanisms

Data encryption and authentication processes

Firewalls

For many reasons, protecting information assets tied to networks, either public or private, requires special attention. Although the Business Vignette describes an extreme threat to networked systems, this incident teaches several important lessons.

First, a small number of people prefer to intrude on your network and invade your systems and applications for pleasure or for profit. They are intelligent and persistent; you must protect your network, systems, and applications from them. Second, networks and systems connected to them can be secured from most intrusions; the degree of protection depends on the resources expended and the type of protection obtained. Third, in spite of obvious threats, many organizations still fail to take simple precautions.[13] Managers are totally responsible for the organization's assets. They must take whatever action is needed to protect the assets entrusted to them.

Network control and security directly relate to controls and audits in applications because networks are sources and sinks of most application data. Network control and security directly relate to other network management disciplines such as problem, change, and recovery management because effective network control reduces network problems and network security reduces damage to information assets. Network management disciplines are an integral part of IT management systems.

ADDITIONAL CONTROL AND PROTECTION CONSIDERATIONS

In addition to those discussed above, other IT manager security and control responsibilities concerning applications and networks must be addressed. In particular, physically protecting major processors, controlling and protecting critical applications, and securing unusually important datasets are issues that IT managers must handle.

Data centers containing central CPUs, server clusters, information libraries, network gateways and bridges, communication controllers, and telephone equipment require extraordinary physical protection. Some practices help data center managers secure their operations:

1. Only people who work in the data center are allowed routine access to the facility.

2. Data center workers must wear special, visible badges that identify them.

---

3. The identity and authorization of all visitors to the center must be validated; they must sign in and out.

4. Duties within the center should be separated so that operators who initiate or control programs cannot access data stores.

Under some circumstances, securing the data center may demand additional actions. For example, in some critical centers, visitors are prohibited entirely and specially constructed floors, walls, and ceilings protect the center. In all cases, however, protection must be consistent with the value of assets housed in the data center.

Downsizing mainframe applications requires IT managers to apply these considerations to distributed operations. Users who operate applications on their own processors must take many of the security and control precautions required of mainframe operators. Thus, managers of distributed operations necessarily assume considerable control and security responsibilities.

Systems programs such as operating systems, file handling utilities, password generation programs, and data management systems must also be specially protected. Managers must take careful precautions with system programmers and network specialists who can access restricted utility programs and control elements like the files of authorized users and their passwords. System support personnel can access utilities to copy or rename data sets or alter executable modules or dataset labels. Most systems also have a superuser capability—one or more privileged passwords that system programmers use to access any system dataset. System programmers need these capabilities to do their jobs.

IT managers must develop individualized security measures for the few data center individuals who have these special privileges. To meet this need, managers can take several routine precautions. Individuals in these responsible positions should rotate or change duties frequently. Their actions should be routinely recorded, and managers should review the records frequently. When system programmers use privileged passwords or access restricted utility programs, they should obtain advance clearance from the center manager. To maintain security and reduce errors to critical information, the manager should validate their actions upon completion.

Privileged network technicians have access to codes, keys, utilities, and passwords for many remote operations. Their work must be handled in the same manner as system programmers' activity.

Managing Sensitive Programs

In addition to the controls and procedures required for operating routine applications, some programs require special handling. Applications that permit or authorize the transfer of cash or valuable inventory items must have specialized controls. Payroll, accounts payable, and inventory control programs are examples of applications needing special protection. Managers must identify and maintain an inventory of them. Their owners must prescribe protection and security conditions covering storage, operation, and maintenance, and must ensure the satisfactory implementation of these special considerations.

First, program source code, load modules, and test data must be classified as sensitive information and protected accordingly. In most cases, the protection should be the highest available on the system. For instance, the owner of the accounts payable program may classify the source code as confidential and restrict access to a single maintenance programmer. Load modules used for program execution may be restricted to another individual. This ensures that maintenance programmers cannot operate the program with live data. The test data should be entrusted to yet another individual who operates the modified program against it and delivers results to the owner. When program testing is complete, the executable load modules are updated and protected. Change control for these special applications must be carefully managed.

Second, these critical programs usually operate differently from routine applications. For example, control over input and output documents is tighter for accounts payable than for most other applications. Checks are hand carried to the computer center after the sequence numbers have been recorded and verified. When processing is complete, the output is returned to the accounting manager who verifies the check count, returns unused checks to the safe, takes the checks and stubs to the distribution center, and gives the check register and other control information to the appropriate accounting manager. The accounting department verifies these operations and records them for later reference.

Some datasets for these applications are also highly sensitive. The vendor name and address file for accounts payable is an example of one such dataset. Anyone scheming to create a fictitious vendor to whom payments will be sent fraudulently must access it. With appropriate controls, firms can prevent this fraudulent act. Accounting managers must validate all changes to this file to prevent possible security breaches.

Application owners must be especially vigilant during maintenance or enhancement. The owner ensures that only authorized changes are made and then reviews all modifications. Owners must control maintenance through close supervision and through documentation and testing procedures. It is important that routine audit and control features in these programs function flawlessly without modification. Additions and changes to these applications must incorporate additional audit and control features. These sensitive applications must be guarded more carefully than the assets they control.

When firms decide to relocate sensitive applications to distributed departmental systems, they must first ensure that all precautions noted above are in place. Facing the daunting task of securing distributed systems, some firms decided to keep critical systems and sensitive programs centralized. In one instance, an organization that owned and operated a midrange computer for financial applications requested to move it to the centralized secure facility rather than tackle the formidable task of securing it.

KEYS TO EFFECTIVE CONTROL

To operate under control, managers must completely understand their control responsibilities. They must know which assets are theirs and the assets' value, and must classify and protect them accordingly. Managers must be actively involved in control processes. Their involvement must be timely and responsive to changing conditions; they must follow up to ensure the effectiveness of their actions.

Operating information assets like systems and applications must routinely produce measurements and reports that reveal the state of control. Managers must review these reports frequently. They must be able to determine that specific operating procedures have been followed and that the operation complies with defined control practices.

Managers should separate duties to disperse information access and handling, and frequently rotate employees in critical positions to different jobs. IT managers must audit their operations periodically, and they must use their findings to improve their business controls. Information assets are usually very valuable. Their owners must protect them carefully according to their relative worth.

Business controls in application systems are of interest to user managers, IT managers, and to the firm's senior executives, too. The capstone discipline—management reporting—must function well. IT managers, their peers, and their superiors periodically require knowledge of correct application performance. Astute IT managers take extra steps to inform all interested individuals of the sound controls applied to application assets. For example, IT managers may provide summaries of problem management actions for senior executives or trend information from internal audits and reviews of their operations.

These periodic reports should highlight the major routine actions that ensure correct and valid performance of application programs. The report must summarize operational deficiencies, if any, and explain the manager's corrective actions. The report should include routine tests and audits of control mechanisms performed since the last report. Additional steps taken to augment existing controls or to make the development or operations areas more secure is important to these executives. Comprehensive reporting is an offensive tactic because, without it, senior executives are likely to seek information through outside audits, independent reviews, or other less welcome means.

SUMMARY

Computer crime has reached epidemic proportions, costing business and industry between $500 million and $5 billion per year, according to *Forbes* magazine.[14] These figures may be low because most computer crime goes unreported and some goes undetected.[15] Numerous incidents prove that almost no organization is safe from computer crime, yet some fail to take even rudimentary precautions. Computer hackers have bulletin boards and several magazines, and meet regularly to exchange new information. These activities should make corporate executives shudder. If they aren't enough, laws in Eastern Europe and Russia legitimize the export of software virus programs. Virus factories are as common as software publishers in the U.S.[16]

Keeping these astonishing facts in mind, remember that business controls and asset protection are fundamental to business operations. They are part of every manager's primary responsibilities. Changing business conditions, new technology, and the increased attention demanded by law now make controls more important. IT controls are especially important because firms depend on computerized systems for control throughout their operation.

Managers must know their assets and their estimated value. Assets must be classified and protected in keeping with their relative worth. For application program assets, individual responsibilities are assigned to owners, users, programming managers, and providers of computing environments. IT managers must ensure that these individuals effectively discharge their responsibilities.

IT managers must exert control over data handling and computer processing; reports must provide evidence that programs operate as specified. They must control the production environment, whether centralized or distributed, through operational disciplines and by separating duties within the center. Physical network elements and data in storage and in transit must be protected. User identification and passwords, dataset classification and access protection, and data encryption all help secure the network from intrusion and protect data from unauthorized viewing or use.

System control programs, utility and data management programs, and asset dispensing programs must be tightly controlled during storage, operation, and maintenance. The manual operations surrounding these applications are also critical and must be carefully controlled. Corporate information of all kinds is valuable: IT managers must protect it against loss and damage.

Review Questions

1. What lessons did you learn from the Business Vignette?

2. What is the first thing that managers must know to establish effective controls? What other things must they know?

3. Why are business controls important to IT and user managers?

4. Why does new technology usually require new and different business controls? What new controls might a firm introducing office automation require?

5. What staff responsibilities do IT managers have regarding business controls?

6. What are some physical information assets that must be controlled? What are the most important intangible assets that must be controlled and protected?

7. Intangible assets usually are very valuable to the firm. Can you give examples of firms whose intangible assets are more valuable than all other assets?

8. Define the concept of separation of duties. How might this concept work in managing physical inventory?

9. What are the characteristics of efficient and effective controls?

10. Identify participants needed for complete control over the development and use of application programs.

11. What are application owners' security and control responsibilities? How are these related to user responsibilities?

12. How do IT manager's duties relate to the duties of other participants in the business control process?

13. Identify the control points in computerized data processing.

14. What are the control features of transaction origination?

15. What are the similarities and differences in controls between batch and interactive data input?

16. Who specifies control features in applications? What management processes ensure correct and complete implementation of these features?

17. Identify the main elements in computer processing designed to provide protection and control.

18. The final data processing control point is data output. What control activities take place there?

19. Describe system auditability. Why do application owners require auditability features as part of system controls?

20. Why are disciplined production processes essential to well-controlled application portfolios? What constitutes a disciplined production process in a client/server operation?

21. To what kind of threats are networks exposed? How can managers act to minimize these threats?

22. What is a firewall? What is its purpose? Why is it needed?

23. System programmers must have almost unlimited access to information system assets. How do managers control this situation to protect system assets?

24. What special precautions must be taken for critical programs?

25. What are the keys to effective system controls and asset protection?

Discussion Questions

1. In what instances do you think application business controls should be listed as a critical success factor in Chapter 1?

2. Why are control issues in applications more important now than they were 10 years ago?

3. The Business Vignette illustrated some of the many things that can go wrong in networked systems. Considering all that you learned in this chapter, discuss actions you would take to protect your networked system.

4. Discuss control issues that accompany the introduction of employee-operated facsimile machines. Who is responsible for resolving these issues?

5. Suppose you were the Merrill Lynch manager controlling the name and address file for the Cash Management Account program. Discuss how you might protect and control this asset from loss and damage.

6. Along with separation of duties, the text discussed the need to rotate employees regularly. Discuss how these two actions work together to improve security and control.

7. In some firms, IT managers own and control application programs and databases. Discuss advantages and disadvantages of this arrangement.

8. Discuss the changes in control responsibilities accompanying end-user and client/server computing. What changes occur in the provider-of-services role? With end-user computing, the owner, users, development manager, and provider of services are all in one department. What are the implications of this staffing arrangement?

9. Describe the IT manager's responsibilities to departments that have adopted end-user computing.

10. Discuss the special control precautions needed when applications are enhanced.

11. If you were the payroll manager, what control actions might you take when the payroll program is being altered to handle tax law changes?

12. System control and auditability must exist across the boundary dividing manual and automated processes. Using the accounts payable program as an example, discuss the shifts in responsibility at this interface.

13. Discuss the reasons why the disciplines of problem, change, and recovery management are business controls issues?

14. Discuss ethical issues that you think might arise in testing business controls in applications.

15. Discuss conflicting issues surrounding ease of use and business controls. How can risk analysis help resolve these issues?

16. Develop a list of critical applications you believe exist in many firms. Why are these applications critical? How should they be protected?

17. What special control considerations are involved when considering purchased applications? How do you think purchased applications should be tested?

18. Discuss important business controls issues arising when a service bureau is employed to process payroll.

19. Describe the role the firm's controller plays in controlling and auditing application systems. If you were the controller in a firm considering moving mainframe applications to departmental systems, what would you want to know before you approve the plan?

Assignments

1. Assume that you are manager of the workstation store in a firm starting to implement end-user computing. Develop the list of business control actions for the store's operation. As store manager, what reports and audit trails do you think are necessary, when should they be produced, and how would you handle separation of duties among your three employees?

2. Suppose a new program developed using the prototyping methodology is replacing one of the applications serving your department. Devise a process by which you, the manager, can feel confident that business controls issues receive sufficient attention during development.

ENDNOTES

[1] Mitch Betts and Gary H. Anthes, "FBI Nabs Notorious Hacker," *Computerworld*, February 20, 1995, 4; John Markoff,"To Catch A Cyberthief," New York Times News Service to the *Gazette Telegraph*, February 18, 1995, A1.

[2] *The Wall Street Journal*, November 7, 1988, 1.

[3] The book, *Cyberpunk: Outlaws and Hackers on the Computer Frontier*, by Katie Hafner and John Markodd, Simon and Schuster, 1991, describes the activities of Mitnick and his accomplice against DEC, a West Berlin cracker called Pengo who sold loot to the Soviets, and a lot more about Robert T. Morris.

[4] Thomas Hoffman, "Oxford Health Plans CFO Resigns," *Computerworld*, November 10, 1997, 110.

[5] Ameritech 1997 *Annual Report*, 34.

[6] Greene, Richard, "Inequitable Equity," *Forbes*, July 11, 1988, 83.

[7] Violation of this principle at the CIA was at the heart of the Aldrich Ames case.

[8] Walter B. Wriston,. *The Twilight of Sovereignty*. New York: Charles Scribner's Sons, 1991, 59.

[9] According to federal investigators, U.S. Department of Defense computers were attacked some 250,000 times in 1995; two-thirds of the attacks were successful—meaning the attacker penetrated the system—but less than one percent were detected and reported. Gary Anthes, "U.S. Easy Target for Cyberattacks," *Computerworld*, May 27, 1996, 7.

[10] Gary H. Anthes, "Firms Seek Legal Weapons Against Info Thieves," *Computerworld*, May 27, 1996, 72. Survey based on 428 respondents at U.S. organizations.

[11] Nikhil Hutheesing and Philip E. Ross, "Hackerphobia," *Forbes*, March 23, 1998, 154. See also "PC Manager Charged in $2M Scam," *Computerworld*, March 30, 1998, 1.

[12] See note 11 for some internal security experiences and perspectives on protection actions.

[13] Gary H. Anthes, "Is Your Data Secure?" *Computerworld*, November 28, 1994, 65. In a survey of 1271 corporate officials—61 percent of whom were IS managers—42 percent said senior management in their organizations believes that information security is "somewhat important" or "not important."

[14] William G. Flanagan and Bregid McMenamin, "The Playground Bullies Are Learning How to Type," *Forbes*, December 21, 1992, 184.

[15] According to Gary Anthes. Eighty-three percent of those who experienced network intrusions didn't report them to law enforcement. (See note 9.)

[16] Peter Stephenson, "Beefing Up your Anti-Virus Strategy," *Lan Times*, June 28, 1993, 62. According to Stephenson, about six new viruses are created and released every day and, for amateurs, do-it-yourself virus-creation tools are available.

Part
Six

Preparing for IT Advances

Information technology is changing people's work and their work environment. Astute managers deal skillfully with people issues as transitions occur for individuals, departments, and organizations. High-performance organizations and effective leaders facilitate people working together to achieve new technology's many benefits. Topics covered in this part include People, Organizations, and Management Systems, and The Chief Information Officer's Role.

Directing human resources skillfully is the most important success factor for managers. Effective information technology managers exhibit extensive people-management skills.

18 *People, Organizations, and Management Systems*

In a deal intended to help both companies move into the era of deregulated utilities, IBM's subsidiary, Integrated Systems Solutions Corporation (ISSC), entered a 10-year outsourcing arrangement with Denver-based Public Service Company of Colorado.

The deal's core is a classic information technology operations contract for ISSC to operate Public Service's Denver data center, its workstations, help desk operation, and intrastate network. ISSC will also acquire the company's application development. The $500 million technology project saves Public Service of Colorado $190 million over 10 years, according to Del Hock, the utility's chairman.

In a separate but related contract, Public Service will perform energy consulting, helping IBM trim its gas and electric costs at 16 U.S. plants. The utility provides these services to IBM and markets them to other companies through a new subsidiary called E Prime. Completing the partnership circle, E Prime and IBM formed an alliance for IBM to develop applications that help E Prime customers manage energy procurement and consumption.

Analysts note that the exchange of services between IBM and E Prime illustrates an emerging trend in which outsourcers act as both vendors and customers with their clients. In another recent example, Electronic Data Systems (EDS) Corporation signed a 10-year outsourcing deal with this Lake Forest, Illinois-based Moore Corporation Ltd., in which Moore supplies EDS with business forms and commercial printing.

And like IBM's alliance with E Prime, AT&T Global Information Solutions and Delta Air Lines are jointly marketing services to the airline industry as part of AT&T GIS's 10-year, $2.8 billion outsourcing deal with the airline. In this case the companies established a 50/50 jointly owned venture.

"These relationships are extending themselves with real interesting nuances," said Allie Young, an analyst at Dataquest, Inc., based in San Jose, California. "These types of contracts show a clear direction for the future of outsourcing. The key is the strategic involvement of the two companies."

The Public Service contract is one of many outsourcing contracts expected as utilities try to blast out of their complacent environments into a competitive market. Recent energy industry outsourcing deals include the following:

Energy Company	Outsourcer	Length of Contract/Amount
Philadelphia Electric	ISSC	10 years/ $450 million
San Diego Gas & Electric	Computer Sciences Corp.	5 years/ $60 million
Halliburton Energy Services	Andersen Consulting, Power Computing, I-Net	10 years/ $500 million
Public Service of Colorado	ISSC	10 years/ $500 million

For Public Service, the $190 million cost savings will be vital in the new competitive environment, where utilities vie to offer services at the lowest price.

The Federal Energy Policy Act of 1992 requires utilities to allow other power companies into their power distribution grid so that wholesale customers can shop for the lowest rates. This policy encourages market competition, so utilities are striving to lower costs to retain their large customer base. Outsourcing is expected to be very popular with utilities.

Philadelphia Electric outsourced its routine mainframe operations, help desk, and LAN and desktop administration to ISSC so it could concentrate on the future. PECO plans to connect its meters with wireless and fiber-optic networks, replacing manual meter reading, and permitting it to record electricity consumption from its offices. The utility believes that using these connections may permit PECO to cycle home appliances on and off, thus achieving cost savings from low-rate, off-peak electricity.

Further strengthening its position, Public Service of Colorado merged with Southwestern Public Service, one of the regions lowest-cost producers. New Century Energies, the company created by the 1997 merger, is expected to generate savings of at least $770 million over the next 10 years. As competition in the electric utility industry increases, mergers, acquisitions, joint ventures, and outsourcing activities will rise as firms strive to lower their costs and their customers'.

However, outsourcing is not confined to the utility industry or the U.S. In 1994, billion-dollar outsourcing deals included Xerox and British aerospace; in 1995, Dutch National Railroad, Hughes Aircraft, and UK Social Security Agency; and in 1996, Lucent Technologies, DuPont, Ameritech, CNA, J.P. Morgan, and GM Plant Floor, among many others. Personnel from these prominent companies also move to outsourcers like EDS, CSC, and IBM when work is transferred. Thus, outsourcing trends are especially relevant to IT employees and their organizations.

INTRODUCTION

This text emphasizes managing information technology assets, focusing on tools, techniques, processes, and procedures needed to control and use these assets most effectively. Strategizing and planning processes set the stage for understanding and embracing software and hardware technology trends. Managing and controlling application assets, and the production facility in which applications are processed, follow technology trends. This book turns now to the most important IT assets, IT people and their organizations and structures.

People and their organizations are critical to the successful functioning of the modern business firm. But employing people effectively requires an environment or corporate culture in which they can thrive and be highly productive for themselves and their organizations. Employees and managers need to know "how we do things around here." To a large extent, how we do things around here is a function of the firm's management system. The management system provides an intellectual framework, not only for management actions, but for employee actions, too.

Peter Drucker's quote in Chapter 1 and the Business Vignette beginning this chapter indicate that dramatic changes are occurring in conventional business organizations. These changes are not confined to the firm, but involve its customers and its suppliers: many entire industries are undergoing rapid and dramatic change. As the Business Vignette noted, deregulation in the utility industry, and, often changes within the firm itself result from collaboration with others. In all cases, information technology is an important change agent supporting and fostering new business activity and structures.

TECHNOLOGY SHAPES ORGANIZATIONS

Today we are witnessing fundamental, pioneering changes in the business environment. The environment is rapidly becoming increasingly complex, driven by tremendous advances in communication and information processing capability. These driving forces significantly affect our business structures. "There is no question that the new information technologies (IT) are having a major impact on the range of strategic options open to an organization. IT is not only creating an environment that is making it imperative for organizations to evolve, it is also helping them adjust to the enormous changes outside their own boundaries," states Michael S. Scott-Morton.[2]

Organizational Transitions

During the 1980s in the U.S., mergers, acquisitions, and leveraged buyouts involved shuffling assets worth $1.3 trillion. The trend accelerated during the 1990s. In 1997, mergers and acquisitions by U.S. firms approached $1 trillion, passing the 1996 record of $650 billion. Activity in the telecommunications sector is especially interesting with SBC Communication's $16.7 billion takeover of Pacific Telesis, Bell Atlantic's $22 billion merger with NYNEX, and US West's purchase of Continental Cablevision for $10.8 billion.

But the grandaddy of all telecomm mergers came when WorldCom, the fourth largest U.S. long-distance company, announced its intention to buy MCI for $37 billion in cash and stock, topping GTE's offer of $28 billion in cash.[3] This announcement came shortly after WorldCom reached an agreement to buy MFS Communications in a $14.4 billion stock deal. With telecommunications deregulation and privatization on the upswing worldwide, and with global communication systems so important to worldwide commerce, major firms are positioning themselves to participate in this exploding market.

Hoping to capitalize on new opportunities through mergers and acquisitions, industry executives are causing turbulence for themselves and their employees. In addition, firms downsize, outsource, and reengineer business processes as they digest consolidations, business partnerships, or joint ventures. Sometimes, corporate reorganizations generate unemployment—in many cases workforce reductions motivate restructuring. Information systems professionals are not immune to this threat.

Through purchases, joint ventures, and planned internal expansion, many firms attain significant international presence. Some of the world's largest, most successful corporations lose much of their national identity when they become truly international businesses. Some examples illustrate the point:

- Honda manufactures in several countries and sells in many.

- IBM manufactures in many countries and sells globally.

- Phillips, the huge Dutch electronics firm, claims 10 percent of the world's color TV market and is the world's largest manufacturer of light bulbs.

- Royal Dutch-Shell, among the world's largest industrial corporations, has Dutch and British origins, but is very cognizant of many nations' interests in its worldwide operations.

- The trend toward dispersed national, transnational, and international business enterprises is certain to accelerate, aided by communication technology and geopolitical considerations.

Centralized Control—Decentralized Management

As firms grow through mergers and acquisitions, alliances between firms become increasingly common. Many examples of alliances exist in business today, especially within semiconductor, computer, and telecommunications industries. Once considered a sign of corporate weakness, alliances and partnerships are now needed to speed new products to market, improve operational efficiency, and obtain skills needed to meet business goals.

Adopting and using alliances and joint ventures to capture the advantage of time, employ critical skills, and access distribution channels or new markets adds complexity to the firm's value chain. Inbound logistics (the firm imports raw materials or subassemblies from suppliers), internal operations (assembly, manufacturing, or process activities), outbound logistics (distribution activities), sales, marketing, and service add value to the firm's product. Activities such as procurement and technology development support these processes, which depend on human resources and the firm's infrastructure.[4] For international firms or those with alliances, one or more parts of the product value chain may lie outside the firm itself.

For example, consider the five firms in five countries—mentioned in Chapter 15—that manufacture jet engines. The five manufacturers produce engine subassemblies, develop engineering and technical documentation, and produce maintenance spare parts for delivery to final-assembly locations, adding value to the final product. Many other examples exist in diverse industries: the auto and computer industries are well-known examples. In all these industries, managers rely upon the firm's infrastructure and information technology architecture to add value to their final product.

Information technology is essential to the success of firms engaged in partnerships and alliances or in developing their businesses internationally. The Business Vignette noted that Public Service of Colorado and IBM shared skills, technology, and resources to help each other solve problems. Their partnership enabled them to share resources used in helping others. Information and guidance from the corporation's hub are intellectual resources that enable firms in five countries to develop and build jet engines. Similarly, these assets enable others to harvest oil on several continents and refine, market, and distribute it in many countries. Executives rely extensively on information technology to operate their networked businesses cohesively.

Centralization *vs.* decentralization is a longstanding corporate strategic issue with firms taking sides. Hewlett-Packard prefers numerous, widely scattered facilities and decentralizes IT. On the other hand, large, highly IT-intensive banks usually prefer centralized IT operations. This issue looms larger now as recent trends toward corporate restructuring and global strategies grow. As firms grow larger, they tend to decentralize operational control, placing their decision-making capabilities closer to operational centers. For many firms, particularly conglomerates, growth via acquisition accompanied by decentralized profit centers is normal. Decentralized operations with limited centralized control are popular.

However, today's executives can capture the best of both worlds by using information technology and telecommunication systems. They can delegate decision making and operational control to operational units while obtaining real-time control information at headquarters.

Telecommunication's Impact

The paradigm of today's information technology is the union of information systems with rapidly advancing telecommunication technology. This gives business executives enormous capability to expand their operations, form alliances and joint ventures, and restructure their assets for greater effectiveness. Using telecommunication systems to shrink time and distance and help establish new relationships, organizations improve efficiency and effectiveness, and promote innovation.[5] Several examples illustrate how companies capture these advantages.

Volvo uses an international computer-aided-design network to design and manufacture heavy trucks. IBM uses its engineering design system and programming support systems through national and international networks to design and manufacture computer systems for a global market. Network systems allow GM to design cars in Germany, build them in Korea, and sell them in the U.S. Thousands of companies in North America now use Electronic Data Interchange (EDI) to transmit information from one computer system to another. With e-mail or Internet technology, electronic communication takes place between the decentralized operations of the firm or from the firm to supplier or customer offices. Just-in-time manufacturing and paperless factory concepts critically depend on telecommunication systems.[6] For most firms today, using sophisticated telecommunication technology to improve business results is a critical success factor.

Span of Communication

Telecommunication systems greatly increase the span of communication and make filtering layers of management unnecessary. Span of communication facilitated by information systems such as intranets makes the traditional span of control irrelevant. By communicating throughout the organization electronically, executives can comprehend operational details needed to maintain control. In today's restructured organizations, middle managers must also expand their spheres of influence to attain corporate goals and objectives. They, too, must use advanced communication techniques to increase their effectiveness over a broader range of organizational activity. Their effectiveness as well as their survival is at stake.

Span of communication can be very large in modern industrial firms. For example, technicians in California work with technologists in New York to help manufacturing engineers in Germany solve difficult production problems on advanced disk drives for European customers. On a smaller scale, application software specialists diagnose customer problems and provide solutions electronically. Customers reside throughout North America, but specialists are in Colorado. On a smaller scale, employees at one computer development and manufacturing facility unite product designs with manufacturing technology and procurement operations to build customer solutions when the customer needs them: just in time. Telecommunication technology enables these operations. Indeed, telecommunication enables parallel processing at the firm level, whether the firm is local, national, or international.[7]

Organizational impacts of technology are not limited to user organizations; IT organizations themselves are undergoing substantial change in most firms. IT organizations must adapt even further to remain relevant. In many firms, the importance of centralized data-processing organization declines and, in its place, new architectures emerge. New structures recognize that performing some information-processing functions centrally is best, while they must perform a great many others locally. New IT organizations require strong leadership to develop, operate, and maintain systems vital to the firm's centralized operations. But more importantly, IT leaders must play strong roles in policy issues (governance) and must ensure that all IT activities, whether centralized or dispersed, conform to and support the firm's strategic goals and objectives.

New IT structures must also form partnerships and develop alliances with users throughout the firm to ensure effective use of technology at all levels of the firm. IT must facilitate user adoption of appropriate information systems and must support their growth and development. Important user tools such as computer-aided-design systems, computer-integrated-manufacturing systems, office systems, Web authoring programs, and end-user systems for marketing, sales, or service are all part of the firm's information infrastructure. IT managers are partly responsible for these user systems' success. IT people and client organizations must be tightly coupled, sharing values, goals, and objectives.

Organization and structure facilitate people working together to achieve personal goals and the firm's goals. Cooperative, innovative endeavors between individuals or groups are the lifeblood of business enterprise. Improved technology, sophisticated management systems, and effective people management enable employees, managers, and the enterprise itself to accommodate the steady stream of change required for success. But the following passage illustrates that an innovator's difficulties are an age-old problem:

> It should be borne in mind that there is nothing more difficult to arrange, more doubtful of success, and more dangerous to carry through than initiating changes. The innovator makes enemies of all those who prospered under the old order, and only lukewarm support is forth-coming from those who prosper under the new. Their support is lukewarm, partly from fear of their adversaries, who have the existing laws on their side, and partly because men are generally incredulous, never really trusting new things unless they have tested them by experience.
>
> In consequence, whenever those who oppose the changes can do so, they attack vigorously, and the defense made by the others is ineffective.
>
> So both the innovator and his friends are endangered together.
>
> —A. Machiavelli, *The Prince*, 1513

PEOPLE—THE ENABLING RESOURCE

Managers universally recognize that human resources are the most important assets contributing to business success. Most CEOs acknowledge this by noting the importance of people resources in letters to stockholders or in annual reports. Commonly, the CEO's letter to stockholders contains a paragraph thanking employees for contributing to the business' success and acknowledging that "people are our most important asset."[8]

When R. J. Stegmeier assumed the position of CEO at Unocal, he published the following statement in his first letter to employees and stockholders:

> In my view, this company's greatest asset has always been its human resource. Without the talent and teamwork of Unocal's 18,000 employees, our oil would stay in the ground and our facilities would stand idle. We face some tough challenges in the months and years ahead, but together—by our creativity and hard work—I'm certain that we can meet those challenges and turn them into opportunities for productivity and growth.

This is a good example of the thought CEOs often express regarding the enormous value of human resources.

Although most corporate vitality originates with skilled employees, the ability to recruit, train, and retain skilled individuals may limit some organizations' performance. Overall unit performance usually relates directly to individual performance. For example, the availability of people trained to sell the firm's products may limit marketing revenue. New product development depends directly on the creativity and ingenuity of the engineering force engaged in development activities. The firm's ability to grow and prosper hinges on the skills, abilities, and energy of the management team in establishing and implementing strategies and plans for success.

It is generally agreed that human resources may limit unit performance, but, in many instances, very significant contributions by a few individuals, or by an individual working mostly alone, materially improve an organization's performance. An engineer whose brilliant idea spawns one or more new products, or a salesman who clinches a big sale is very important to the firm. Thomas Musmanno, who invented the Cash Management Account, and the managers at Merrill Lynch who implemented the CMA are truly valuable human resources.

A section of Boeing's 1997 annual report, entitled *Our Strength Is in Our People*, states "Our people are the real source for our competitive advantage. We will continually learn, and share ideas and knowledge. We value the skills, strengths, and perspectives of our diverse team. We will foster a participatory workplace that enables people to get involved in making decisions about their work that advance our common business objectives." [9] Most exceptional, high-performance firms feel similarly about their people.

People and Information Technology

Since electronic data processing began in the 1950s, shortages of highly skilled people have somewhat limited effective implementation of information systems in business and industry. The scarcity of skilled employees and experienced managers limits the introduction of complex and rapidly evolving technology. Opportunities for programmers and systems technicians have remained firm for thirty years; these skills are still critical today and predicted to remain critical in the future. Numerous articles detail the catastrophes of inexperienced or ill-prepared individuals attempting to increase automation in corporations with high expectations. Clearly, twenty years' experience taught us much about coping with these difficulties, but many current examples remind us of the many struggles ahead.

Not only does the information-handling skills gap remain open, but current activities in telecommunications, client/server computing, operating system development, Internet technology, and outsourcing increase demands for skills in short supply. Talented individuals who can capitalize on advances in hardware, software, and telecommunications are in great demand. Additionally, skills and experience needed to implement strategies in these growing and potentially profitable areas differ in many ways from the currently available skill base. For example, object-oriented programming in C++ differs significantly from programming in COBOL 74, and writing mainframe batch systems differs greatly from designing and implementing Web applications. Managing IT people skillfully has always been difficult. New and emerging challenges promise to keep it that way.

The era of personnel shortages in the information arena is far from over. Prudent IT managers and their superiors must plan to cope with skills shortages lest they find themselves straying from the critical path of their parent organization's roadmap to success. Some alternatives exist. Managers must place the highest priority on effectively managing present employees familiar with the organization and its mission. Organizations that fail to manage their present staff in a superior manner stand little chance of obtaining and retaining outstanding individuals.

ESSENTIAL PEOPLE-MANAGEMENT SKILLS

Managers' ability to employ and retain skilled individuals is key to using information-processing technology effectively. Individuals skilled in technology must also creatively support the firm's technology use. To capture advanced technology's benefits and capitalize on its opportunities requires talented people. Employing skilled people, sensitively managing them, and providing effective leadership are necessary conditions for success. Managers who fail to lead their people skillfully and effectively usually achieve mediocre or marginal results in most endeavors. Successful managers employ solid people management skills: they consider these skills the cornerstone of their success.

What do we mean by solid people-management skills? How do employees perceive good people-management traits? What can employees reasonably expect of their managers? What distinguishes managers possessing these skills from those who do not?

Answers to these questions depend directly on assumptions about organized human effort and the values individuals derive from participating in organized activities. Abraham Maslow attempted to explain individual needs as a pyramidal hierarchy with basic physiological needs such food and drink at the bottom and self-actualization needs at the apex. The most essential needs like physiological and safety needs are followed in order of importance by social needs and esteem needs, which precede self-actualization needs. Individuals try to satisfy their most basic needs first. When they satisfy these, they try to satisfy the next-most-important needs. For example, individuals whose physiological and safety needs are satisfied, try to satisfy social needs, such as love or a sense of belonging. When they meet these needs, individuals attempt to attain esteem and recognition. According to Maslow, when individuals fulfill all other needs, they strive for self-actualization.[10]

Most IT employees strive to fulfill higher needs because they have satisfied their physiological and safety needs. Typical organizational activity lets managers appeal to employees' higher needs and to accomplish objectives for the firm. Skillful managers believe that most employees prefer to engage in meaningful, self-satisfying work that offers them a chance for self-development. Congruent individual and organizational goals is the ideal situation. Managers who understand their employees as individuals and arrange for this goal congruence greatly improve morale and increase productivity.

Effective People Management

Respect for individual dignity forms the basis of effective people management. Good managers recognize the enormous, frequently unrealized potential in each individual. Effective people managers strive diligently to understand each individual's preferences and motivations. They use this knowledge to display respect for each employee's unique personal characteristics. Talented professionals expect to be treated as special individuals; good managers meet this expectation.

Good people managers make appropriate assumptions about individuals and act according to these beliefs until proven wrong. These assumptions form the basis for managers' behavior toward employees and, in turn, influence employees' attitudes toward managers.[11] The assumptions are very important, not only because they govern attitudes, but because they can become self-fulfilling. Therefore, the assumptions and attitudes greatly influence individual and group performance.

Good People Managers' Attitudes and Beliefs

Good people managers believe that employees are honest and industrious and act with the firm's best interests in mind. Good managers also believe that employees are intelligent, willing to learn, and desire self-fulfillment.

In keeping with these beliefs, effective managers establish environments of high but reasonable expectations for their employees. Believing that professionals want to contribute meaningfully to the organization, managers work with individual employees to establish challenging goals. The manager and employee agree on objectives aligned with the organization's needs that also lead to employee self-fulfillment. Employees who want to achieve challenging, self-fulfilling goals congruent with organizational goals are ideally positioned to contribute significantly to the firm. And, in the process, they achieve self-satisfaction. They are productive and self-fulfilled.

Management equals leadership in the most successful organizations. Employees expect their managers to lead with clear goals and objectives for the organization based on a consistent vision of the future. Good managers clearly articulate this vision to their employees. They establish organizational goals and objectives, provide pathways, and set directions that lead employees toward meeting these challenging goals, for the organization and themselves. The best managers lead their employees with clear vision and motivate them with desires to achieve challenging corporate and personal objectives.

Managers must also set high and challenging performance standards for themselves. By setting good examples for their employees, effective managers establish high performance and highly productive environments for the organization. Managers who lead high-performance organizations have high expectations for themselves and have attained agreement on challenging goals for their employees. Managers achieve self-satisfaction and self-fulfillment from attaining difficult goals just as employees do, and their morale and productivity rises with that of their employees.

When managers and employees reach mutually acceptable standards of accomplishment, employees should assume responsibility for meeting those standards. They should have authority to accomplish the associated tasks. Managers must also provide tools for employees to do their jobs and train them to use the tools effectively. In addition, good managers delegate responsibility to individual employees and grant them the requisite authority to accomplish objectives. Through delegation, managers empower employees to meet organizational objectives. Managers must also pinpoint accountability and responsibility so employees clearly know what is expected. In this way, employees earn rewards for their individual accomplishments.

All organizations face obstacles and obstructions that impede progress. Frequently, bureaucracy appears to impede progress because meaningless rules make work difficult. Sometimes bureaucracy is an excuse for lack of progress. Good people managers' task is to remove or reduce barriers to accomplishment. Managers should remove or revise truly meaningless rules. Managers should explain the need for meaningful and required rules and assist employees in complying with them. Managers represent the firm to employees; they should enforce effective regulations and eliminate ineffective ones.

Good managers expect employees to be thoughtful about their work, but they also expect action. In most situations, a balance exists between exhaustive analysis and thoughtless action. Frequently, hoping to obtain all facts surrounding a decision, employees and managers spend far more time analyzing problems than solving them. Understanding the minute consequences of certain actions may be impossible, but, occasionally, employees and managers must take action despite incomplete information. That's the essence of decision making, after all. Effective managerial and staff decision making means striving for the middle ground, balancing expended effort against potential risk.

Effective managers expect employees to solve problems for the firm, and they also expect employees to prevent problems. Often, solving problems requires more time and energy than preventing them. Good people managers encourage and reward problem prevention. In many cases, preventing problems is difficult. Although visible external signs, such as distressed operations and frenetic activity, accompany problem solving, problem prevention usually consists of thoughtful actions taken without fanfare while operations proceed smoothly. Good managers must take special care to acknowledge and reward the subtle but valuable efforts of those who work to prevent problems or who exert effort to keep small problems from growing.

Through their actions and attitudes, superior people managers establish environments of high productivity and good morale. In these environments, individuals accomplish challenging objectives and experience high degrees of self-fulfillment. These environments encourage creativity, innovation, and invention. Managers who provide an atmosphere of productivity filled with opportunities for self-satisfaction foster innovation. (For many, innovation is the highest form of self-fulfillment.) But managers' actions must also support those whose innovations fall short of expectations or whose attempts at invention failed. Managers know that not all innovative endeavors succeed. They must also remember that the surest way to stifle innovation is to belittle or criticize individuals whose innovative ideas were not totally successful.

Achieving High Morale

Studies show that employee's morale and their opinions of their immediate manager's performance strongly correlate with their level of trust and confidence in their manager.[12] Managers whose behavior is reliable, based on predictable organizational norms, and consistent across similar situations gain employee trust and confidence. Managers who maintain steady, consistent behavior patterns inspire employee trust and confidence. Employees view them more favorably. Employee morale is also predictably high in the departments these managers lead.

High employee trust and confidence in management stems from many managerial behavior factors that they can control and incorporate in their management practice. Managers can take actions to increase employee trust and confidence:

1) Maintain two-way communication with employees to understand their needs and desires and to share company information.

2) Provide training and complete information so that employees can work effectively and efficiently.

3) Inform employees of promotional and career advancement opportunities.

4) Listen to employee suggestions for improving the work environment and respond encouragingly to all suggestions.

5) Sponsor teamwork and cooperation among department members.

6) Be available when employees seek consultation.

7) Understand the amount and quality of each employee's work.

8) Use the knowledge of each employee's work to grant fair salary increases and promotions.

9) Lead the department enthusiastically to achieve its goals and objectives.

Good people managers inspire employee trust and confidence and communicate effectively. They share information candidly and consistently, and welcome dialogue intended to improve business operations. Effective managers treat individuals with respect and dignity. They understand individual wants, needs, and contributions. Good people managers understand and publicly and privately reward good performance. Good people management is the mark of effective executives and a critical success factor for IT managers.

Ethical and Legal Considerations

Organized human activities involve individual and group interactions implicitly governed by rules of conduct. These rules or disciplines of what is good, bad, or morally required are termed *ethics*. Ethics are important for all organizations and their managers, who represent the organization to employees and help establish rules of conduct. Through their words and actions, managers govern the implementation of these rules and obligations.

Culture, law, religion, nationality, and other factors form the basis for individual ethical standards. In most firms, wide differences exists in individual standards, even among cohorts. These differences stem from parentage, heritage, religious beliefs, education, training, and experience. Managers must assume that employees intend to behave ethically, but that they may not come to the firm with a sense of ethics consistent with the firm's or fellow employees'. Managers have special responsibilities toward the company and its employees regarding ethical considerations.

Well-managed organizations act to ensure that managers and employees understand the basic beliefs and policies governing behavior within the organization and in external business relationships.[13] They conduct training sessions regularly and provide specially trained people to answer specific questions about the organization's policies and practices. These organizations establish a code of conduct and ensure that all employees follow it.

Table 18.1 lists actions that managers can take to foster ethical behavior.

TABLE 18.1   An Ethics Guide for Managers

1. Develop a statement of ethical principles for the firm.

2. Establish employee and manager rules of conduct.

3. State and enforce penalties for rule violations.

4. Emphasize ethics as a critical factor in the firm.

5. Inform employees of applicable laws and regulations.

6. Recognize ethical behavior in performance evaluations.

7. Maintain business controls, thus removing temptation.

8. Support a confidential forum for answering ethical questions.

9. Lead by outstanding example.

Managers must understand the firm's code of conduct and present it to employees unambiguously. Managers themselves must develop a code of behavior that encompasses the firm's policies and leaves no doubt that the firm values ethical behavior. Managers must answer employee questions thoughtfully; there may be subtleties. When in doubt, seek guidance from senior executives; discuss ethical dilemmas with them and seek their wisdom. An organization cannot build standards for ethical behavior on a manager-by-manager basis without creating serious internal inconsistencies that undermine the firm's rules of business conduct and behavior.

Ethical issues raised by the information age are numerous and increasing; information systems managers must be especially concerned about them. Consider privacy: is it ethical, for example, for firms to read employees' outgoing e-mail? Or consider property: Is employee use of the organization's resources such as e-mail for personal business ethical?[14] Firms must answer these questions and myriad others thoughtfully and clearly to free managers and employees from paranoia and guilt. For some situations, the firm establishes rules of conduct; for others, sources outside the firm establish laws and regulations.

Information technology diffusion in organizations and society raises many issues that have both ethical and legal overtones. Laws and regulations that individuals and firms are expected to know and follow govern issues like intellectual property rights, licensing, copyrights, and royalty payments. Although gray areas exist, especially regarding liability, employees and managers must remain informed. Desktop software and hardware are especially critical areas for most organizations.[15] Established by the firm and implemented by the information center and workstation store, business controls and other management practices are critical in managing intellectual and physical property, and in fulfilling license and royalty payment agreements.

Periodically, the legal and audit departments should review the organization's position and recommend improvements. The firm's managers, including IT managers, should seek advice and counsel from the firm's legal and human resource staffs. These specialized departments support managers just as some IT departments do. Individual managers should verify the firm's legal and personnel policies with the firm's experts and not attempt to chart these difficult waters unaided.

THE COLLECTION OF MANAGEMENT PROCESSES

Operating the IT management system effectively is an IT manager critical success factor. The management system contains tools, techniques, and processes, exercised periodically, that furnish a framework or background to guide employees' and managers' actions. For example, the strategic planning process establishes long-term corporate directions for using information technology. Likewise, the problem management system validates managers' intentions to achieve service levels. To be effective, the IT management system must align with the firm's management system, support and augment it, and embrace the firm's values and basic beliefs.

Systems guiding IT managers toward critical goals are essential for them, their organizations, and their firms. These critical factors (discussed in Chapter 1) consist of business issues, strategic and competitive issues, planning and implementation concerns, and operational items. This text presents management systems that help managers achieve their critical goals and deal effectively with issues they and their firms face.

Strategizing and Planning

The management tasks of building IT strategies and developing long- and short-range plans to implement IT's strategic direction are critical first steps for IT managers. Strategizing and planning are the IT management system's cornerstone. These activities link the IT organization to the firm's management system and align IT's strategic direction with the firm's business direction. These linkages and alignments are critical to IT and the firm. Figure 18.1 shows the relationship between these activities.

Information technology planning and control begin with the firm's business strategy. The firm's business strategy builds its foundation: the firm's mission, goals, and objectives. The firm's business strategy and IT strategy intrinsically link through shared goals, objectives, and processes. Business objectives for the firm in need of information technology resources and actions translate into IT strategy directions. Thus, IT strategy development supports the business strategy development process. Because the IT strategy is the basis for IT plan development, the firm develops shared strategic directions and plans, forging links between IT and the firm's many parts. Interweaving and sharing goals, objectives, and processes ensures alignment between IT long-range plans and the business strategy.

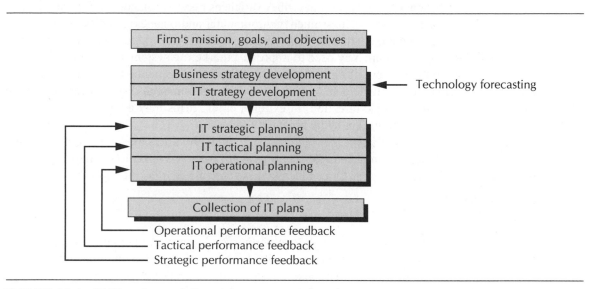

FIGURE 18.1   IT Planning and Control

These cohesive, interrelated strategizing and planning activities deal directly with many issues facing senior executives, IT managers, and their peers throughout the firm. Specifically, these processes align IT and corporate objectives, educate senior managers about IT's role and potential, and demonstrate IT's business contribution. Properly implemented, these processes eliminate strategic planning issues. They provide mechanisms for the firm to systematically exploit information technology for competitive advantage.

IT strategizing and planning processes offer excellent opportunities to coach others in the firm, including executives, about using technology to achieve objectives, including attaining competitive advantage. These processes must consider forecasts of technology capabilities. The firm matches technology capability with its needs for new technology and develops plans to incorporate it into the business. Planning for improved use of current technology and, when necessary, adoption of new technology can reduce costs, improve efficiency, and provide or sustain competitive advantage. When handled skillfully, these actions also ensure realistic, long-term technological expectations.

For several reasons, the process described above may not succeed, and the resulting IT strategic plan may not align with the firm's business plan. For example, the process succeeds only when the firm has well-defined mission statements and business strategies. Without these, IT strategies may not correlate well with intended business directions. Some firms do not include IT in the planning process because they consider IT unessential to success. These firms do not need strong information technology support for corporate goals and objectives. In other firms, IT managers share in the process but lack needed business skills and knowledge to construct strategies and plans aligned with the firm's plans. To mitigate these difficulties, IT managers must ensure that they understand their role and contribute substantially to the firm.

Controlled plan execution stems from operational, tactical, and strategic performance information. Performance information describes variances between planned and actual results. It also includes variances between assumed environmental and business conditions and actual business and environmental conditions. For example, if competition threatens the firm's markets, the IT development team may need to improve a marketing system's development schedules. Depending on the nature of plan variances, feedback may generate course corrections in operational, tactical, or strategic planning. Also, major environmental disturbances may cause the firm to adjust its strategic direction. In extreme cases, firms may adjust their mission to capture perceived opportunities or avoid potential threats.

The critical activities of strategizing and planning direct the firm's exploitation of its current and future assets and lay foundations for future growth. The management systems described in this text are excellent frameworks for conducting these crucial activities successfully.

Portfolio Asset Management

IT managers must deal intelligently with many important issues surrounding application portfolio asset management. Earlier chapters described methods for prioritizing application backlog, managing development processes, and finding local application development alternatives. Figure 18.2 illustrates the relationships between these activities.

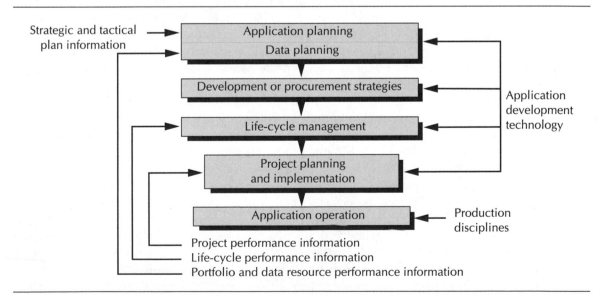

FIGURE 18.2 Application Management

Most firms spend significant portions of their IT budget maintaining and enhancing application programs and managing associated data resources. Management systems to handle these challenges begin with IT strategizing and long-range planning. These processes establish directions for enlarging and enhancing the application portfolio and lay the foundation for incorporating technology advances. They help managers optimize the portfolio and add value to its assets.

Strategy and plan development activities generate information needed for application and data planning. Managers compare application performance and capability with the firm's requirements and analyze data resources' sufficiency. When prioritizing, managers use this information to identify applications qualifying for investment and specify the preferred method for acquiring them. This decision-making process combines computer experts' technical knowledge with senior managers' business knowledge and vision.[16] From this analysis, IT managers glean information to build acquisition plans and establish installation schedules.

Life-cycle management, project planning, and implementation govern resource investment decisions for the applications selected. These management-oriented processes include business case development, phase reviews, resource allocation and control, risk analysis, and risk reduction. These processes ensure the controlled and disciplined achievement of objectives of the portfolio's strategic and tactical plans. Skillfully executing these management processes improves applications to satisfy the firm's functional and business objectives.

Frequently the firm learns about new, valuable technology when planning and managing applications. For example, executives may learn more about Web technology, imaging systems, client/server implementations, CD-ROM data storage technology, or important data transfer technology like fiber optics or wireless LANs. New application development technology such as advanced languages, methodologies like object-oriented design, and new development tools and methods may also gain prominence. Introducing new technological developments is essential to application management systems.

Application asset managers must consider alternative acquisition methods like alliances, joint ventures, and purchased application packages, and alternative strategies like client/server systems, intranet computing, and outsourcing application development and operation. Because businesses need modern applications and rapid solutions to business problems, IT managers who bring the firm new solutions must use various techniques to improve productivity and increase cost effectiveness.

Management information available during these processes helps assess whether activities proceed satisfactorily and provide data for needed course corrections. Information derived from phase reviews, for example, helps keep development projects on schedule or validates needed schedule corrections.

Application and data planning combined with strategic and tactical planning lay foundations for developing the firm's information architecture—an important concept for all organizations. A coherent information architecture develops from application and data planning when strategic plan objectives, combined with assessments of present application and data performance, are measured against needs. In most firms today, developing and maintaining effective information architectures is both difficult and critical.

Systematically applying portfolio management processes causes executives to focus on application issues and provides tools and techniques to address their concerns. By using portfolio management systems effectively, firms acquire needed applications and functions, on schedule, and within budget, based on sound business cases. New and enhanced applications are available for the firm's productive use in a controlled and optimal fashion.

Production Operation Disciplines

Sound portfolio management delivers new and enhanced applications that improve the firm's business. These applications codify the firm's internal functions and link it to customers and suppliers through telecommunication systems. In many cases, they provide competitive advantage. Operating applications smoothly and incident free is a critical success factor for the firm's managers.

Service-level agreements, batch operations, and online operations relate production operations departments to IT customers. Figure 18.3 illustrates these relationships.

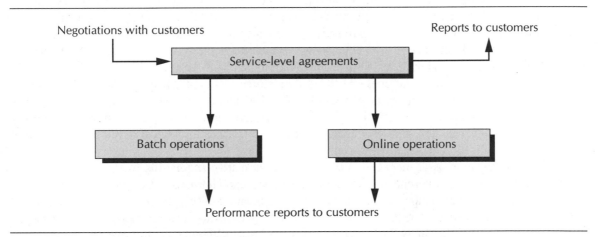

FIGURE 18.3   Production Operations—Customer Performance System

Service-level agreements between service users and service providers develop and document customer expectations. Parties negotiate technically achievable and financially sound service levels that meet business needs. Batch processing, online processing, or a combination of both batch processing, online processing, or a combination of both deliver production service. Customers' service requirements described in service-level agreements drive batch or online process performance.

To understand this, consider inventory management. Routine inventory transactions entered throughout the day via client/server applications accumulate information for the nightly batch run. The applications transmit results from the nightly inventory processing, a scheduled batch operation, to the inventory department's server before 7:00 a.m. daily. In each activity, the inventory department reached agreement with service providers on needed and affordable service parameters. IT's performance in meeting committed service levels is measured and reported regularly.

These processes form the basis for achieving customer satisfaction in production operations. They establish requirements rigorously and provide systematic means to meet them. These management systems meet many, but not all, necessary conditions for successful production operation.

Sometimes service problems occur during program execution that require changes to applications or their operating environment. Disciplined processes for managing these activities are mandatory. Achieving service levels also depends on hardware capacity, software systems, and their performance. Figure 18.4 shows the management systems for dealing with these complexities.

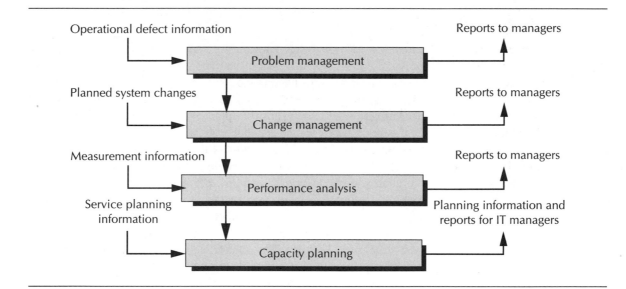

FIGURE 18.4   Production Operations—Internal Systems Management

Production operations occasionally experiences problems or defects that actually or potentially affect service. Problem management handles these incidents and reports significant activities to managers for monitoring and control purposes. Customer problem resolution is also confirmed with the management team.

Controlled system changes to complex data processing or telecommunications operations originate from two primary sources: planned changes and changes required to correct problems. The change management discipline obtains input from these sources, manages changes through implementation, and reports results to appropriate managers. Together, problem management and change management provide systematic ways for service providers to correct inevitable operational faults and implement changes needed to correct problems or implement planned system alterations.

Performance management and capacity planning complete the production operation management system. Because service levels are based on known or anticipated performance factors, ongoing performance measures must be available to validate system productivity. Service providers receive reports of productivity analysis for their action, if required. Thus, capacity planning receives input from performance analysis and service-level planning. Plan input, customer needs, and performance analysis generate future capacity requirements for tactical and operational plans.

Managing production operations is a critical success factor for service providers. Management systems for operating applications attack operational issues and help managers deal with them successfully. This operational framework, combined with tactical and strategic management systems, forms an important base for management success.

Network Management

Advances in various technologies for processing, transmitting, and storing binary data lead to systems that handle voice, data, and image information interchangeably. The consolidation of supporting organizations usually follows consolidating hardware and software for processing all information types. In most cases, merged organizations find consolidating previously separate management systems advantageous, too. This means that management systems for centralized or distributed operations also support network management.

For example, the problem management system processes network faults (defects that impair or may impair network service levels), and the change management system handles network changes. Many organizations consolidate network performance and capacity planning with application systems performance and capacity planning. This works well for IT organizations and also helps IT customers. Because customers are much less able than IT people to distinguish network components from application components, they are much less likely to care about these distinctions. They want to obtain needed services; they appreciate simple management systems that provide services without bureaucracy.

In addition to network management systems discussed so far, IT needs network configuration management to provide data for other disciplines. Figure 18.5 illustrates sources and uses of data that configuration management develops and stores.

Configuration management maintains several databases needed to manage problems, changes, and plan recovery actions. Managers can also obtain network and component status information from these databases. Automated network management systems use configuration data to optimize network operations and update other relevant databases. Systems for managing networks critically depend on this information.

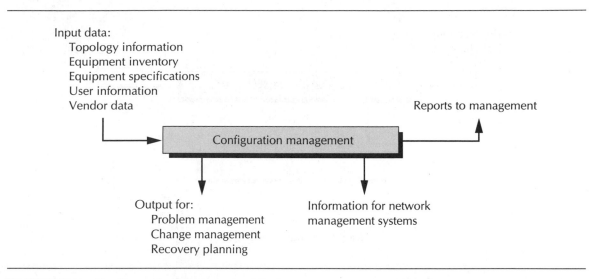

Input data:
    Topology information
    Equipment inventory
    Equipment specifications
    User information
    Vendor data

Configuration management

Reports to management

Output for:
    Problem management
    Change management
    Recovery planning

Information for network
management systems

FIGURE 18.5   Configuration Management System

Financial and Business Controls

Effectively operating IT organizations means that managers exert control over all IT business aspects. They must understand the organization's objectives and manage IT to meet them systematically and regularly. Managers must implement formal strategizing and planning processes that meet established objectives and install disciplines to manage application acquisition and operation. In addition, IT managers must control financial and business aspects of their operations through appropriate systems and business controls.

Processes to monitor IT finances and maintain financial control begin with tactical planning and budgeting. Whether IT uses cost centers or profit centers, or operates as corporate overhead, annual planning generates an IT operating budget. Usually the firm's controller produces monthly financial statements describing the actual financial position *vs.* the budgeted or planned position. Using these statements, IT managers review and compare actual expenditures with planned expenditures and resolve variances. Figure 18.6 displays this process.

If IT operates cost centers, its financial plan contains planned cost-center support and expenses, both subject to variances. Large variances or unsatisfactory trends require managers to analyze information and make corrections. In some cases, negotiation with clients causes changes to planned rates or service levels for some service classes.

If IT operates a profit center, both revenue and expense must be analyzed for discrepancies. Changing revenue, altering expenses, or permitting profit margin changes can correct variances. Increasing or decreasing business volumes or changing prices can alter revenue. Because IT profit centers operate as businesses within businesses, managers have many opportunities to make adjustments affecting revenue, expenses, or margins.

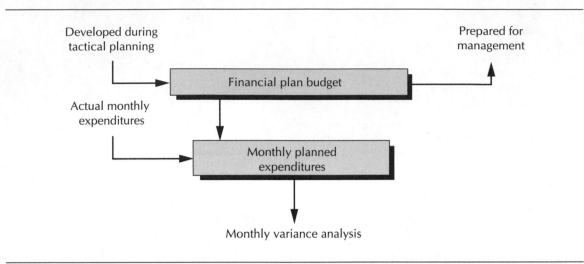

FIGURE 18.6 Budget Management

Sound financial management systems help solve many top management issues. Budgets resulting from thoughtful tactical and operational plans that account for user services and recover costs help reduce conflict between users and service providers. IT organizations that use sound financial management techniques operate in a businesslike manner and spend precious resources on activities that the firm's managers deem needed.

Business Management

Sound management systems are essential because they provide frameworks for accomplishing the firm's important objectives. Properly functioning systems guide and help managers discharge their responsibilities in a logical, organized manner. But systems operate through people; good people relationships are needed to support effective systems. To achieve maximum effectiveness, IT managers must execute their management systems harmoniously with others.

IT managers can foster improved interaction with clients, increase coupling between customers and suppliers, and provide input for improving service through various means. Many firms form an IT steering committee, composed of client organizations' executives, to consult with key IT executives. The steering committee, a high-level sounding board, helps establish IT strategies and develop IT initiatives. In many ways, steering committees sponsor IT to users and increase executives' awareness of IT benefits. Steering committees' actions and deliberations influence service levels by recommending and approving spending levels. In firms that are culturally biased toward participative management, steering committees can improve communication between top management and functional managers on important IT matters.

Another effective mechanism for linking IT with clients is to assign individual IT managers liaison responsibility to client organizations. For example, an application programming manager's assignment may be to represent IT to manufacturing, or an operations manager's assignment may be product development. The manager represents IT to the function and the function to IT.

Service representative's responsibilities include:

1) Obtaining information on new or changing client needs

2) Broadening communications between clients and IT

3) Answering user's questions about IT procedures

4) Facilitating problem resolution for clients

5) Advising on client-developed programs

In short, service representatives serve everyone's mutual advantage by providing high-speed communication links between IT and other functions. In large, diversified firms, many service representatives interact with product development, manufacturing, marketing, sales, service, finance, administration, and possibly other departments, too.

Effective IT executives use every means at their command to maintain and improve communication between IT and the rest of the organization. They understand the value of effective communication and strive to maintain close contact with their peers throughout the firm. Formal and informal communication is an important ingredient of successful IT management systems.

THE IT MANAGEMENT SYSTEM

The text describes a management system that focuses managers' attention on activities that improve both their effectiveness and the firm's. To succeed, managers must accomplish what the firm wants. This system focuses managers' attention on activities that yield the greatest payoff for the firm. IT improves the firm's competitive posture through effective operations.

Management systems also help IT conduct activities efficiently. They force managers to focus on critical operational, control, and financial details. This focus improves efficiency and increases productivity. When businesses are able to use internal information more effectively, decision making improves and IT and firm performance increase dramatically.

SUMMARY

Today's business environment is undergoing rapid and fundamental changes and becoming increasingly complex. Information systems—especially telecommunications—are a major factor driving these changes because they provide the means for altering competitive firms' positions. Information systems furnish opportunities to restructure in response to environmental changes. Restructuring enables firms to reduce costs, improve their value chain, and compete globally. They also present both problems and opportunities.

Organizations are comprised of people working together toward common goals directed by managers who provide leadership. People are the firm's most important assets; they must be managed skillfully and sensitively. This means that effective people managers respect their employees' dignity and make assumptions that reflect positive beliefs about them. Productive organizations operate in environments filled with high expectations, supported by managers and employees who encourage innovation and recognize and reward accomplishment. High performance organizations are filled with people who achieve self-satisfaction and self-actualization by accomplishing challenging tasks for the firm. Good people management is a skill effective IT executives cultivate and a critical success factor for them.

Management systems support IT managers and guide them and their organizations to achieve corporate and organizational objectives within the firm's cultural norms. IT management systems assist managers to develop strategies and plans supporting the firm's strategic direction. IT planning processes ensure alignment between IT and firm objectives. Portfolio management supplies sound applications supporting the firm's operations.

Management systems for operations and networks establish environments fostering careful planning, service level achievement, and cost control. Financial management and control are accomplished through planning, budgeting, and review activities. Although these activities form a sound basis for IT managers, open communication channels increase their effectiveness. Steering committees and IT service representatives play important communication roles and promote understanding and cooperation. Effective IT and client managers champion these communication techniques.

Review Questions

1. What role does information technology play in the dramatic changes occurring in today's business environment?

2. How are alliances or joint ventures important to firms' success?

3. How is telecommunication technology important to centralization and decentralization in business operations?

4. How do Volvo, IBM, and GM use telecommunications to improve their businesses?

5. Define the span of communications. What is its importance to middle managers today?

6. What current IT trends increase the importance of people skills?

7. Describe Maslow's hierarchy of individual needs. In their working environment, where in the hierarchy do most IT people exist?

8. What assumptions do good people managers make concerning their employees?

9. Why should managers and employees establish high expectations for themselves?

10. How can managers and employees balance thoughtfulness and action when making decisions?

11. The text states that managers expect employees to solve and also prevent problems. What possible difficulties do these expectations create?

12. What should managers do when employees present them with ethical dilemmas?

13. Why are strategizing and planning activities so important for IT managers?

14. Why does IT strategy and plan development fail in some firms? What are the consequences of this failure?

15. Why is the application portfolio management system important? What critical issues does it address?

16. What are the critical steps in application life-cycle management?

17. What role do service-level agreements play in production operations?

18. How are IT budgets developed?

19. What activities do steering committees perform? Why are these activities valuable for IT and users?

Discussion Questions

1. Describe the arrangement between Public Service of Colorado and ISSC. Could part of this arrangement be called co-sourcing? If so, explain why.

2. In the Business Vignette, Allie Young commented that contracts like that between Public Service and ISSC show a clear direction for outsourcing's future. The key is the two companies' strategic involvement. Discuss these ideas' significance.

3. Describe how firms use alliances or joint ventures to improve their value chain or expand their operations.

4. Discuss how the issue of centralization *vs.* decentralization relates information technology's use and application.

5. Discuss how partnerships and alliances apply to IT organizations today. Describe their relevance to IT downsizing and outsourcing.

6. Describe how effective people managers' knowledge of human needs improves their managerial performance.

7. In what ways do you think that managers' assumptions about people impact employees' performances?

8. Review the characteristics of good people managers. Which characteristic is most important in your opinion? Which characteristic do you think is most difficult to attain?

9. Describe the environment that good people managers attempt to create. What advantages accrue to managers and employees working in this environment?

10. What managerial behavior patterns improve employees' perceptions of managers' performance and inspire their trust and confidence?

11. What actions should firms take to develop and foster ethical behavior among employees and managers?

12. Describe how management systems help manage IT functions. How do management systems relate to the corporate culture?

13. How might IT implement controls to ensure that plan implementation proceeds correctly?

14. Describe the system for portfolio asset management. What processes provide input data for it? What are desired outcomes of effective portfolio asset management?

15. How can application management systems help manage the large data stores associated with the firm's portfolio?

16. Describe the management system for computer operations, and relate it to other management systems.

17. Discuss the advantages and disadvantages of separating application and network operational processes.

18. How do IT profit centers maintain financial control?

19. How do steering committees and service representatives augment IT management systems?

20. How does the service representative concept differ between organizations operating cost centers and those selling IT services to other firms?

Assignments

1. Read "Ethical Standards for Information Systems Professionals: A Case for a Unified Code," by Effy Oz in *MIS Quarterly*, December 1992, 423, and prepare a two-page summary for the class.

2. Interview your university's human resources manager and discuss his or her views on people management. Compare and contrast your interview observations with the ideas this text advocates.

3. Visit a computer center near your school, and review its written operational procedures. Compare them to the management systems this text describes. Did you find any processes or procedures unique to the center you visited? If so, why are these important?

[1] Mark Helper, *Computerworld*, February 6, 1995, 14. Copyright 1994 by Computerworld, Inc., Framingham, MA 01701. Reprinted from *Computerworld*.

[2] Michael S. Scott-Morton, "Information Technology and Corporate Strategy", *Planning Review*, September-October, 1988, 28.

[3] On April 6, 1998, Citicorp and Travelers Group agreed to an $83 billion merger, the largest to date in the U.S. The merger created a financial giant with $697.5 billion in assets and annual revenue nearing $49 billion.

[4] Michael Porter defines and discusses the value chain in detail in *Competitive Advantage*: *Creating and Sustaining Superior Performance* (New York: Free Press, 1985), 12.

[5] Machael Hammer and Glenn E. Mangurian, "The Changing Value of Communications Technology," *Sloan Management Review*, Winter, 1987, 39.

[6] In 1997, total telecommunications traffic was roughly 80 percent voice and 20 percent data. Because data transmission is rapidly growing, these numbers will reverse by 2004. "Telecommunications Survey," *The Economist*, September 13, 1997, 27.

[7] The "virtual office," wherever individuals and their personal, portable office equipment is, and "virtual corporations," dispersed, sometimes temporary units electronically tied to corporate headquarters, are embryonic concepts likely to gain importance.

[8] For example, see Exxon's 1997 *Annual Report*, 5. Chairman Lee Raymond writes, "Exxon is the successful enterprise it is today because of *many strengths, the greatest of which is our people.* To them I extend special thanks for turning in another record year."

[9] Boeing's 1997 *Annual Report* to stockholders, 15.

[10] Abraham H. Maslow, *Motivation and Personality* (New York: Harper & Row, 1954), 80.

[11] Douglas McGregor, *The Professional Manager,* ed. Warren Bennis and Caroline McGregor (New York: McGraw-Hill Book Company, 1967), 16.

[12] M. L. Pesci, personal communication.

[13] Ethical standards vary among organizations within a country and between countries. See Effy Oz, "Ethical Standards for Information Systems Professionals: A Case for a Unified Code," *MIS Quarterly*, December, 1992, 423.

[14] One firm's employee handbook states its policy unambiguously. "There is no personal privacy when you use Salomon Smith Barney's equipment and services." The firm "may monitor, copy, access, or disclose any information or files that you store, process or transmit." Patrick McGeehan, "Two Analysts Leave Salomon in Smut Case," *The Wall Street Journal*, March 31, 1998, C1.

[15] "Get Legal on the Desktop," *Datamation*, May 1, 1995, 56.

[16] Thomas H. Davenport, Michael Hammer, and Tauno J. Metsisto, "How Executives Can Shape Their Company's Information Systems," *Harvard Business Review*, March-April 1989, 131.

19

The Chief Information Officer's Role

Caterpillar Uses IT to Remain Competitive

Tens of thousands of Caterpillar jobs in the U.S. have been saved due partly to this leading construction equipment manufacturer's impressive injection of information technology. Facing strong foreign competition, Caterpillar invested in new computer and telecommunication technology to increase its global competitive position. Thanks largely to technology investments, Caterpillar's sales and profits are surging. "To survive in this new globally competitive world, we had to modernize," says vice chairman James Wogsland. "Information technology is the glue for everything we do."[1]

To improve its competitiveness, Caterpillar launched two massive initiatives. The first initiative, termed "Plant with a Future," aimed to increase plant floor automation, reduce manufacturing costs, improve marketplace responsiveness, and retain Caterpillar's leading market position. Led by chairman Donald Fites, the second initiative broadly restructures Caterpillar to improve internal and external customer responsiveness. Information technology plays an important role in both undertakings.

As part of the restructuring, Robert Hinds, formerly director of computer-integrated manufacturing, was named CIO (Chief Information Officer) and corporate IT director. Hinds manages an annual budget of $180 million, a 1300-person corporate IT staff, and a centralized data center in Peoria, Illinois, home to many Caterpillar facilities. In addition to the corporate IT staff, operating divisions employ another 1300 IT people to manage divisional information systems. "Our corporate systems make us look like an integrated company—because we are—but giving autonomy to the business groups gives us a competitive edge," Hinds says.[2] Using workstations or PCs, 90 percent of Caterpillar's employees can access corporate data, validating Hinds' assertion that technology is truly part of the job regardless of where you work.

Finding or creating superior support systems was plant and division managers' first step in reengineering the company. In many instances, Caterpillar's plant and divisional IT teams built their own systems, while Hinds' corporate IT staff rebuilt critical corporate systems, including purchasing, logistics, and MRP II (manufacturing resource planning) systems. Hinds also established company-wide program development standards to ensure compatibility among these dispersed activities.

In addition, Caterpillar developed factory integration specifications that equipment suppliers must build into their controllers. Because nearly every manufacturing equipment vendor supplies proprietary, built-in software, the specifications forced them to meet Caterpillar's manufacturing system interface. This strategy eliminates the "islands of information" problem at the outset.

Another critical IT activity supports Caterpillar's dealer network, the company's only direct link with customers. Cat's 70 dealers operate more than 300 outlets in the U.S. alone. "Our goal is to give the customer 100 percent of the parts he needs over the counter when he needs them," says Larry O'Neill, a $100 million Elmhurst, Illinois, Caterpillar dealer. "We can't carry every part, but by jointly sharing information, we can get any part in here by 6:30 a.m. the next day."[3]

Under Hinds, corporate IT built a tri-level telecommunication system to support Caterpillar's worldwide operations. Digital microwave links back up 70 miles of fiber-optic cable supporting company activities in the Peoria area. Leased fiber-optics capacity and satellite links support North American dealer operations and other company activities beyond Peoria. Leased fiber and satellite services link operations outside North America to Caterpillar systems.

To advance its automation efforts, Caterpillar developed proprietary EDI software its suppliers use. More than 1000 key suppliers, providing about 80 percent of Caterpillar's parts, exchange data electronically with Cat.

Caterpillar has completed several other important IT initiatives. Its corporate IT staff developed and built a financial modeling system that helps executives perform what-if analyses and estimate future project profitability. IT implemented an activity-based cost accounting system to help managers obtain production cost contributions by plant for particular products. And, Cat engineers now use a computer-aided design system that reduces product development time by up to four years.

"Plant with a Future" is a great achievement. It reduced work-in-process inventory by 60 percent, decreased product build time from 45 days to 10 days or less, and increased on-time customer delivery by 70 percent. Caterpillar's IT and reengineering initiatives reduce inventory carrying costs, save millions on Cat's annual $5 billion purchases, improve customer satisfaction, and provide competitive leadership. Caterpillar's 1996 revenue was $16.5 billion—Value Line projects $19.8 billion in 1998 revenue with earnings per share growing 11 percent.[4]

Caterpillar is an outstanding example of IT and corporate leaders combining technology and reengineering activities to create superior systems and processes and greatly improve business effectiveness. It serves as a model for firms that have yet to capitalize on IT's enormous potential for organizational improvement.

INTRODUCTION

Information technology managers in most modern firms wield considerable responsibility and authority. Power and responsibility flow from the line organization they head, and, because the firm widely deploys information technology, IT executives exercise additional authority and shoulder considerable staff responsibility. As organizations increasingly depend on IT, they also depend more on senior IT managers. Consequently, they enjoy a position of opportunity for great success or substantial failure. To achieve success, IT managers must operate so their superiors view their accomplishments positively.

This text's premise is that superior managers are made, not born, and that superb management skills are learned and developed, not the result of genetics.[5] Therefore, studious managers can learn and apply this book's management systems and practices to defuse the many management issues arising in practice, only some of which are discussed here. CIOs and senior IT managers are well-positioned to observe management issues, follow emerging trends, and implement well-designed management systems.

However, do senior IT executives enjoy success commensurate with their responsibilities? Can well-trained individuals consistently succeed in the job? Or is the evolving nature of technology and its influence on organizations so dominant that success is mostly a matter of chance? Given the difficulties of IT management and the changing character of its responsibilities, what are the CIO's critical success factors? This chapter explores these and other topics.

SENIOR IT EXECUTIVES' CHALLENGES

The CIO or senior IT executive's role in modern firms captures the opportunities and pitfalls inherent in the technology, embraces the ebb and flow of managers' and employees' aspirations and motivations, and symbolizes the powerful effect information technology has on business and industry worldwide. The role is extremely difficult. Unfortunately, the job's realities differ markedly from the philosophical foundation of the CIO title. Consequently, the title and position are alternately praised and cursed.

Numerous articles chronicle the dilemma of the CIO. In 1986, for example, Gordon Bock writing for *Business Week* stated, "There's a new breed of manager surfacing in the executive suite. Some members of this new information elite sit behind such recognizable nameplates as senior vice president, vice president for information services, or information resources manager. Others are beginning to get a higher-sounding title to reflect their new status: chief information officer, or CIO."[6] For some, the CIO position was destined to be prominent indeed. In 1990, for example, Thomas Friel, managing partner for information technology at Heidrick & Struggles predicted that "In five years, virtually every major company will have a CIO who's a peer to the CEO."[7]

Contrary to predictions, current literature suggests that business firms have yet to define fully the CIO's role and position. For example, in nearly 80 percent of *Computerworld's* Premier 100 companies, one or more management levels separated the CIO from the CEO. Analysis of the premier 100 companies also reveals that a CIO's average tenure exceeds six years, whereas, in CIO ranks generally, the annual turnover approaches 40 percent, a rate significantly greater than that of top executives in general.[8] Literature also suggests that part of CIOs' identity problems result directly from their behavior in the firm. In addition to their actual performance, their perceived performance in relation to expectations may be the critical aspect for evaluating the CIO position and title.

In some firms, the title itself created difficulties. In some cases, according to Professor Wetherbe at the University of Minnesota, the title inspired such animosity that its bearers happily abandoned it. But the title does not necessarily make the job risky for the incumbent—the task itself is hazardous. Individual performance deficiencies caused some failures, but others stemmed from organizational consolidations, cost reduction exercises, or bad times for the industry or firm. Peter Keen, however, quickly gets to the subject's essence when he says, "The CIO position is a relationship, not a job. If the CIO/top management team relationship is effective, the title doesn't matter. If it is ineffective, the title doesn't matter."[9]

Today, many more firms employ an information chief than did 10 years ago; however, very few CIO's possess the skills and the opportunity to advance in the corporation. And,

---

many companies still view information technology staffers with skepticism or apprehension. "With most of my peers, there tends to be a distrust with the information services organization and with the CIO for that matter," says Bob McLendon, CIO of Texas Instruments.[10] His remarks characterize the feelings of some regarding the CIO title and position; nevertheless, many superior individuals have achieved great success as CIOs of important organizations.

Some exemplary CIOs include DuWayne Peterson, formerly at Merrill Lynch; Max Hopper, earlier at AMR and Sabre; Ron Ponder, previously at FedEx and AT&T and now CEO at Beechwood Data Systems; Bob Martin of Wal-Mart; Patricia Wallington at Xerox; and Paul Strassmann at the Defense Department and now a respected researcher and writer of today's use of IT.[11] These individuals share one characteristic—a firm conviction that organizations must use IT, not only as a supporting service, but to drive business success. In one way or another, all successful CIOs have contributed importantly to their organizations' prosperity.

THE CHIEF INFORMATION OFFICER

CIO's Organizational Position

Whether they are called CIOs, VPs of information systems, or information resource managers, senior IT executives have line and staff responsibilities for the firm's information technology resources. Today, the coalescence of telecommunication and information processing technologies leads to consolidating these technologies' formerly separate organizations under the senior IT executive's leadership. Because information technology is now widely dispersed throughout organizations, senior IT executives gain extensive staff responsibility, too. Just as chief financial officers are responsible for their firms' expenditures although they do not spend most of the money, the firm's chief information officer is responsible for IT expenditures although the IT line organization does not consume most resources.

In leading firms, with critical and well-developed information technology, chief information officers or CIOs are responsible for managing its corporate information systems function, information infrastructure, and their operation. CIOs make technology investments in these facilities and recommend or approve other IT investments; they develop and implement IT strategies to increase the firm's revenue and profits; they set standards for information or telecommunication operations in the firm not under their direct control; they recommend and enforce corporate policy on IT matters including procurement, security, data management, personnel, cost accounting, and others.[12] However, exact responsibilities differ markedly among organizations.

For example, a survey of 137 health-care CIOs revealed that 99 percent were responsible for information systems, 65 percent for telecommunications, 31 percent for management systems, 15 percent for medical records, and 10 percent for admitting services.[13] They believe their most important role is to integrate information systems, telecommunications, and management systems. According to the respondents, knowledge of hospital systems ranked fourth among the top attributes CIOs need. They placed more importance on leadership ability, vision, imagination, and business acumen.

Reporting relationships between senior IT executives and CEOs vary considerably—from direct reporting to three levels removed. Many CIOs report to chief financial officers, executive or senior VPs, VPs, or division heads. Fewer than 20 percent report directly to CEOs. Many times, the firm's culture or the industry it's in influences reporting relationships. At information-intensive firms, CIOs tend to earn higher salaries and occupy prominent positions.

At National Car Rental, for example, where information technology is critically important to its strategy, the CEO and CIO share many responsibilities.[14] In this unusual situation, the CIO is National's second highest paid executive. "Any CEO who doesn't have the CIO as a direct-report is absolutely crazy," says Vincent Wasik, National's CEO. This unusual partnership pairs a business-wise CEO with a technology-smart CIO, both committed to using automation fully to revolutionize the car rental business.[15]

In addition to reporting relationships, many CIOs exert significant influence through committee appointments and staff assignments. Regardless of their organizational level, however, to be effective CIOs must closely communicate with top executives or be an integral part of the executive management team. Highly effective CIOs consider themselves corporate executives, not mere IT functional managers.[16]

However, close contact with executive managers is not enough; the executive management team must recognize information technology's value and the importance of making sound decisions regarding its use. Top managers sometimes fail to appreciate IT because they lack awareness of IT's strategic importance or recognize only the operational importance of computers and automation. Although most senior executives consider information a resource, some find credibility gaps in IT's direction. CIOs must overcome these difficulties by educating top management and marketing IT accomplishments to them. Satisfied IT user-partners—the beneficiaries of technology investments—help greatly in selling and promoting IT's business image with senior executives and others.

CIO's Career Paths

The most popular route to the CIO position goes through information systems, although some CIOs combine their IS experience with that from other disciplines. Consulting and telecommunications backgrounds combined with IS experience is also common among CIOs. When telecommunication organizations merge with IS, former IS managers usually head the consolidated organization because they developed or managed critical organizational systems and are believed to have broad management skills. Telecommunication managers, on the other hand, are usually seen as merely managing the phone system. Many other managers, however, move to the top IT position after serving in line or staff positions elsewhere in the company. Engineering, product development, or general management backgrounds are also common.

According to a Coopers & Lybrand survey, more than 75 percent of CEOs, COOs, and CFOs prefer that CIOs have both an IT and a general or line management background. Only two percent indicated that background didn't matter.[17]

Firms are more likely to appoint CIOs from outside IT organizations when the IT function needs revamping. IT organizations that resists downsizing, outsourcing, or decentralization, or that fail to support rapidly changing business conditions are likely to be reorganized under former line managers. In many cases, the IT culture is more easily changed by replacing the CIO than through other, less drastic actions. Newly appointed CIOs from outside the IT fraternity usually carry a strong mandate for change.

Abrupt changes in the top IT position are quite common. A 1994 survey of 400 CIOs revealed that nearly one-third of their predecessors had been dismissed or demoted, and that nearly one in five left voluntarily for other positions. According to some studies, CIOs' average tenure is less than three years. The turmoil at the top is partially attributed to the IT profession's relative newness and to frequent redefinitions of roles and corporate relationships; part is just normal turnover or job rotation within the organization.[18] Reengineering business processes and adopting new operational modes strain financial, technical, and personnel resources. Difficulties in recruiting, training, and retaining competent IT professionals who can implement complex IT systems on new platforms with constrained, no-growth IT budgets also fuel the turmoil.

Successful IT executives, however, enjoy considerable mobility. Careers of some top CIOs indicate great demand for exceptional individuals, and that technology skills and general business management skills are largely transferable across industries. The position of their superior, however, is usually beyond their reach.

Performance Measures

Regardless of the top IT executive's title, the firm's top executives have clear expectations of the position and the IT organization. The CIO's highest priority is satisfying these expectations. What is expected of CIOs? What must they do for success?

CIOs' performance is measured by their success in applying information technology cost effectively, achieving goals and objectives for the firm, and bringing value to it. Organizations expect CIOs to identify technological opportunities and provide leadership in capitalizing on them for the firm's advantage. To accomplish this, CIOs must educate senior managers on appropriate opportunities and enlist their agreement to incorporate these opportunities into the firm's strategies and plans. Simultaneously, management measures CIOs on their ability to develop planned courses of action that balance opportunities, expectations, and risks.

CHALLENGES WITHIN THE ORGANIZATION

Businesses evolve by changing their business practices or adopting new practices to improve their competitiveness or to respond to changes in the external environment. Mergers, acquisitions, alliances, joint ventures, and new business formations are external indications of such evolution. Firms also evolve internally by restructuring, downsizing, and reorganizing as they seek to upgrade internal operations, streamline business processes, and boost their financial results. Functions like IT supporting the firm's operations must be flexible enough to respond smoothly to evolving corporate structures and changing business requirements.

Organizational changes and new business methods impact senior IT managers heavily. IT executives must use current information systems innovatively and adopt new computing and telecommunication systems and products to facilitate organizational transitions and deal with competitive threats. Deeply concerned about competition, corporate executives must consider computer and telecommunication systems vital elements in their competitive struggles. To succeed in today's dynamic environment, business managers must understand technology and develop visions of how they can use it to improve competitiveness. CIOs are primarily responsible for making this happen.

Information technology advances continually create new opportunities for alert CIOs and new pitfalls for complacent ones. Today, advanced technology leads to product development, marketing, sales, and service innovations. IT organizations must support cost-effective technological innovations so their firms can maintain future competitiveness. IT executives become change agents when firms explore technological menus of opportunity, make selections, and introduce new technology.

Changing and improving the business is obligatory for most firms in nearly all industries today. Fundamental business forces compel CIOs and IT organizations to find new and better ways to operate. Competitive pressures demand improved performance. Most corporations mandate CIOs to reexamine the firm's operations, searching for improved productivity. CIOs should strive to reduce cost, improve quality, or enhance responsiveness through reengineering and technology applications. Many top executives expect information technology to add substantial value to the firm's output, thus improving its financial performance.

Some employees, however, may respond unfavorably to changes needed to maintain the firm's health and viability when responding to competitive threats. Employees perceive IT managers who encourage and facilitate changes accordingly. In addition, using present technology innovatively and adopting advanced technology may significantly alter IT organizations. Some IT professionals may resist these changes, and others may resent them.

THE CHIEF INFORMATION OFFICER'S ROLE

Formulating Information Policy

CIOs are the firm's senior executive responsible for formulating and obtaining approval of information policies. They help the firm's executives establish organizational objectives for using information technology and monitor progress toward achieving them. These duties include developing and approving IT strategies, approving IT resource allocation, establishing IT cost accounting methods, developing policy instructions for IT procurement, overseeing outsourcing contracts, ensuring quality IT personnel hiring and training policies, and establishing standards for data security, disaster recovery, and IT business controls.[19] When the firm's executives approve new IT policy instructions, CIOs ensure that employees throughout the organization understand and follow them. In addition to policy development, CIOs establish the firm's information processing standards and ensure that all systems acquisition and operation activities adhere to those standards. Policies and standards are essential because they establish rules governing acquisition and use of IT assets to meet the firm's strategic and operational goals and objectives. Well-formulated policies and standards greatly increase opportunities for positive returns on IT assets. Chief information officers are responsible for recommending these policies and standards and for ensuring they are followed.

CIOs' most critical task is to ensure that IT and business strategies and plans are tightly coupled and approved by the firm's senior management team. Chapters 3 and 4 provide guidance in how to accomplish this. After formulation of a basis for congruent strategy and plan development, CIOs must ensure its effective operation and IT's role as a working partner in the firm's major plans.[20] Building sound relationships for achieving this goal is CIOs' top priority, particularly recently appointed CIOs.

CIOs must ensure the financial integrity of business investments in systems and technology, whether in corporate systems or local development and operations. Company policy must establish this authority. CIOs must evaluate, review, and approve major hardware, telecommunications, and application software investments. Using techniques like those explained in Chapter 16, CIOs must analyze costs, compare them to those of outside providers (benchmarking), and achieve agreement on expenditures.

CIOs must ensure cost-effective IT operations. They must ensure sound procurement decisions, improve productivity in local development, extend hardware and software life, and increase operational efficiency. CEOs expect improvements in productivity, performance, schedules, and quality. CIOs should keep expectations within achievable ranges and then help achieve them.

CIOs must promote and enforce business control, security, and asset protection policies and standards. CIOs must establish guidelines for problem, change, and recovery activities in all telecommunications and computing areas. Preserving the firm's hardware, software, and data assets must be high priority for CIOs.

Evaluating Technology Futures

CIOs are responsible for providing technological leadership to the firm for information handling. Firms' executives depend on CIOs to forecast technology trends and assess their significance to the firm. CIOs must assemble forecasts and develop assessments useful to the firm's senior officers in strategic planning. Because information technology advances so rapidly, evaluating technology futures is difficult; however, the importance of information technology makes technology evaluation critical.

Evaluation is difficult because technology advances rapidly, hype and fads abound, and potential changes tend to be large. For example, analysts predict an increase in computer processing power of two orders of magnitude over the next two decades without cost increases. Likewise, enormous bandwidth increases occur as utility companies, railroads, pipelines, and traditional communication companies install fiber-optic cables at breakneck speed. These trends, together with dozens of others like new application development methodologies, advanced Internet applications, and vast storage increases, make technology forecasting both difficult and risky.

Nevertheless, technology forecasting and evaluation is critical for the organization. CIOs must make sure that those who need the information understand technology and technology trends. In particular, they must apprise executives responsible for strategic decision making of technology futures. Strategic decision making attempts to shape or define the future, including future deployment of high technology. Technology innovation's increased pace and industry's strong propensity to adopt it make technology forecasting and evaluation increasingly critical.

CIOs can obtain expert advice on technology trends from many sources, but most offer vague or incomplete information about using advanced technology. Commonly available forecasts of exponential increases in circuit density per chip, for example, or in bits stored per cubic centimeter offer little help in forecasting these capabilities' usefulness to individual businesses. In fact, the chief difficulty in evaluating future technology is the abundance of promising new advances. Andrew Grove puts it best, "I have a rule, one that was honed by more than thirty years in high tech. It is simple. What can be done, will be done."[21] According to Grove, technology is like a natural force, impossible to stop.

Although Grove's rule makes eliminating any future technology difficult, forecasting its uses involves much more than just extrapolating past trends forward. Because new technology creates new uses, extrapolating present use is of limited utility. The Internet provides a good example. Ten years ago, foreseeing Web technology's tremendous value would have been difficult. Even today, predictions for 10 years hence are likely to miss widely, failing to foretell important developments. Nevertheless, predicting intranet's and extranet's capabilities is important. Setting difficulties aside, business managers and technologists must evaluate emerging technologies and attempt to forecast innovative uses for existing and developing technologies.

In the late 1990s, some trends will become more important than others because of the large sums devoted to their development. Most significant is the global information infrastructure's explosive development, in which personal workstations increasingly are less computing machines than communication devices. Corporations connect thousands of servers to the Internet monthly, and this trend is just the beginning. During the next five years, telecommunication companies plan to spend upwards of $100 billion on infrastructure in the U.S. and billions more linking the U.S. to the rest of the world. Many organizations plan to use the emerging information infrastructure to provide global access for their services.[22] Business executives here and abroad plan to exploit these trends for improved national and global business effectiveness.

Most experts believe that technologies supporting data transfer and personal interactions will be highly important. Voluminous information from many sources will be easily available to the firm's knowledge workers. Communication technology and appropriate software (groupware and Web technology) remove barriers of time and distance and allow cooperative work groups to form, perform, and disband readily and rapidly. For the next several years IT spending on communication infrastructures, Internet and online systems, and EDI applications will take top priority.[23] Improved human-computer interfaces and expanded communication capability form the technological foundations for Drucker's information-based organization.

CIOs and executive managers must evaluate technology directions, select appropriate technology and products, develop strategies, and establish plans to adopt them to improve business processes and operations. Developing strategies that improve firms' economic contribution is critical. So is avoiding the temptation to spend energy and resources chasing each new technology development.

Introducing New Technology

Organizations' adoption and acceptance of new ideas results from many individual decisions to use the innovation or adopt the new product. Individuals become adopters through communication-based processes including 1) innovation awareness, 2) becoming interested and seeking information, 3) evaluation considering needs, 4) trying the product or innovation, and 5) adopting the producter innovation if conditions are favorable. First described by Everett Rogers, this process in which new ideas permeate an organization is called *innovation diffusion*.[24]

Individuals' willingness to accept new ideas or innovations differs significantly. Those few individuals eager to embrace new ideas become champions of innovations or *pioneers*. A somewhat larger group, *early adopters,* accepts innovations readily. For successful innovations, adoption processes continue until the majority accepts the innovation. The last individuals to adopt, those most resistant to new ideas, are called *laggards*. Table 19.1 shows adoption propensity and the percentage of individuals typically in each class.

TABLE 19.1   Innovation Adoption Processes

Category of Adopters	Percent in Class	Cumulative Percent
Pioneers	2.5	2.5
Early adopters	13.5	16.0
Early majority	34.0	50.0
Late majority	34.0	84.0
Laggards	16.0	100.0

CIOs must understand individual adoption differences because they are fundamental to successful technology introduction. When introducing new ideas, perceptive managers search for pioneers and early adopters who are opinion leaders. Opinion leaders effectively establish awareness and encourage others' interest. They influence the majority to adopt the innovation.

Pioneers in one area are not necessarily pioneers in others. Some managers, for example, readily adopt and implement new organizational structures but resist changes in personnel policies. Some managers readily accept product strategy changes but keep personal computers out of the executive suite. Skillful CIOs understand that all employees, from top executives to nonmanagerial employees, have adoption propensities. Wise CIOs acknowledge individual differences and use their knowledge to advantage when implementing organizational changes, new business methods, and other innovations.

CIOs varying propensities to accept change enhance or inhibit their effectiveness. Surrounded by a sea of change, CIOs must remain flexible and adaptable to lead effectively through turbulent alliances, restructuring, and new business processes. Although CIOs must exhibit caution and recognize risk, they must not be laggards in today's rapidly evolving environment.

FACILITATING ORGANIZATIONAL CHANGE

To streamline business processes and improve efficiency and productivity, managers frequently reorganize and introduce new information technology. Likewise, adopting and using new information technologies brings steady, unrelenting change to the enterprise and the IT organization. For example, microcomputers, Web technology, and networked systems alter work patterns, improve communication, streamline organizational processes, and improve competitive positions. Indeed, IT managers justly deserve their reputation as change agents.

Most modern firms operate large numbers of individual workstations, exploit some form of intranet communication, and support end-user computing widely. Most use employee workstations for database query, statistical analysis, decision support, and many other applications. Properly implemented systems and new applications from word processing to executive support tools improve productivity. New computer applications enable employees and managers to become more effective and efficient.

Work once performed on large centralized computers moves even closer to its using departments by linking more powerful personal workstations to larger systems. Downsizing mainframe-based systems and installing new network applications on client/server systems is an important current trend in business and industry.

Downsizing presents many problems for IT professionals and managers, particularly CIOs. Strong, but frequently overlooked, financial returns from new ways of doing business are sometimes mostly illusion. Two-year studies of top information technology users reveal that 59 percent of the most effective technology users continue to operate centralized IS functions, whereas bottom-ranking firms were largely decentralized.[25] Although many firms claim decentralization improves productivity, solid evidence supporting their claims is scarce.

Earlier sections noted that decentralized computing increases total costs, largely because hidden personnel costs increase greatly. Although employees usually favor new tools and organizations claim improved responsiveness, many studies reveal little if any improved business health. Obviously, decentralization poses significant challenges for managers at all levels. In a decentralized environment, CIOs have special responsibilities to ensure that reengineering with new IT capabilities translates into measurable business results.

Some firms today tend to accept benefits from technology as fact, although they are mostly intangible, imprecise, and unmeasurable. For example, a recent survey by Meta Group found that most assumed benefits would follow directly from intranet investments and other such projects.[26] Although installing systems based on directives from the CEO is one thing, assuming that flashy new technology will benefit the firm is quite another. CIOs who make such assumptions without performing benchmark studies, prototype installations, or rigorous analysis operate recklessly. New technology costs are real and tangible; if benefits don't translate into financial returns, how can one assume they generate any?

Today, capitalizing on new system approaches and improving the firm's competitiveness requires new IT structures. But, structure also relates to management style. Centralized companies with more autocratic management styles tend to adopt conservative competitive strategies—more aggressive companies tend to be less centralized. Highly competitive, high-performance companies need flexible structures, especially in such critical areas as IT. Thus, CIOs and IT organizations must remain flexible enough to respond quickly to changing business conditions.

IT leadership and expertise is critical when rebalancing workload between centralized and distributed operations. After decentralization, IT professionals are needed to preserve centralization's good features like data security, operational integrity, and business controls. Assuming that downsizing yields people savings in IS may be valid, but people reductions in IS may coincide with increases in user departments.

CIOs must recognize that some IT professionals may resent plans to downsize or adopt new distributed systems because they believe their mainframe experience will become unusable. They may reluctantly abandon their COBOL skills for new languages—some may leave the organization. Others relish the opportunity to acquire new skills and work in a changing organization. They know firms need their application skills and organizational knowledge, although, applications that reside on LAN-based workstations and intranets are growing profusely. They also know that their knowledge of database operations, application controls, and business procedures remains valid despite changing environments. For many reasons, CIOs must remain alert to numerous people issues when restructuring information processing.

Acquiring or building distributed systems, moving applications from mainframes to LAN-based systems, and building intranets and extranets on mainframe enterprise servers requires outstanding planning and implementation skills like those Caterpillar exemplifies in the Business Vignette. Its corporate IS group established companywide rules, standards, and policies defining an information architecture. Caterpillar also defined vendor specifications so vendors' equipment could be integrated into its established factory automation architecture. Cat's planned evolution avoids isolated, nonintegrated systems and databases and forms an architectural base for future growth and development.

Decentralization can sometimes be carried too far; sometimes, centralization is the more appropriate strategy. In Cambridge, Massachusetts, separate data processing systems in 50 city agencies combined in a centralized system yielded an 11 percent reduction in the data processing budget over three years. With annual savings exceeding $800,000, the payback period was less than two years.

Sometimes firms adopt strategies to centralize. When companies centralize their corporate structure, IT organizations usually follow the corporate strategy. Trailer Train Company of Chicago decided to consolidate divisions, once wholly-owned subsidiaries, and centralize corporate structure. The MIS organization replaced small distributed computers in the divisions with large central mainframe support and also reduced costs.

Other reasons to centralize exist: some firms experience security or control problems; others find functional areas preferred not to manage complex computer operations. Some simplify their operations by returning applications to centralized facilities and data-processing professionals. In other cases, the decision does not hinge directly on out-of-pocket costs.

Top executives must consider many factors in striking a balance between centralized and decentralized processing. Although technology is available, critical business factors usually drive decentralizing decisions. These factors include improved customer response, higher quality service, and lower costs for the firm and its customers. To strike the proper balance, CIOs must carefully blend business, organizational, and technical skills.

"Most companies run their data-processing operations the wrong way," according to John Singleton, President, Security Pacific Automation Co. "DP is not about being centralized or decentralized. It's more that all managers have to be business executives first and technologists second."[27] The Chapter 12 Business Vignette describes John Singleton's approach to IT business management and shows him to be an extraordinarily successful executive.

Finding Better Ways to Do Business

Maintaining the status quo can be fatal to the CIO's position and career. The old adage "if it ain't broke, don't fix it" is definitely invalid for CIOs and their organizations. CIOs must constantly seek to improve the organization's productivity and create or sustain business advantage. They must control costs and risks, secure returns on IT investments, and find innovative ways to improve organizational effectiveness.

You are already familiar with many means for effective CIOs to reduce costs and improve performance. Among them are using cost-effective hardware, adopting alternative application acquisition methods, downsizing and reorganizing IT operations, and using effective disciplines to manage production operations and networks.

However, as competition intensifies, businesses strive to reach ever higher levels of effectiveness. Marketing sharpens its skills and refines its messages for specific segments. Customers demand unblemished product quality, reduced product life-cycle costs, and individualized service support. Many firms find that Internet technology supporting restructured processes helps satisfy these demands.

Because Internet technology is important and so potentially valuable, business enterprises spend large sums on hardware, software, and networks supporting e-mail, intranets, extranets, and Web technology. Many firms believe Internet technology will deliver large future benefits by greatly increasing communications within and between the firm and its suppliers, partners, and customers. Many CIOs see these new communication capabilities as an unprecedented opportunity to use information technology for competitive advantage—an opportunity not to be missed.

Outsourcing

In addition to internal reorganizations, many firms believe that subcontracting some IT operations to third parties is cost effective. Although service bureaus and subcontract application development have always been available, outsourcing—moving people and work segments to specialty vendors—claimed the spotlight when Kodak and Merrill Lynch subcontracted major operations. Kodak negotiated with IBM to manage Kodak's data center operation, and Merrill Lynch gave its network management to MCI and IBM. One of many examples, EDS, partly owned by General Motors, performs most of GM's processing. "Within the next twenty years, you will see over half of the information technology assets moving into the hands of commercially run service utilities," claims Paul Strassmann, who sees outsourcing trends accelerating.[28]

Reasons to outsource vary, but strategic, technical, financial, or scale considerations motivate most firms. For example, a manufacturing firm, determined to concentrate on its core strengths, may find network management a distraction, needing critical, but unavailable skills. By outsourcing its network activity, the manufacturer avoids investing resources in complex, rapidly evolving telecommunication systems. Similarly, many firms prefer to concentrate on their core strengths and buy IT capability from others.

Firms currently using service bureaus for payroll and related financial applications find that the outsourcers' economies of scale provide cost savings. Seeking to capitalize on this finding, they discover additional applications to offload. By moving additional work outside, firms can sometimes reduce even further the costs of work already outsourced through volume discounts. Cost reductions in outsourcing decisions are important motivators, but outsourcing economics remain controversial; each case must be decided on its merits.[29]

By performing data processing for many firms, outsourcers do achieve economies of scale. By combining workloads, they keep large, efficient processors fully loaded. They use people and supporting functions efficiently too. For instance, operating large production facilities, they can afford to hire talented system-support programmers and network specialists to support their systems. Large operations fully utilize these highly skilled people. Smaller firms may not, so outsourcing offers skill and cost advantages.

Whether outsourcing grows in the future as predicted depends on more than economics. Other, major non-economic advantages and disadvantages play significant parts in outsourcing decisions.

Outsourcing centralized mainframe operations permits IT organizations to concentrate on other important issues like strategic applications, client/servers, or intranet development instead of computer center processes. IT can concentrate on user support and business objectives, rather than hardware, systems software, and associated personnel. Even as hardware costs decline, acquisition, training, and retraining costs for hardware, operating system, and network technicians rises rapidly. More importantly, finding and retaining people with the needed technical skills is increasingly difficult for small operators.

Outsourcing has its critics. Although some consider computer hardware and networks mundane, those who prefer retaining firm control consider them vital assets. Subcontracting data center and network operations to others, however capable, raises long-term issues like operational quality, strategic direction, and control of critical production activities, critics claim. Can the vendor provide long-term, high-quality service? Can the vendor cooperate with the firm on problems, changes, recovery, and other operational issues? Because reversing outsourcing decisions is difficult, firms may find themselves locked into computer architectures and management systems ill-suited to their long-term goals and objectives.

Nevertheless, CIOs must consider outsourcing alternatives for parts of their operations. In most cases, financial issues are not the primary consideration; however, finances are always important. Although outsourcing decisions are usually long term and strategic, they probably depend heavily on subjective factors like the corporate culture, IT competence, and industry conditions in addition to strictly financial considerations.

Professionally-managed information utilities gain popularity because, in many cases, local IT craftsmen fall behind IT utility professionals in cost and risk containment. Computer or information utilities have grown slowly during the past 20 years, rapidly gained importance recently, and promise to increase in importance in the next decade. CIOs and senior executives must prepare themselves and their organizations to consider and capitalize on outsourcing if appropriate.

SUCCESSFUL CIOS ARE GENERAL MANAGERS

Successful CIOs demonstrate skills commensurate with the general management positions they hold. Additionally, skills they learned as IT general managers help them advance to more responsible positions beyond IT. Katherine Hudson is CEO of an international manufacturing company. Her career is an example of these ideas.

A Business Vignette

Exceptional Executive Advances

Katherine Hudson joined Kodak in 1969 and held various positions in its legal, public affairs, finance, and investor relations departments. In 1987, she was promoted to vice president and named head of the newly formed corporate IS department overseeing a $500 million budget and more than 3000 employees.[30] In this position, she reported directly to Kodak's president.

As director of corporate IS, Ms. Hudson strived to implement productive new technology in operating departments. Under her direction, Kodak installed local networks in its plants, reducing inventory by 90 percent and improving on-time delivery performance to 98 percent. Ms. Hudson also established Centers of Excellence in the business units that worked with senior IS managers to identify and understand technology trends useful to Kodak. She considers IS alignment and technology forecasting to be the most critical aspects of her job.

In a pioneering effort, she upset the status quo by outsourcing significant portions of Kodak's data-processing operations to IBM, Digital Equipment Corp., and Businessland, Inc. This bold, successful endeavor won praise both inside and outside Kodak. She gained more responsibility when she was named to head Kodak's professional printing and publishing imaging division, a position she left in late 1993.

Ms. Hudson is now president and chief executive of W. H. Brady company, a Milwaukee firm manufacturing more than 20,000 different industrial products. She believes that insight she gained into multifunctional operations through her IS experience at Kodak is valuable in her new role. Ms. Hudson is truly an exceptional executive.

WHAT CIOS MUST DO FOR SUCCESS

CIOs critically depend on many people throughout the organization. In particular, people at the top are extremely important to CIOs. CIOs must gain top executives' confidence by understanding the business from their vantage point and must identify with the business and be sensitive to its priorities. CIOs must be fully contributing members of the executive suite, providing leverage to senior executives through initiative, creativity, and vision. Their role includes executive education and salesmanship. CIOs must not assume immediate adoption of complete understanding of their designs.

CIOs must develop visions and strategies for the firm's use of information technology that add substantial, tangible value to the firm. Reducing costs and headcount, and avoiding expenses are excellent activities, but the firm's executives expect much more. They expect IT to contribute substantively to the firm's value chain. They expect CIOs to contribute to bottom-line results and share a vision for improving the firm's results that extends beyond IT activities; they expect the vision to attract advocates. Advocates and stakeholders must reside at all organizational levels, not just at the top.

CIOs are action oriented; they want to make things happen. They are impatient and intolerant of mediocrity, and they set high expectations for themselves and their organizations. At the same time, they have the patience to work with the current situation. They recognize that progress usually occurs in many small increments. They realize that effective executives continually strive to improve their operations across a broad front. They know that business success rarely comes in one grand stroke.

Corporate culture is a powerful force in nearly all organizations: the CIO must understand the culture and must work within it. Realistically, CIOs cannot single-handedly change firms' accounting procedures or corporate personnel policies. A conservative firm will probably not accept radical infusions of high technology. Firms expect CIOs to operate the IT function within the larger organization's norms. Therefore, CIOs should pay careful attention to the corporate culture.

CIOs must manage expectations within the firm regarding information technology. Unfulfilled expectations are a leading cause of failure for managers at all levels. Unfortunately, IT managers suffer from this difficulty more than most. The problem arises naturally because many communication channels to the firm's executives, managers, and employees inflate information technology benefits and understate its costs or implementation difficulties. Additionally, lack of discipline within the IT organization itself frequently creates unrealistic expectations.

CIOs depend on many people for success. Effective CIOs cultivate good human relations and encourage innovation. They withhold criticism for well-intentioned but flawed inventions, and freely praise accomplishments. Exceptional CIOs are unselfish. They believe that the amount of good they can do is almost unlimited—if they don't care who gets the credit.

Four kinds of people work in most firms: those who make things happen, those who prevent things from happening, those who watch things happen, and those who don't know or don't care what's happening. Astute CIOs have an ability to distinguish between them and an approach for coping with them. CIOs must make things happen and must develop support from those of similar persuasion. CIOs should ignore those who prefer the status quo, concentrating instead on motivating onlookers to get involved. For those who don't know or don't care what's happening, CIOs should prescribe education. With effort, most people buy into an idea that benefits the firm and, therefore, probably benefits them.

CIOs must understand the firm's position in its information technology use and its ability to assimilate new technology. CIOs must balance the availability of new technology with the firm's need for it and the firm's propensity to adopt it. Effective CIOs believe that the best place to work is where they are with what they have. They build on current capabilities. They understand the systems that run the firm, and they know the capabilities and limitations of the people who run the systems. Effective CIOs always try to improve the environment, but they use the present environment as the base for improvement within the culture.

SUMMARY

The chief information officer's position is precarious in both theory and practice. Derived from the accepted CFO title, the CIO position stands on shaky ground because, unlike money, information cannot be quantified or measured. Information is created, used, and discarded without the CIO's knowledge or approval. Therefore, identifying a firm's executive as chief information officer is somewhat misleading.

Nevertheless, regardless of the title, the senior IT executive's job is large and important. The nature of the work performed, however, makes the job hazardous. CIOs are expected to be the firm's technological leaders, providing business direction for new technology selection and introduction. Information technology's adoption and implementation frequently cause fundamental changes within organizations. Changes are difficult to manage; their success is sometimes doubtful, and CIOs are vulnerable when things go astray.

Chief information officers face many challenges as their firms alter strategic direction, reorganize, or adopt new business methods. CIOs must assess technology changes and provide guidance to the firm on technology adoption and introduction. They must facilitate internal organizational changes and find more effective ways of operating the firm and the IT organization. They must play a strong general management role.

CIOs depend on many people for success; they must be astute managers of people. CIOs are action-oriented and must surround themselves with similarly inclined individuals. To be effective, they must study the convictions of the firm's executives and also comply with the corporate culture.

Successful CIOs learn the practice of management. They learn to develop and manage IT management systems, and to manage themselves. People and technology are incredibly complex. After extensive study, considerable introspection, and prolonged experience, wise managers maintain a respectable level of humility.

Review Questions

1. Why do myth and confusion surround the CIO title? Does your university have a CIO?

2. Why is the senior IT executive in a position of power and responsibility? Distinguish between the CIO's line and staff responsibilities. Give examples of each type of responsibility.

3. Discuss the reasons why CIOs have difficult jobs. List six challenges for CIOs.

4. To whom do most CIOs report? What entrees to the executive management team besides direct reporting do CIOs have?

5. What career paths do most CIOs follow? Why are CIOs in dead end jobs in most firms? What, if anything, can CIOs do to improve this situation?

6. What backgrounds do most CIOs have? What background and experiences do most executives prefer in CIOs?

7. When are firms most likely to import a CIO from outside? What do they hope to accomplish by doing this?

8. Enumerate CIO performance measures. What can CIOs do to improve these measurements?

9. What changes occur within and outside firms to challenge the CIO?

10. What is the CIO's role in policy matters?

11. What responsibilities do CIOs have regarding technology? Why are these tasks difficult?

12. Why is technology forecasting risky, difficult, *and* important?

13. Why is forecasting technology use more difficult than forecasting its availability?

14. What technologies promise to be important in the near future?

15. Describe how the firm incorporates technology forecasting results into its business.

16. Describe innovation diffusion processes. How would you use these ideas if you wanted to install a new user-driven application?

17. What does outsourcing mean? What motivates firms to consider outsourcing?

18. What are the disadvantages of outsourcing? Which disadvantage is most significant in your opinion?

19. What must CIOs do to succeed? What role does corporate culture play in CIOs' success plans?

20. What philosophical difficulties surround the title of chief information officer?

Discussion Questions

1. Discuss conceptual difficulties with the CIO title.

2. Discuss the significance of Keen's comment that the CIO is a relationship, not a job.

3. Why do you think Bob McLendon of TI finds users distrust information-services organizations? Do you think this attitude is widespread? Why?

4. Discuss CIOs' line and staff responsibilities. How have these responsibilities changed over time? In large corporations, CIOs may have no line responsibility but considerable staff responsibility. What does this mean organizationally and for the CIO?

5. Where do individuals gain the experience necessary to become CIOs? Sketch the path that an MBA might take to become a CIO. What business experiences would you recommend for this individual?

6. Discuss the balance CIOs need between technical skills and general or line management skills. How would you recommend an aspiring CIO obtain these skills?

7. CIOs must respond to changes initiated by organizations, and they must take actions, which also create changes. Discuss the interplay between these two activities.

8. Why it is critically important for CIOs to engage in policy formulation and enforcement? Discuss the connection between policy formulation and the CIO's line and staff roles.

9. If you were CIO of a major firm, how would you obtain technology trend information? Describe the process you would use to ensure that the appropriate people evaluated trends.

10. How might astute CIOs use innovation diffusion theory when dealing with top executives? How would innovation diffusion apply during the development and installation of an executive information system? How might firms wanting to install EDI linking them to suppliers apply innovation diffusion theory?

11. Discuss the consequences of not using strict ROI calculations for IT projects. What risks does this involve for CIOs? How can they mitigate these risks?

12. Discuss advantages and disadvantages of outsourcing the firm's telecommunication system. Under what conditions would firms outsource their telecommunication system and not mainframe systems?

13. What considerations are important in deciding to outsource application development? What industry factors might influence this decision? What strategic considerations are important? What role does corporate culture play in these decisions?

14. Discuss the balance CIOs must achieve among technology availability, the firm's need for it, and the firm's adoption propensity. What actions can CIOs take to alter the balance among these variables?

15. Describe Robert Hinds' general management responsibilities.

Assignments

1. "Conclusion: Effectiveness Must Be Learned" is the title of the final chapter in Peter Drucker's book, *The Effective Executive*, 166-174. Read this chapter and write a summary for class.

2. Analyze IT factors contributing to Caterpillar's success as discussed in the Business Vignette. Categorize them as IT line responsibilities, IT staff responsibilities, or client department responsibilities. Compare Caterpillar's actions with those you learned to be valuable in this text. How would you describe the relationships and roles of CIO and CEO as compared with what you consider desirable?

3. Interview a senior information executive in your community. Determine that person's job responsibilities and what he/she must do for success. How does this person balance technical and business skills in performing the job? Compare and contrast this person's view of the CIO title and job with that developed in the Business Vignette.

[1] Doug Bartholomew, "How IT is Helping Caterpillar Fend off Japanese Competition and Keep Jobs in the U. S.," *Information Week*, June 7, 1993, 36.

[2] See note 1.

[3] See note 1.

[4] *The Value Line*, November 11, 1997, 1347.

[5] Speaking about executive effectiveness, Drucker said "effectiveness, in other words, is a habit; that is, a complex of practices. And practices can always be learned." Peter Drucker, *The Effective Executive* (New York: Harper & Row, Publishers Inc., 1986), 23.

[6] Gordon Bock, "Management's Newest Star," Business Week, Special Report, October 13, 1986, 160. I prefer the CIO title because it is widely understood and applied in writings and in practice.

[7] Jeffrey Rothfeder, "CIO is Starting to Stand for 'Career is Over'," *Business Week*, February 26, 1990, 78.

[8] See Paul Strassmann's articles, "The Price of Uncertain Leadership," *Computerworld*, November 10, 1997, 72, and "The Dangers of CIO Turnover," Computerworld, June 10, 1996.

[9] Peter Keen, *Every Manager's Guide to Information Technology* (Boston, MA: Harvard Business School Press, 1991), 55.

[10] "Information Chiefs Get Plaudits but Rarely Promotions," *The Wall Street Journal*, November 10, 1994, B1.

[11] For an informative discussion of outstanding CIOs, see "The CIO Hall of Fame," CIO, September 1997.

[12] For a complete discussion of CIO roles and responsibilities, see Paul Strassmann, Chapter 26, "Roles," and Chapter 27, "Charter for the CIO," *The Politics of Information Management* (New Canaan, CT: The Information Economics Press, 1995), 315-350.

[13] "Trends," *Computerworld*, March 5, 1990, 118.

[14] Dan Richman, "National's Steering Committee", *CIO*, June 1991, 74.

[15] Surprisingly, Strassmann found that firms are more likely to use IT effectively when the CIO does not report directly to the CEO. *The Squandered Computer* (New Canaan, CT: The Information Economics Press, 1997), 115-116.

[16] Charlotte S. Stephens, William N. Ledbetter, Amitava Mitra, and Nelson F. Ford, "Executive or Functional Manager," *MIS Quarterly*, December 1992, 449.

[17] Allan E. Alter, "Good News, Bad News...," *CIO*, January, 1990, 18.

[18] For example, according to one report, fifty-one percent of CFOs held their positions less than 36 months. D. McCorkindale, "Is Everybody Happy?" *CFO Magazine*, September 1996.

[19] See Paul Strassmann, Chapter 27, "Charter for the CIO," in *The Politics of Information Management* (New Canaan, CT: The Information Economics Press, 1995), 335-350.

[20] John C. Daily, "What It Takes To Be CIO," *Datamation*, November 1, 1995, 61.

[21] Andrew Grove, "What Can Be Done, Will Be Done," *Forbes ASAP*, December 2, 1996, 193.

[22] Fiber-optic cable capacity across the Atlantic and Pacific oceans will increase by more than a factor of eight from 1995 to 2000, according to Telegeography, Inc. "Telecommunications Survey," *The Economist*, September 13, 1997, 27.

[23] Source: Xplor International, Torrance, CA, *Computerworld*, December 1997, 33.

[24] Everett M. Rogers, *Diffusion of Innovations* (New York: Free Press, 1962), 79.

[25] Paul Strassmann, *The Squandered Computer* (New Canaan, CT: The Information Economics Press, 1997), 115-116.

[26] Joseph E. Maglitta, "Beyond ROI," *Computerworld*, October 27, 1997, 77.

[27] Deborah Cooper, "High Five," *CIO*, December 1988, 55.

[28] Paul Strassmann, *The Politics of Information Management* (New Canaan, CT: The Information Economics Press, 1995), 334. Outsourcing trends are international. In 1997, non-U.S. companies performed 40 percent of outsourcing activity, according to Dun & Bradstreet. *Computerworld*, March 30, 1998, 41.

[29] Paul Strassmann presents an excellent discussion of outsourcing in *The Squandered Computer* (New Canaan, CT: The Information Economics Press, 1997), Part V, "Outsourcing," 179-213.

[30] "Former Kodak IS director to Head Global Plastics Company," *Computerworld*, December 13, 1993, 10. Katherine Hudson was elected to the board of Apple Computer in 1994 but left when Steve Jobs returned.

Bibliography

PART ONE   FOUNDATIONS OF IT MANAGEMENT

Bower, Marvin. *The Will to Lead*. Boston, MA: Harvard Business School Press, 1997.

Boar, Bernard H. *The Art of Strategic Planning for Information Technology*. New York: John Wiley & Sons, 1993.

Campbell, Andrew, and Marcus Alexander. "What's Wrong with Strategy." *Harvard Business Review*, November-December, 1997.

Drucker, Peter F. "The Coming of the New Organization" *Harvard Business Review*, January-February, 1988, 45.

Earl, Michael J. "Experiences in Strategic Information Systems Planning." *MIS Quarterly*, No. 1, 1993, 1.

Gasman, Lawrence. *Telecompetition*: *The Free Market Road to the Information Highway*. Washington, DC: The Cato Institute, 1994.

Hills, Mellanie. *Intranet Business Strategies*. New York: John Wiley & Sons, Inc., 1996.

Hope, Jeremy, and Tony Hope. *Competing in the Third Wave*. Boston, MA: Harvard Business School Press, 1997.

Hopper, Max. "Rattling SABRE: New Ways to Compete on Information." *Harvard Business Review*, May-June, 1990, 118.

Jarvenpaa, S. L., and Blake Ives. "Executive Involvement and Participation in the Management of IT." *MIS Quarterly*, No. 2, 1991.

Kaplan, Robert S., and David P. Norton. *Balanced Scorecard*: *Translating Strategy into Action*. Boston, MA: Harvard Business School Press, 1996.

Koelsch, Frank. *The Information Revolution*: *How It Is Changing Our World and Your Life*. New York: McGraw-Hill Rynerson, 1995.

Maitra, Amit K. *Building a Corporate Internet Strategy*. New York: Van Nostrand Reinhold, 1996.

Montgomery, Cynthia A., and Michael E. Porter, eds. *Strategy*. Boston, MA: Harvard Business School Press, 1991.

Porter, Michael E. *Competitive Advantage*. New York: Free Press, 1985.

Strassmann, Paul A. *The Politics of Information Management*. New Canaan, CT: The Information Economics Press, 1995.

Sullivan, Cornelius H. "The Changing Approach to Systems Planning." *Journal of Information Systems Management*, Summer, 1988, 8.

Tappscott, Don. *The Digital Economy*. New York: McGraw-Hill, 1996.

Wiseman, Charles. *Strategic Information Systems*. Homewood, IL: Irwin, 1988.

Wriston, Walter B. *The Twilight of Sovereignty*. New York: Charles Scribner's Sons, 1992.

Zuboff, Shoshana. *In the Age Of the Smart Machine*. New York: Basic Books, Inc., 1988.

PART TWO   TECHNOLOGY AND INDUSTRY TRENDS

Auletta, Ken. *The Highwaymen*: *Warriors of the Information Superhighway*. New York: Random House, 1997.

Cairncross, Frances. *The Death of Distance*. Cambridge, MA: The Harvard Business School Press, 1997.

Clarke, Arthur C. *How The World Was One: Beyond the Global Village*. London: Victor Gollancz LTD, 1992.

Hudson, Heather. *Global Connections*. New York: Van Nostrand Reinhold, 1997.

Hyman, Leonard S., Edward di Napoli, and Richard Toole. *The New Telecommunications Industry: Meeting the Competition*. Arlington, VA: Public Utilities Reports, Inc., 1997.

Maney, Kevin. *Megamedia Shakeout*. New York: John Wiley & Sons, Inc., 1995.

Martin, Chuck. *The Digital Estate*. New York: McGraw-Hill, 1997.

Morley, John C., and Stan S. Gelber. *The Emerging Digital Future*: *An Overview of Broadband and Multimedia Networks*. Boston, MA: Course Technology, 1996.

Silberschatz, Abraham, and Peter Baer Galvin. *Operating Systems Concepts*. Reading, MA: Addison-Wesley Publishing Company, 1997.

"Premier 100, The Productivity Payoff." *Computerworld*, Special Issue, September 19, 1994, 4-55.

Shelly, Gary B., Thomas J. Cashman, and Judy Hill. *Business Data Communications*: *Introductory Concepts and Techniques*, 2d ed. Boston, MA: Course Technology, 1998.

Stallings, William. *Data and Computer Communications*, 5th ed. Englewood Cliffs, NJ: Prentice-Hall, Inc., 1996.

Stamper, David A. *Business Data Communications*, 4th ed. Redwood City, CA: The Benjamin/Cummings Publishing Company, Inc., 1996.

"The Computer in the 21st Century." *Scientific American*, Special Issue, Vol. 6, No. 1, 1995.

"The Global 100, Outstanding Users of Information Technology From Around the World." *Computerworld*, Special Issue, May 1, 1995, 6-64.

Terplan, Kornel. *Communications Network Management*, 2nd ed. Englewood Cliffs, NJ: Prentice Hall, Inc., 1991.

White, Curt M. *Data Communication and Computer Networks*: *An OSI Framework*. Boston, MA: Course Technology, Inc., 1996.

PART THREE   MANAGING APPLICATION PORTFOLIO RESOURCES

Bacon, C. James. "The Use of Decision Criteria in Selecting Information Systems/Technology Investments," *MIS Quarterly*, No. 3, 1992, 335.

Benjamin, Robert I., and Jon Blunt. "Critical IT Issues: The Next Ten Years. *Sloan Management Review*, Summer 1992, 7.

Bernard, Ryan. *The Corporate Intranet*. New York: John Wiley & Sons, Inc., 1996.

Brooks, Frederick P. *The Mythical Man Month—Essays On Software Engineering*. Reading, MA: Addison-Wesley, 1974.

Burch, John G. *Systems Analysis, Design, and Implementation*. Danvers, MA: boyd & fraser publishing company, 1991.

Clemons, Eric K. "Evaluation of Strategic Investments in Information Technology." *Communications Of the ACM*, January 1991.

Davenport, Thomas H., and James E. Short. "The New Industrial Engineering: Information Technology and Business Process Redesign." *Sloan Management Review*, Summer 1990.

Davenport, Thomas H. *Process Innovation*. Boston, MA: Harvard Business School Press, 1993.

Humphrey, Watts. *Managing the Software Process*. Reading, MA: Addison-Wesley Publishing Company, 1993.

"IS Shops Form Alliances as Development Costs Rise," *Datamation*, September 1, 1988, 19.

Keen, Peter G.W. *Shaping the Future*: *Business Design Through Information Technology*. Boston, MA: Harvard Business School Press, 1991.

Linthicum, David. *Guide To Client/Server And Intranet Development*. New York: John Wiley & Sons, Inc., 1997.

Martin, James, and Carma McClure. *Structured Techniques*: *The Basis For CASE*. Englewood Cliffs, NJ: Prentice-Hall, Inc., 1988.

Swider, Gaile A. "Ten Pitfalls of Information Center Implementation." *Journal Of Information Systems Management*, Winter, 1988, 22.

Verity, John, and Evan I. Schwartz. "Software Made Simple." *Business Week*, September 30, 1991.

Weinberg, Randy S. "Prototyping and the Systems Development Life Cycle," *Journal of Information Systems Management*, 1991.

Wojtkowski, W. Gregory, and Wita Wojtkowski. *Applications Software: Programming with Fourth-Generation Languages*. Boston, MA: boyd & fraser publishing company, 1990.

Yourdon, Edward. *Modern Structured Analysis*. Englewood Cliffs NJ: Yourdon Press, 1991.

PART FOUR   TACTICAL AND OPERATIONAL CONSIDERATIONS

Borovits, Israel. *Management of Computer Operations*. Englewood Cliffs, NJ: Prentice-Hall, 1984.

FitzGerald, Jerry. *Business Data Communications*, 3rd Ed. New York: John Wiley & Sons, 1990.

Jenkins, George. *Information Systems Policy and Procedures Manual*. Englewood Cliffs, NJ: Prentice-Hall, 1997.

Loyd-Evans, Robert. *Wide Area Networks*: *Performance and Optimization*. Reading, MA: Addison-Wesley Publishing Co., 1996.

Nemzow, Martin. *Enterprise Network Performance Optimization*. New York: McGraw-Hill, 1995.

Shafe, Laurence. *Client/Server*: *A Managers Guide*. Reading, MA: Addison-Wesley Publishing Company, 1995.

Singleton, John P., Ephraim R. McLean, and Edward N. Altman. "Measuring Information Systems Performance: Experience with the Management by Results System at Security Pacific Bank," *MIS Quarterly*, June, 1988, 325.

Stallings, William. *Business Data Communications*. New York: Macmillan Publishing Company, 1990.

Toigo, Jon William. *Disaster Recovery Planning: Managing Risk & Catastrophe in Information Systems*. Englewood Cliffs, NJ: Prentice Hall, 1989.

Van Schaik, Edward A. *A Management System for the Information Business: Organizational Analysis*. Englewood Cliffs, NJ: Prentice-Hall, Inc. 1985.

PART FIVE   CONTROLLING INFORMATION RESOURCES

Bysinger, Bill, and Ken Knight. *Investing In Information Technology*. New York: Van Nostrand Reinhold, 1996.

Deardon, John. "Measuring Profit Center Managers." *Harvard Business Review*, September-October, 1987, 84.

FitzGerald, Jerry, and Ardra FitzGerald. *Fundamentals of Systems Analysis*, 3d ed. New York: John Wiley & Sons, 1987. See Chapter 9, "Designing New System Controls," and Appendix 2, "Controls for Inputs, Data Communications, Programming, and Outputs," A15-A55.

Hoyt, Douglas B., ed. *Computer Security Handbook*. New York: John Wiley & Sons, 1995.

Keen, Peter G.W. *The Process Edge*. Boston, MA: Harvard Business School Press, 1997.

Mason, Richard O. "Four Ethical Issues of the Information Age." *MIS Quarterly*, January, 1986.

Roach, Stephen S. "Services Under Siege." *Harvard Business* Review, September-October 1991, 82.

Strassmann, Paul. *The Squandered Computer*. New Cannan, CT: The Information Economics Press, 1997

PART SIX   PREPARING FOR IT ADVANCES

Brancheau, James C., and James C. Wetherbe. "Understanding Innovation Diffusion Helps Boost Acceptance Rates of New Technology," *Chief Information Officer Journal*, Fall 1989.

Clermont, Paul. "Outsourcing without Guilt," *Computerworld*, September 9, 1991.

Christensen, Clayton M. *The Innovator's Dilemma: When New Technologies Cause Great Firms to Fail*. Cambridge, MA: Harvard Business School Press, 1997.

Emory, James. "What Role for the CIO?" *MIS Quarterly*, No. 2, 1991, vii.

Feeny, David F., Brian R. Edwards, and Keppel M. Simpson. "Understanding the CEO/CIO Relationship," *MIS Quarterly*, No. 4, 1992, 435.

Ives, Blake. "Transformed IS Management," *MIS Quarterly*, No. 4, 1992, iix.

Keen, Peter G. W., Walid Mongayuer, and Tracy Torregrossa. *The Business Internet and Intranets: A Manager's Guide to Key Terms and Concepts*. Boston, MA: Harvard Business School Press, 1998.

Meyer, Christopher, and Stan Davis. *Blur: The Speed of Change in the Connected Economy*. Reading, MA: Addison-Wesley, 1998.

Pensouneault, Alain, and Kenneth Kraemer. "The Impact of IT on Middle Managers." *MIS Quarterly*, No. 3, 1993, 271.

"The Big Issue: Where Do We Go From Here?" *Forbes ASAP*, December 2, 1996.

Walton, Richard E. *Up And Running: Integrating Information Technology And the Organization*. Boston, MA: Harvard Business School Press, 1989.

OTHER REFERENCES AND READINGS

Brynjolfsson, Erik, and Loren Hitt. "The Big Payoff from Computers." *Fortune*, March 7, 1994, 28.

Davis, Stan, and Bill Davidson. *2020 Vision*. New York: Simon & Schuster, 1992.

Dertouzos, Michael. *What Will Be: How the New World of Information Will Change Our Lives*. San Francisco: HarperEdge, 1997.

Fried, Louis. *Managing Information Technology in Turbulent Times*. New York: John Wiley & Sons, Inc., 1995.

Gates, William H., III. *The Road Ahead*. New York: Viking Penguin, 1995.

Grove, Andrew S. *Only the Paranoid Survive*. New York: Doubleday, 1996.

Gilder, George. *Microcosm, The Quantum Revolution in Economics and Technology*. New York: Simon and Schuster, 1989.

Hammer, Michael, and James Champy. *Reengineering the Corporation*. New York: Harper Collins, 1993.

McKinnon, Sharon M., and William J. Bruns, Jr. *The Information Mosaic*. Boston: Harvard Business School Press, 1992.

Negroponte, Nicholas. *Ones And Zeros*. New York: Knopf, 1995.

Rocklin, Gene. *Trapped in the Net: The Unanticipated Consequences of Computerization*. Princeton, NJ: Princeton University Press, 1997.

Stoll, Clifford. *Silicon Snake Oil*: *Second Thoughts on the Information Highway*. New York: Doubleday, 1995.

Stout, Rick. *The World Wide Web*: *Complete Reference*. Berkeley, CA: Osborne McGraw-Hill, 1996.

Wang, Charles B. *Techno Vision*. New York: McGraw-Hill, Inc. 1994.

Womack, James P., Daniel T. Jones, and Daniel Roos. *The Machine That Changed the World*. New York: Rawson Associates, 1990.

Wriston, Walter B. *The Twilight Of Sovereignty*. New York: Charles Scribner's Sons, 1991.

Index

---

response time, 371, 372
  as measure of online activities, 325
  measuring, 327–328
  network service-level agreement requirements, 394
  online transaction processing, 376
  workload volume and, 393–394
restructuring
  networks and, 389–390
return on investment (ROI), 239, 415, 416, 434–435
return on investment in information technology (ROIT), 415
review. *See* phase review process
Rhapsody, 112, 128, 129
risk
  absolute levels, 250
  aversion, 208
  management, 207–208, 355
  reduction, 250–251
  sources of, 247–250
  strategy statement and, 73–74
risk analysis
  change management and, 346
  defined, 247
  determining trends, 250
  phase review process, 241, 245–250
  subcontract development and, 274
Rmon2 (Remote Monitoring 2), 402
roaming, 170
Rockart, John F., 21, 98
Rockwell International Corporation, 6
Rogers, Everett, 508
ROI (return on investment), 415, 416
ROIT (return on investment in information technology), 415
routers, 126, 162
Royal Dutch Shell, 474
RPG, 264

S

Sabre system, 40–41, 51, 53, 54, 347, 502
sampling, 146
Samsung, 179
San Diego Gas & Electric, 472
SAS, 264
satellites
  capacity, 403
  cellular networks, 171–172, 194–195
  low-earth-orbit (LEO), 171, 194–195
satisfaction
  analysis, 219–221
  measurements, 327–328
  user surveys, 328
SBC Communications, 177, 187, 188, 190, 193, 473
schedule
  management, for early computers, 128
  in service-level agreements, 324
scheduling programs, 366
Scientific American, 133
Scott-Morton, Michael S., 473
Searle, G. D., 386
secondary storage controls, 452–453
second-generation programming, 263
secret clearance, 446
security
  active threats, 456
  of client/server systems, 456
  controls, 453

of data, 457–459
data encryption, 456–457
of load modules, 461
network controls and, 456–459
passive threats, 456
of physical facilities, 456
of program source, 461
of test data, 461
user identification and verification, 457
Security-General, 389
"Security of Federal Automated Information Systems," 348
Security Pacific Automation Co., 279, 315–316, 316, 511
self-actualization, 479
Semi-Automated Business Research Environment. *See* Sabre system
semiconductor technology, 114–119
  chip density, 114–115
  chip design and manufacturing, 117–118
  chip lines, 116
  chip performance, 116–118
  consumption of, 179
  suppliers, 179
  trends, 114–119
senior managers. *See also* chief executive officer (CEO), chief information officer (CIO)
  responsibility for IT strategy development, 63
  strategic issues for, 35–37
sensitive programs, 460–461
separation of duties, 446
servers
  in client/server systems, 124–127
  for distributed computing, 287–288
  failure, 318
  hardware suppliers, 180
  Web, 127
service bureaus, 279, 280
service criteria, 320
service-level agreements (SLAs), 320–322
  administrative information, 323
  availability, 324, 325
  capacity management, 378
  contents, 322–328
  document, 321
  goals of, 329–330
  Management by Results system and, 315–316
  network management and, 392, 393–395
  objectives, 364, 365
  planning, 94–95
  production operations, 488–489
  satisfaction measurements, 327–328
  schedule, 324
  service specifications, 323–324
  systems management and, 379
  timing, 324–325
  value of, 318–319, 320
  workload forecasts, 326–327
service-level attainment, 94–95, 371
service-level negotiations
  capacity management and, 378
  process, 321
service providers
  contingency planning responsibilities, 350
  satisfaction measurements, 327–328
  service-level negotiations, 321
service representatives
  to client organizations, 492–493

strategic value of, 49–51
telephone system, 139–152
third-party, as risk source, 248
value of, 403
wireless systems, 169–172
Tele-Communications, Inc., 185, 190
telecommunications-based enterprises, 389
Telecommunications Competition Act of 1996, 131–132, 183, 184–185, 192, 390
programmers and, 261
telecommunications industry, 182, 185–190
business developments, 186–189
cost recovery, 432
international, 404
local service providers, 185–186
regulation of, 183–185
U.S., 178–179
telecommuters, 168
teleconferencing, 168
Teledesic, 195
Telefonos de Mexico (Telmex), 177
telematique, 113, 139
telephone helpline
information center, 299
telephone signals, 143–145
analog, 142–143
digital, 142–143
telephone systems, 139–152
access lines, 140
customer premise equipment, 140
digitizing voice signals, 146–147
disaster recovery and, 398
growth of, 14
international distribution, 403–404
network architecture, 139
network capacity and performance, 145–146
network transmission media, 150–152
switching systems, 147–148
Telecommunications Law (1996) and, 184–185
telephone network characteristics, 142–143
time-division multiplexing, 148–150
traffic, 139
television
analog broadcast, 169
broadcast systems, 169
channel capacity, 168–169
TeleWest Communications plc, 189
Tel-Save, 191
Terplan, Kornel, 391, 392, 401, 406
test cases
in change management, 346
test data
protection of, 461
Texas Instruments, 179, 502
Tflop, 123
third-generation programming, 263
thousand line of code (KLOC)
defects per, 272–273
throughput measurements, 372
ticketless-travel systems, 43
Time Assignment Speech Interpolation (TASI), 155
time dimension
agile operations, 36
as an asset, 48–49
critical applications and, 350
in service-level agreements, 324–325

as strategic thrust, 40, 52
strategic time horizon, 74–75
Time Division Multiple Access (TDMA), 171
time-division multiplexing, 148–150
Time Warner (TW), 184–185, 189
CNN, 134
T nets, 386
tones (telephone system), 141
Toshiba, 179, 180
TPS (telecommunications-based transaction processing systems), 37
TPS (transaction processing systems), 12, 37
traffic measuring tools, 398–399
Trailer Train Company, 510
training
computer science programs, 261
for distributed computing, 306
information center, 299
retraining programmers, 262
user, 243, 276
transaction controls, 450–451
transaction processing systems (TPS), 12, 37
transistors, 114, 116
transitions
organizational, 473–474
transmission
accuracy, 399
availability, 399
bandwidth requirements, 399
channels, multiplexing, 144–145
error rates, 399
media, 150–152
network availability, 399
rates, 149, 150, 159
Transmission Control Protocol/Internet Protocol (TCP/IP), 163
trend analysis
for problem management, 341
trillion flops (Tflop), 123
trunk lines, 166–168
TRW, 195
T services, 154–155
T-1C service, 154
T-1 service, 154–155
T-2 service, 154
T-3 service, 154
T-4 service, 154
turn-around time, 321–322, 371
as measure of batch production runs, 324–325
twisted pair cable
telephone system, 140
two-channel voice grade line, 145

U

UK Social Security Agency, 472
unclassified information, 446
uninterruptible power supplies (UPS), 349, 355
uniprogramming operating systems, 129
Unisys, 180, 181
United Airlines, 42
United Parcel Service (UPS), 385–386
UNIX, 130, 163
Unocal, 477
upper CASE tools, 266
UPSNet, 386
UPS (uninterruptible power supplies), 349, 355
UPS (United Parcel Service), 385–386
U.S. Commerce Department, 348

U.S. Defense Department, 163, 267
U.S. Justice Department, 34, 187
U.S. Postal Service, 385–386
USAir, 42
USA Today, 40
user classes
    cost recovery for dedicated services, 430–431
user documentation. *See* documentation
user interface
    control point, online systems management, 368–369
    object-oriented programming for, 268
users
    control responsibilities, 448
    distributed computing and, 294
    identification and verification, 457
    response time, 327–328, 371, 373
    satisfaction surveys, 328, 373
    spending by, 304
user training
    phase review process, 243
    purchased applications and, 276
US Media Group, 189
US West, 184–185, 187, 189, 190, 473
US West Cellular, 190

V

validation, 446, 452
value-added network firms
    contingency planning and, 353
Van Dyke, Robert, 329
Van Heusen, 386
variances
    analysis of, for operational plans, 90
    in change management, 346
    cost centers, 427–428
VDSL, 154
vendor network architecture, 163
verification of users, 457
Very High Data Rate Digital Subscriber Lines (VDSL), 154
video signal compression, 169
VisualAgeC++, 269
Visual C++, 269
voice signals. *See also* telephone system
    bandwidth requirements, 398
    compression, 155
    digitizing, 146–147
    telecommunications transmission, 143
Volvo Corp., 133, 403, 475

W

Waldeck, James, 416
Wallington, Patricia, 502
Wall Street Journal, 40
Wal-Mart, 123, 502
Walt Disney Company, 134, 188, 190
Wasik, Vincent, 503
"waterfall" life cycle, 235, 236. *See also* life-cycle approach
wave-division multiplexing technology, 403
Web sites. *See also* World Wide Web
    addresses, 163
    costs of, 385–386
    Fortune 500 companies and, 386
    maintenance of, 385–386
Web technology, 511. *See also* World Wide Web
    distributed computing and, 306–307

network managers and, 406
    programming, 265
    restructuring and, 389
    servers, 127
    trends, 507
WebTV Networks Inc., 33
Well, 441
Western Digital WD4360WC, 119
Windows 95, 34, 129, 181
Windows 98, 129, 181
Windows NT, 128, 129, 130, 181
    follow-ons, 34
Windows Terminals, 34
Wintel systems, 117, 181
wireless systems, 169–172, 194
    land-based, 169–171
    satellite cellular networks, 171–172
wireline
    direct, 164
Wiseman, Charles, 39, 52
WMRM CD-ROM drives, 121
Wogsland James, 499
Wometco Cable Corporation, 189
Working Capital Management Account (Merrill Lynch), 44
workload forecasting, 326–327, 375
workload scheduling systems, 366
workload volume
    response time and, 393–394
workstations, 14, 180
    compatibility policies, 295
    for distributed computing, 287, 288, 290, 291
    growth of, 389
    personal, 124
    purchasing through workstation store, 297–298
    terminals, 124
workstation store, 297–298
World Class Network, 189
WorldCom, 61, 191, 192–193, 474
Worldspan Travel Agency Information Services, 42
World Wide Web. *See also* web sites; web technology
    client/server computing and, 307
    connecting to Federal Express via, 46
    distributed computing and, 291
    importance of, 387–388
    package delivery company use of, 385
    WebTV, 33
World Wide Web news service, 134
WORM CD-ROM drives, 121
Wriston, Walter, 35, 182, 197
write-many-read-many (WMRM) CD-ROM drives, 121
write-once-read-many (WORM) CD-ROM drives, 121

X

Xerox, 268, 502
Xerox Parc, 265

Y

Yankee Group, The, 194
year 2000 problem, 210, 261
Young, John, 49
Yourdan, Edward, 261

Z

Zeglis, John D., 62